AF126621

Extraction of Cellulose-Based Polymers from Textile Wastes

Extraction of Cellulose-Based Polymers from Textile Wastes

Editors

Helena P. Felgueiras
Jorge Padrão
Joana C. Antunes

MDPI • Basel • Beijing • Wuhan • Barcelona • Belgrade • Manchester • Tokyo • Cluj • Tianjin

Editors
Helena P. Felgueiras
Centre for Textile Science and
Technology
University of Minho
Guimarães
Portugal

Jorge Padrão
Centre for Textile Science and
Technology
University of Minho
Guimarães
Portugal

Joana C. Antunes
Centre for Textile Science and
Technology
University of Minho
Guimarães
Portugal

Editorial Office
MDPI
St. Alban-Anlage 66
4052 Basel, Switzerland

This is a reprint of articles from the Special Issue published online in the open access journal *Polymers* (ISSN 2073-4360) (available at: www.mdpi.com/journal/polymers/special_issues/Extr_Cellul_Based_Polym).

For citation purposes, cite each article independently as indicated on the article page online and as indicated below:

LastName, A.A.; LastName, B.B.; LastName, C.C. Article Title. *Journal Name* **Year**, *Volume Number*, Page Range.

ISBN 978-3-0365-4734-3 (Hbk)
ISBN 978-3-0365-4733-6 (PDF)

© 2022 by the authors. Articles in this book are Open Access and distributed under the Creative Commons Attribution (CC BY) license, which allows users to download, copy and build upon published articles, as long as the author and publisher are properly credited, which ensures maximum dissemination and a wider impact of our publications.

The book as a whole is distributed by MDPI under the terms and conditions of the Creative Commons license CC BY-NC-ND.

Contents

Helena P. Felgueiras, Jorge Padrão and Joana C. Antunes
Extraction of Cellulose-Based Polymers from Textile Wastes
Reprinted from: *Polymers* **2022**, *14*, 2063, doi:10.3390/polym14102063 1

Geraldo Cardoso de Oliveira Neto, Micheline Maia Teixeira, Gabriel Luis Victorino Souza, Valquiria Demarchi Arns, Henrricco Nieves Pujol Tucci and Marlene Amorim
Assessment of the Eco-Efficiency of the Circular Economy in the Recovery of Cellulose from the Shredding of Textile Waste
Reprinted from: *Polymers* **2022**, *14*, 1317, doi:10.3390/polym14071317 5

Halimatuddahliana Nasution, Esam Bashir Yahya, H. P. S. Abdul Khalil, Marwan Abdulhakim Shaah, A. B. Suriani and Azmi Mohamed et al.
Extraction and Isolation of Cellulose Nanofibers from Carpet Wastes Using Supercritical Carbon Dioxide Approach
Reprinted from: *Polymers* **2022**, *14*, 326, doi:10.3390/polym14020326 21

Torki A. Zughaibi and Robert R. Steiner
Forensic Analysis of Polymeric Carpet Fibers Using Direct Analysis in Real Time Coupled to an AccuTOF™ Mass Spectrometer
Reprinted from: *Polymers* **2021**, *13*, 2687, doi:10.3390/polym13162687 35

Shengjun Wang, Jiaqi Guo, Yibo Ma, Alan X. Wang, Xianming Kong and Qian Yu
Fabrication and Application of SERS-Active Cellulose Fibers Regenerated from Waste Resource
Reprinted from: *Polymers* **2021**, *13*, 2142, doi:10.3390/polym13132142 47

Maha Mohammad Al-Rajabi and Yeit Haan Teow
Green Synthesis of Thermo-Responsive Hydrogel from Oil Palm Empty Fruit Bunches Cellulose for Sustained Drug Delivery
Reprinted from: *Polymers* **2021**, *13*, 2153, doi:10.3390/polym13132153 59

Samsul Rizal, Abdul Khalil H. P. S., Adeleke A. Oyekanmi, Olaiya N. Gideon, Che K. Abdullah and Esam B. Yahya et al.
Cotton Wastes Functionalized Biomaterials from Micro to Nano: A Cleaner Approach for a Sustainable Environmental Application
Reprinted from: *Polymers* **2021**, *13*, 1006, doi:10.3390/polym13071006 81

Binwei Zheng, Weiwei Zhang, Litao Guan, Jin Gu, Dengyun Tu and Chuanshuang Hu
Enhanced Water Resistance of Recycled Newspaper/High Density Polyethylene Composite Laminates via Hydrophobic Modification of Newspaper Laminas
Reprinted from: *Polymers* **2021**, *13*, 421, doi:10.3390/polym13030421 117

Samsul Rizal, Funmilayo G. Olaiya, N. I. Saharudin, C. K. Abdullah, Olaiya N. G. and M. K. Mohamad Haafiz et al.
Isolation of Textile Waste Cellulose Nanofibrillated Fibre Reinforced in Polylactic Acid-Chitin Biodegradable Composite for Green Packaging Application
Reprinted from: *Polymers* **2021**, *13*, 325, doi:10.3390/polym13030325 129

Neha Tavker, Virendra Kumar Yadav, Krishna Kumar Yadav, Marina MS Cabral-Pinto, Javed Alam and Arun Kumar Shukla et al.
Removal of Cadmium and Chromium by Mixture of Silver Nanoparticles and Nano-Fibrillated Cellulose Isolated from Waste Peels of Citrus Sinensis
Reprinted from: *Polymers* **2021**, *13*, 234, doi:10.3390/polym13020234 145

Omid Yazdani Aghmashhadi, Lisandra Rocha-Meneses, Nemailla Bonturi, Kaja Orupõld, Ghasem Asadpour and Esmaeil Rasooly Garmaroody et al.
Effect of Ink and Pretreatment Conditions on Bioethanol and Biomethane Yields from Waste Banknote Paper
Reprinted from: *Polymers* **2021**, *13*, 239, doi:10.3390/polym13020239 . . . . **159**

Editorial

Extraction of Cellulose-Based Polymers from Textile Wastes

Helena P. Felgueiras *, Jorge Padrão and Joana C. Antunes

Centre for Textile Science and Technology (2C2T), University of Minho, Campus de Azurém, 4800-058 Guimarães, Portugal; padraoj@2c2t.uminho.pt (J.P.); joana.antunes@2c2t.uminho.pt (J.C.A.)
* Correspondence: helena.felgueiras@2c2t.uminho.pt

The extraction and exploration of cellulose-based polymers is an exciting area of research. For many years, wood (especially from bleached kraft wood pulp) was considered the main source of cellulosic compounds because of its abundance in nature [1,2]. However, in the past decade, researchers have been devoted to finding alternatives to extract cellulose from byproducts of agricultural crops and/or textile wastes, both highly available at a very reduced raw material cost. Indeed, because of the ever-increasing consumption of cotton-based products, the amount of cotton waste generated, including pre-consumer (fiber linters, yarn slivers, fabric scraps from factory offcuts, unsold brand-new garments) and post-consumer (used and unwanted garments) wastes, has increased substantially in landfills [2,3]. In finished cotton fabrics, cellulose content can be up to 99% since non-cellulose components are eliminated during scouring and bleaching, which are routine preparation procedures. Considering the urgent demands for a circular economy and sustainable actions, researchers have been taking the first steps towards finding new and greener extraction systems for agricultural and textile wastes to endow the raw materials present within those wastes with a second life. This Special Issue brings together 10 original articles that detail the recent progress and new developments in this field.

Rizal et al., in a very compelling and critical overview of the worlds' current situation regarding cotton waste fibers and their ineffective processing mechanism to mitigate their environmental impact, provided evidence of new work being conducted to employ these wastes in functional products. Indeed, different pre-treatment techniques were identified for their efficiency in extracting cellulose nanocrystals from cotton wastes, and many applications in the packaging and biomedical fields were highlighted [3]. Neto et al. explored and assessed the economic and environmental gains from the mechanical shredding of cellulose in cotton fabrics in a textile company, identifying the circularity associated with the adoption of such methods. Data suggested the existence of opportunities for the circular economy by resorting to mechanical recycling, even though there are still some limitations related to the consumption of electrical energy and amount of lubricants employed [4]. Nano-fibrillated cellulose extracted from waste peels of citrus sinensis by a chemical method involving alkali and acid hydrolysis and modified with silver nanoparticles also synthesized using extract of citrus sinensis skins as a reducing agent were engineered for heavy metal sorption. Data found these isolates and nanoparticles especially effective in removing cadmium and chromium particles from pharmaceutical effluent samples [5]. Rizal et al., demonstrated that cellulose nanofibrillated fibers isolated from waste cotton fabrics and combined with supercritical carbon dioxide via high-pressure homogenization could be used to enhance polylactic acid/chitin properties, particularly those related with the thermal-mechanical and wettability of the system [6]. Cellulose nanofibers have also been isolated from bleached carpet wastes using an innovative supercritical carbon dioxide treatment approach, which while isolating the materials also removed impurities. This treatment was unveiled as a green approach for enhancing the isolation yield of cellulose nanofibers from textile wastes with improved value and high quality [7]. Additionally, focusing on carpet fibers, Zughaibi et al., divulged the accuracy of a new analysis technique, the Direct Analysis in Real Time (DART™) coupled to an Accurate time-of-flight

Citation: Felgueiras, H.P.; Padrão, J.; Antunes, J.C. Extraction of Cellulose-Based Polymers from Textile Wastes. *Polymers* **2022**, *14*, 2063. https://doi.org/10.3390/polym14102063

Received: 30 April 2022
Accepted: 16 May 2022
Published: 18 May 2022

Publisher's Note: MDPI stays neutral with regard to jurisdictional claims in published maps and institutional affiliations.

Copyright: © 2022 by the authors. Licensee MDPI, Basel, Switzerland. This article is an open access article distributed under the terms and conditions of the Creative Commons Attribution (CC BY) license (https://creativecommons.org/licenses/by/4.0/).

(AccuTOF™) mass spectrometer, for identifying distinct polymeric fibers in crime scenes, and highlighted its potential for forensic sciences [8].

Wang et al., engineered a flexible SERS substrate based on regenerated cellulose fibers, obtained from wastepaper, modified with gold nanoparticles through dry-jet wet spinning method, an approach turned eco-friendly by using ionic liquids. The gold nanoparticles were incorporated on the regenerated fibers through electrostatic interaction. The SERS scaffolding system was found very effective for identifying toxins and chemicals like dimetridazole in aqueous solutions [9]. Recycled newspaper fibers have also been collected and reinforced with high density polyethylene for potential outdoor applications. The composite samples were seen to improve water resistance and reduce loss of tensile strength in wet conditions, depending on the amount of stearic acid used to modify the newspaper fibers [10]. Additionally, Aghmashhadi et al., using waste banknote paper as raw matter, proposed the production of value-added products like bioethanol and biogas. In this study, the authors analyzed the influence of the presence and absence of ink on the fibers in the bioethanol and biogas yields. They determined that waste banknote paper without ink and treated with sulfuric acid and a nitrogen explosive decompression process could be a suitable feedstock for sustainable biorefinery approaches [11].

In the field of drug delivery, Al-Rajabi et al., explored a green synthesis method for developing a thermo-responsive cellulose hydrogel using cellulose extracted from oil palm empty fruit bunches. The thermo-responsiveness was guaranteed by the incorporation of Pluronic F127 polymer onto the hydrogels. A sustained release of silver sulfadiazine was observed over time. In the end, it was seen that the thermo-responsive cellulose-based hydrogel could give rise to cost-effective and sustainable drug delivery systems using abundantly available agricultural biomass [12].

Author Contributions: Conceptualization, H.P.F.; writing—review and editing, H.P.F., J.P. and J.C.A.; funding acquisition, H.P.F. All authors have read and agreed to the published version of the manuscript.

Funding: This research was funded by the Portuguese Foundation for Science and Technology (FCT) grants PTDC/CTMTEX/28074/2017, PTDC/CTM TEX/28295/2017 and UID/CTM/00264/2021.

Conflicts of Interest: The authors declare no conflict of interest.

References

1. Tavares, T.D.; Antunes, J.C.; Ferreira, F.; Felgueiras, H.P. Biofunctionalization of Natural Fiber-Reinforced Biocomposites for Biomedical Applications. *Biomolecules* **2020**, *10*, 148. [CrossRef] [PubMed]
2. Teixeira, M.A.; Paiva, M.C.; Amorim, M.T.P.; Felgueiras, H.P. Electrospun Nanocomposites Containing Cellulose and Its Derivatives Modified with Specialized Biomolecules for an Enhanced Wound Healing. *Nanomaterials* **2020**, *10*, 557. [CrossRef] [PubMed]
3. Rizal, S.; Abdul Khalil, H.P.S.; Oyekanmi, A.A.; Gideon, O.N.; Abdullah, C.K.; Yahya, E.B.; Alfatah, T.; Sabaruddin, F.A.; Rahman, A.A. Cotton Wastes Functionalized Biomaterials from Micro to Nano: A Cleaner Approach for a Sustainable Environmental Application. *Polymers* **2021**, *13*, 1006. [CrossRef] [PubMed]
4. de Oliveira Neto, G.C.; Teixeira, M.M.; Souza, G.L.V.; Arns, V.D.; Tucci, H.N.P.; Amorim, M. Assessment of the Eco-Efficiency of the Circular Economy in the Recovery of Cellulose from the Shredding of Textile Waste. *Polymers* **2022**, *14*, 1317. [CrossRef] [PubMed]
5. Tavker, N.; Yadav, V.K.; Yadav, K.K.; Cabral-Pinto, M.M.; Alam, J.; Shukla, A.K.; Ali, F.A.A.; Alhoshan, M. Removal of Cadmium and Chromium by Mixture of Silver Nanoparticles and Nano-Fibrillated Cellulose Isolated from Waste Peels of Citrus Sinensis. *Polymers* **2021**, *13*, 234. [CrossRef] [PubMed]
6. Rizal, S.; Olaiya, F.G.; Saharudin, N.I.; Abdullah, C.K.; Olayia, N.G.; Mohamad Haafiz, M.K.; Yahya, E.B.; Sabaruddin, F.A.; Ikramullah; Abdul Khalil, H.P.S. Isolation of Textile Waste Cellulose Nanofibrillated Fibre Reinforced in Polylactic Acid-Chitin Biodegradable Composite for Green Packaging Application. *Polymers* **2021**, *13*, 325. [CrossRef] [PubMed]
7. Nasution, H.; Yahya, E.B.; Abdul Khalil, H.P.S.; Shaah, M.A.; Suriani, A.B.; Mohamed, A.; Alfatah, T.; Abdullah, C.K. Extraction and Isolation of Cellulose Nanofibers from Carpet Wastes Using Supercritical Carbon Dioxide Approach. *Polymers* **2022**, *14*, 326. [CrossRef] [PubMed]
8. Zughaibi, T.A.; Steiner, R.R. Forensic Analysis of Polymeric Carpet Fibers Using Direct Analysis in Real Time Coupled to an AccuTOF™ Mass Spectrometer. *Polymers* **2021**, *13*, 2687. [CrossRef] [PubMed]

9. Wang, S.; Guo, J.; Ma, Y.; Wang, A.X.; Kong, X.; Yu, Q. Fabrication and Application of SERS-Active Cellulose Fibers Regenerated from Waste Resource. *Polymers* **2021**, *13*, 2142. [CrossRef] [PubMed]
10. Zheng, B.; Zhang, W.; Guan, L.; Gu, J.; Tu, D.; Hu, C. Enhanced Water Resistance of Recycled Newspaper/High Density Polyethylene Composite Laminates via Hydrophobic Modification of Newspaper Laminas. *Polymers* **2021**, *13*, 421. [CrossRef] [PubMed]
11. Aghmashhadi, O.Y.; Rocha-Meneses, L.; Bonturi, N.; Orupõld, K.; Asadpour, G.; Garmaroody, E.R.; Zabihzadeh, M.; Kikas, T. Effect of Ink and Pretreatment Conditions on Bioethanol and Biomethane Yields from Waste Banknote Paper. *Polymers* **2021**, *13*, 239. [CrossRef] [PubMed]
12. Al-Rajabi, M.M.; Teow, Y.H. Green Synthesis of Thermo-Responsive Hydrogel from Oil Palm Empty Fruit Bunches Cellulose for Sustained Drug Delivery. *Polymers* **2021**, *13*, 2153. [CrossRef] [PubMed]

Article

Assessment of the Eco-Efficiency of the Circular Economy in the Recovery of Cellulose from the Shredding of Textile Waste

Geraldo Cardoso de Oliveira Neto [1,2], Micheline Maia Teixeira [1], Gabriel Luis Victorino Souza [1], Valquiria Demarchi Arns [1], Henrricco Nieves Pujol Tucci [1] and Marlene Amorim [2,*]

1 Industrial Engineering Post-Graduation Program, Universidade Nove de Julho (UNINOVE), São Paulo 03155-000, Brazil; geraldo.prod@gmail.com (G.C.d.O.N.); micheline.maiateixeira@gmail.com (M.M.T.); gabrielvictorino@uol.com.br (G.L.V.S.); valquiria.demarchi@cocamar.com.br (V.D.A.); henrricco@gmail.com (H.N.P.T.)
2 GOVCOPP & DEGEIT, University of Aveiro, 3810-193 Aveiro, Portugal
* Correspondence: mamorim@ua.pt

Abstract: There is a growing demand for the adoption of cyclical processes in the fashion industry. The trends point to the reuse of cellulose from cotton fibres, obtained from industrial waste, as a substitute to the former linear processes of manufacturing, sale, use, and discarding. This study sets up to explore and assess the economic and environmental gains from the mechanical shredding of cellulose in cotton fabrics in a textile company, identifying the circularity associated with the adoption of such methods. The study resorted to a case study methodology building on interviews and observation. For the environmental estimations, the study employed the material intensity factor tool, and for the economic evaluation the study uses the return on investment. The study also offers an estimation of the circularity of the processes that were implemented. The adoption of the mechanical shredding for cotton cellulose generated economic gains of US$11,798,662.98 and a reduction in the environmental impact that amounts to 31,335,767,040.26 kg including the following different compartments: biotic, abiotic, water, air, and erosion. The findings suggest the existence of opportunities for the circular economy in the textile sector of about 99.69%, dissociated to the use of mechanical recycling, while limited by the consumption of electrical energy and lubricants in the recycling process, leading the way to a circular economy.

Keywords: recovery of cellulose; textile fibers; eco-efficiency; circular economy; textile industry

1. Introduction

The elimination of waste and pollution, the circulation of products and materials, and the regeneration of nature, are the key principles underlying the circular economy perspective, which pursues the goal of rebuilding financial, human, and social capital [1]. According to the principles of a circular economy, the production and consumption activities must be organized to preserve the value of the products, components, and materials throughout all of the value chain and the product life cycle [2]. In the textile sector, the circular business models involve the materials, the facilities, the packaging, the garments, and the programs aimed at retuning textile items instead of discarding such items [3]. As such, all of the value chain is included with initiatives in the several phases of design, production, consumption, and waste management [4].

The transfer of linear business models to the circular economy has been gaining room [5], particularly in the last decade as researchers and policy makers have been discussing strategies to meet sustainability goals [6]. The textile industry, and sector specialists, have also been involved in the development of actions to build knowledge for the circular economy while contributing to raising awareness about sustainability [7]. In this context, the circular economy, as a management approach in the food and textile sectors, offers a real opportunity for the reuse of food waste to produce natural fiber that can be

used in the production of new textile products [8]. The research work by Costa et al. [7] also offers an important vision about the valorization of proteins which can be recovered by means of chemical and physical processes from milk byproducts and which can have different applications in the textile industry.

It is noteworthy that demographic growth has influenced the availability of resources globally due to the increase in the consumption of disposable goods, leading to potential excessive levels of consumption and the generation of waste [9]. In the manufacturing context, the textile sector raises concerns about the abundant use of natural resources and of fossil derived materials that raise environmental and social problems due to the manufacturing and discarding of textile clothing [10]. For this reason, organizations are adopting strategies to minimize environmental impacts, such as the use of acetate to transform cellulose waste which is present in cardboard, paper, and cotton (therefore, introducing such cellulose waste into new cellulose textile fiber) [11]. Therefore, the potential for the use of this type of waste as a raw material for conventional processes of fiber spinning holds important contributions for the environment and create opportunities for circular economy [12].

The recycling of textile mixtures has raised the awareness about the development of methods that are safe for the environment. In studies that address fabric made with yarn mixing polyester and recycled cotton, some technical evaluation has been conducted about the mechanic recycling of the fibers. Despite the existing functional differences between virgin and used polyester fibers, recycled polyester fibers can be used in adequate fabric, minimizing the production of virgin fiber and contributing to reducing the environmental impact [13]. The process of enzymatic hydrolysis can also be used as a strategy to recycle the mix of polyester and cotton. It involves two stages. The first stage involves a mechanical pre-treatment by means of grinding with the purpose of obtaining short fibers for the textile structure. Then, an alkaline chemical pre-treatment follows wherein the material is exposed into a sodium hydroxide solution for a given amount of time. Despite the fact such a process had not been implemented at an industrial level, it proves to be a valid opportunity for the recycling of mixed fabrics including polyester and cotton (which is aligned with the goals of the circular economy in the textile sector) [14]. Concerning the mixtures of polyester and wool fiber, some experiments involving enzymatic treatment of the mix have also evidenced promising results for the recovery of synthetic fibers, since the textile waste involving mixed materials have aggravated the economic and environmental impacts in the sector [15].

The environmental impacts of the bio recycling method, which recovers polyester fibers from the mixture of polyester and cotton in fabric waste, have been evaluated using the life cycle assessment method (LCA) for the stages of pre-treatment, melt-spinning, and enzymatic hydrolysis. This allowed for the observation that the pre-treatment process contributes with an impact of 60%, while the melt spinning has smaller impacts for the ecosystems (14%) and for human health (15%), and enzymatic hydrolysis affects the quality of the ecosystem (14%) and human health (12%). The results suggest that the bio recycling of textiles creates opportunities for the reduction of environmental impacts [16].

The evaluation of the environmental impact associated to the use of mechanically recycled cotton fiber instead of using virgin cotton fiber was also conducted using the LCA method. The results observed, in the scenarios that were evaluated, highlighted some key aspects such as the denim fabric using cotton fiber as a raw material (53%) and the energy consumed in the e spinning phase (16%). In terms of what concerns the environmental impact, the use of the 100% recycled cotton fiber has the smaller impact. The results of existing studies might encourage textile companies to use recycled raw materials as well as the use of eco-friendly technologies for energy production [5].

It stands out that very often, textile materials are discarded in landfill sites or incinerating plants [17], leading to serious environmental risks [18]. Meanwhile, to reduce environmental impacts, some companies have been adopting strategies to reuse the textile waste [19]. The implementation of recycling reutilization practices for textile waste following the consumption can reduce pollution and the volume of discards in landfills, reducing both the environmental impacts and the costs associated with the discarding activities [20]. Likewise, the separation of large volumes of textile materials through automated processes can also contribute to minimize the production of virgin textile fibers and to the increase in the levels of reutilization and recycling levels [21].

According to studies addressing the use of cotton fiber recycled through mechanical processes instead of virgin cotton fiber, the economic evaluation performed using relative costs led to different results for each scenario. When using 100% recycled cotton fiber, instead of virgin fibers, the cost variation was only of 1%. The utilization of the combined heat and energy plant led to the observation of a reduction in cost around 35 to 40% for each meter of fabric. However, the increase in the consumption of electricity to produce thread with recycled fiber led to an increase in the energy costs of 8% [5].

As such, several studies have demonstrated the existence of efforts to change the perception about the textile waste which is generated throughout the production chain, addressing several good practices that are environmentally friendly [20]. Such practices are related to the development of materials from waste [5,14,19,22–24], the recycling of textiles [12,20,25–27], and the development of textile fibers from non-textile waste, among others [7–9,28]. Table 1 offers an overview of the literature addressing the textile industry and the approaches related to economic and environmental aspects that hold opportunities for the circular economy.

Table 1. Overview of existing literature focusing on the textile industry and the approaches for economic and environmental gains, and opportunities for the circular economy.

			Environmental Approach		Economic Approach	
References	Year	General Approach and Opportunities for the Circular Economy	Qualitative	Qualitative	Qualitative	Qualitative
[27]	2016	Production of synthetic geotextiles non weaved with recycled textile waste	X			
[12]	2019	Utilization of chemical processes for separating cotton and polyester from textile waste	X			
[29]	2019	Methods for textile recycling	X		X	
[21]	2020	Automatic separation of large volumes of textile waste	X		X	
[13]	2020	Evaluation of the properties of recycled thread made of polyester and cotton	X			
[16]	2020	Evaluation of the lifecycle of textile bio recycling	X	X		
[15]	2020	Enzymatic processes for the selective digestion of wool fibres and mixed wool and polyester fibre	X			
[17]	2020	Project TEX2MAT for the recycling of materials from textile flows and selected multi-materials	X			
[10]	2021	Enzymatic and biological processes	X			
[14]	2021	Removing cotton from textile mixture of cotton and PET through enzymatic hydrolysis	X			
[9]	2021	Creation of protein fibres regenerated from waste in the food industry	X			
[18]	2021	Incorporation of dyed cotton flakes in the polypropylene	X		X	
[22]	2021	Utilization of hemp as a natural fibre for the development of green label products	X			
[23]	2021	Replacing polyester by three biological alternatives	X			
[5]	2021	Investigating the contribution of using fibre of mechanical recycled cotton instead of virgin fibre	X	X	X	X
[20]	2021	Valorisation of textile waste for end-of-life products	X		X	
[30]	2021	Investigating the consumption of electric energy to produce cotton clothes	X			
[28]	2021	Investigation of the technological innovation in fibres produced from proteins	X			
[25]	2021	Identification and valorisation of solid waste from textiles, pre-treatment methods	X			
[26]	2021	Recovery of waste in the Nazi period with direct influence in industry sectors, including textile	X			
[31]	2021	Use of iron nitrate to modify the leftovers of cotton in nature and to absorb the back reactive colorant	X			

Table 1. *Cont.*

References	Year	General Approach and Opportunities for the Circular Economy	Environmental Approach		Economic Approach	
			Qualitative	Qualitative	Qualitative	Qualitative
[7]	2021	Recovery of milk proteins and by-products by chemical and physical processes for applications in non-food industries	X			
[19]	2021	Scientific research addressing the application of recycled fibres	X			
[8]	2021	Utilization of food industry waste to produce bio-textiles	X			
[24]	2021	Development of t a methodology for the fabrication of textile composites with cellulose regenerated from textile waste	X			
[6]	2022	Influences of digital solutions in the textile industry with opportunities for the circular economy	X			

The research conducted by Fidan et al. [5] and Subramanian et al. [16], describes the conduction of environmental assessments using the LCA method. However, these studies do not distinguish those results across departments. None of the existing studies offers an assessment of the environmental impact using the mass intensity factor (MIF) tool, or present evidence about reductions in the volume of waste that is attributed to the recycling, reuse, or the development of new materials. Concerning the economic evaluation of the results, no studies were identified using an economic evaluation that builds on the estimation of the return on investment (ROI). The existing studies only offer a qualitative description of the economic gains. In the study of Fidan et al. [5] the authors used relative costs to describe the economic gains associated with each of the scenarios addressed in the study.

Therefore, gaps were identified in the literature, which justify the development of this study and to offer a contribution to the following research question: Does the adoption of mechanical splitting of cotton cellulose generate economic and environmental gains and opportunities for the circular economy in the textile industry? This study therefore aims to assess the economic and environmental gains that can be derived from the mechanical splitting of cellulose in cotton fabric in a textile company, and to discuss the opportunities for the circular economy associated with this method (Figure 1).

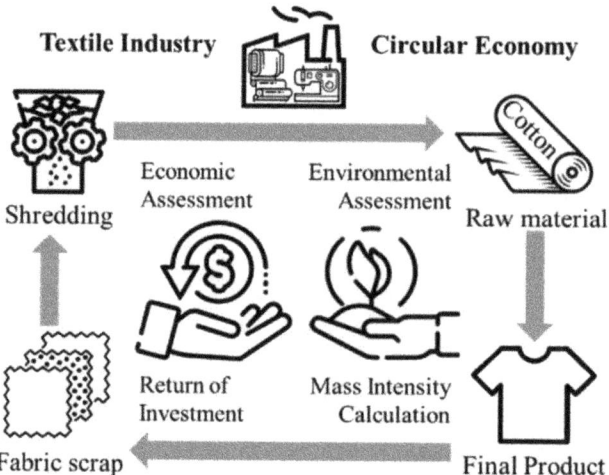

Figure 1. Proposed application of the circular economy in the textile industry.

Among the processes used for recycling of textiles, the mechanical recycling is a process widely adopted. In this process, the grinding or splitting is used to extract the fiber from textile waste either before or after consumption and use [25]. Cotton is one of the most important materials for textile production, but is has raised concerns about its environmental impacts, since its growth and cultivation require large volumes of water and pesticides and its manufacturing involves high levels of energy consumption [5]. Globally the production of fibres has been growing and reaching levels around 100 million of tons per year [17]. A large share of this production is directed to supply the textile industry that has also been registering important levels of production growth [29]. This growth is influenced also by the prevalent fast fashion trends [28]. The fast fashion tendency contributes to the increase in the negative impacts for the environment, which are related to the large volumes of water and energy that are consumed in the production and distribution processes including the extraction of raw materials, the production of textile fibres, the weaving, the dying, washing, and end of life processes [6]. In this context the mechanical splitting of the textile cellulose can offer a promising strategy to obtain environmental and financial gains, contributing to reduce the pressures that the textile sector faces.

2. Materials and Methods

2.1. Data Collection Methods

This study addressed the mechanical spliting of cellullose in cotton textiles in the specific context of a textile company, with the purpose of identifying opportunities for circular economy and attaining economic and environmental benefits that are quantifiable. According to Yin [32], the conduction of case studies is recommended when the researcher wants to explore research questions such as "how" and "why". In addition, according to Eisenhardt [33], the study has characteristics of an exploratory work which investigates a contemporary scenario and addressed a real-world problem. Following the identification of the research gap, the study involved the conduction of a literature review using the following research terms: "circular economy"; "textile industry"; "fiber"; "cellulose"; "shredded" (Figure 2). The databases included in the search where Science Direct, Emerald Insight, Wiley, Taylor & Francis, and Google Scholar. All of the articles were analysed according to the principles proposed by Bryman [34] with the purpose of identifying the theoretical constructs building on content analysis techniques.

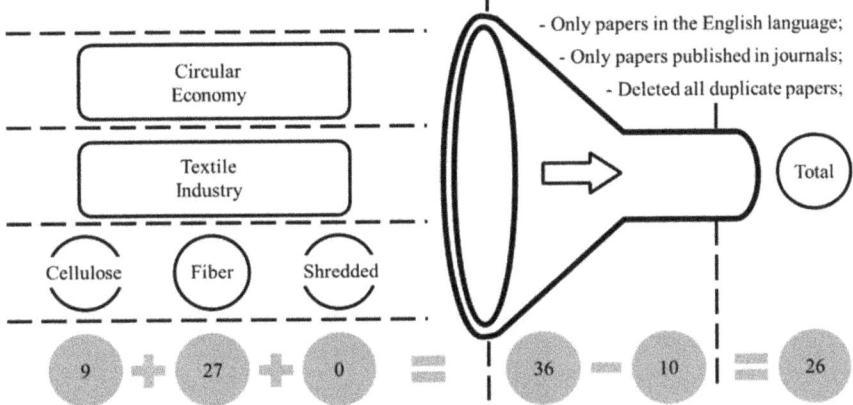

Figure 2. Protocol used for the systematic literature review.

The 26 articles identified supported the observed paucity of research in the context of textile industries in the topics related to circular economy, notably for studies addressing the economic and environmental impacts resorting to quantitative and qualitative methods. Building on the existing sources, the following conceptual model was proposed (see Figure 3).

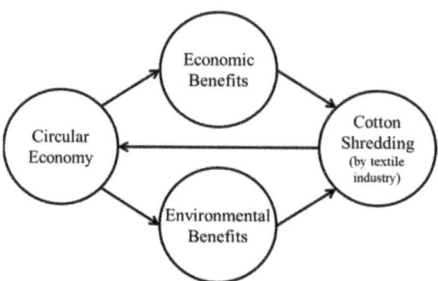

Figure 3. Conceptual model.

The next phase in the study involved the conduction of field study with the purpose of exploring the theories identifies in the literature towards the managerial practices adopted in the textile industry. To this end, the data collection involved the conduction of semi-structured interviews with the technical managers of the company addressed in the study. According to Bryman [34] and Collins and Hussey [35], this is an adequate approach to collect quantitative data. The process also involved the development of an interview script with a key set of questions to support the conduction of the interviews.

2.2. Data Analysis

The data analysis was organized into two steps. First, the environmental aspects were addressed, and next the economic perspective was explored. It is important to highlight that the textile company addressed in the study approached the circular economy by investing in equipment aiming at the reduction of waste. Therefore, it was necessary to consider both the costs and the environmental impacts associated with these changes in the production process following the acquisition of that equipment. To increase the transparency of the impacts being investigated the study adopted a comparative logic approach (i.e., analysing the situation "before" and "after").

The environmental analysis was performed with the calculation of the mass intensity total (MIT), as this is a method that considers biotic compartments (x), abiotic (w), air (z), and water (y) for analysing the environmental impacts of the process focused on in the study [36]. According to Tucci et al. [37], it was necessary to first proceed with the identification of the quantities consumed in the processes and convert these quantities into mass units (M). Next, it was necessary to multiply the quantities by the corresponding intensity factor (IF) (Equation (1)).

$$MIF = M \times IF \quad (1)$$

The values for the IF are released, and updated regularly, by the Wuppertal Institute. As previously mentioned this study considered the elements of cotton, oil, and electrical energy (Table 2).

Table 2. Impact factors.

IF	Abiotic	Biotic	Water	Air	Erosion
Cotton Fiber	8.6	2.9	6.814	2.74	5.01
Energy	2.67	–	37.92	0.64	–
Synthetic Oil	1.22	–	4.28	0.01	–

Source: Wuppertal Institute [38].

Accordingly, it was possible to calculate the mass intensity per compartment (MIC) by adding all together (Equation (2)) the three elements considered in the study for each compartment. In other words, it was possible to identify the shares for environmental impacts that were avoided in each compartment.

$$MIC = IF\ (residue\ A\ compartment\ w) + IF\ (residue\ B\ compartment\ w) + IF\ (residue\ C\ compartment\ w) + \ldots \quad (2)$$

Finally, the sim of the MIC for each of the compartments let to the total environmental impact that was prevented. This estimation was defined as the mass intensity total (MIT) (Equation (3)).

$$MIT = MICw + MICv + MICz + MICn \quad (3)$$

It is noteworthy that the study considered the reduction in the use of cotton as raw material by reusing the production waste, as well as the consumption of energy and lubricant oils in the new equipment (which allowed for the reuse of the materials). As such, it was possible to estimate the level of circularity of the company using the equation below.

$$Circularity\ Index = \frac{Mass\ Intensity\ Total\ Before}{Mass\ Intensity\ Total\ After} \quad (4)$$

The economic analysis was performed by calculating the ROI, which has the objective of determining the amount of time that is necessary to cover the debts that were incurred to support the total investments made. By determining the ROI, companies can analyse the economic viability of each innovation, change, or modernization, regardless of the volume of the investment. This is a standard method that is applied across all industries [39,40].

The study considered in the stage "before" the reusing of the textile wastes the direct costs related to the acquisition of raw materials.

In the "after" stage, other costs such as human resources were considered, which grouped costs such as hiring employees, technical training, overtime to adapt the new workflow and other similar costs. In addition, costs with laboratory tests, food, as well as costs with lubricants, electricity, janitorial, maintenance, storage, and insurance, logically affected by the acquisition of new machines, with emphasis on the large increase in insurance costs.

The data collected allowed for the estimation of the depreciation costs for the equipment acquired and the costs related to taxes and social contributions. As such, it was possible to perform a comparative analysis for each year to determine the moment when the accumulated returns would surpass the yearly depreciation and the return period for the investment.

3. Results and Discussion

The object of this case study is a cooperative company that started its activities in the 1960s, bringing together 46 agricultural producers who wanted to diversify their products. Cotton was selected as the product that would initiate this diversification. However, the success of the cooperative resulted in the expansion of business. Currently, there are 97 operational units and more than 15,000 members.

In the 1980s, the cooperative opened its yarn industry with the aim of adding value to the cotton delivered by the cooperative members. However, it was in 2011 that the entire industrial park was modernized to increase export volumes, improving the quality of the final product and complying with international standards (mainly from the European market, which are stricter than national standards). Since then, the guide to good practices in the textile sector has been followed, with emphasis on waste reduction programs.

The high levels of demand for virgin fibres carry important impacts for the environment, since their cultivation often involves the use of pesticides. Moreover, along all of the manufacturing process, much waste is generated which are considered by-products with low value.

On the contrary, in the context of the textile industry, we have been observing a growth in the studies concerning the reduction of waste aiming at eliminating or reducing them, according to the principles of the circular economy. One of the avenues for that is the reutilization of cellulose by splitting the remainders of textile cuttings, as well as of new textiles or those that are at the end of their lifecycle, by incorporating them into the production of split thread. Moreover, over the last years new industrial equipment in the textile industry, that have higher capacity, demand fibres with higher intrinsic quality, and this strengthens the challenge for pursuing the goals of circular production.

In Figure 4 we present the flowchart for the stages involved in the process of mechanic recovery of cellulose by splitting textile material.

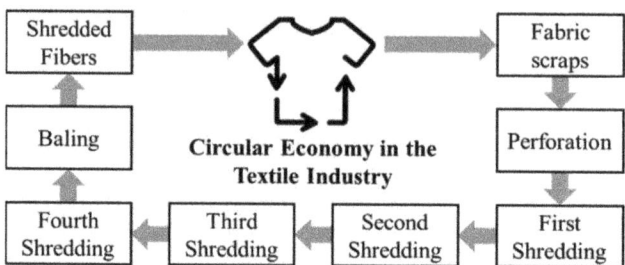

Figure 4. Process for the mechanical recovery of cotton fiber.

The process involves:
- ✓ The separation of the textile material by color: the separation of the textile scraps is made according to the color and is performed in the textile industry with the purpose of obtaining split fibres with solid colors and with the lowest levels of contamination. The index of contamination involves the mix of scraps with different colors from the main colour.
- ✓ Perforation: this stage aims to reduce the size of the scraps to a size of approximately 5 cm, to facilitate the shredding process that will follow.
- ✓ First to fourth shredding: each shredding stage aims to separate the fibres from the scrap. In each step, the separation is improved.
- ✓ Baling: the shredded fibres are compacted in bale format and are conditioned in bags.

3.1. Technical Analysis for the Recovery of Textile Fibres

In the context of the thread industry, which uses shredded fibres as a substitute for virgin cotton fibre, some control points are important such as the measurement of the composition of the mixture bale and the identification of the contamination and the intrinsic quality of the fibres and its thread. Since they originate from the scrap from textiles used to produce clothing, the risks of mixing materials in the moment of separation that is performed in the textile company is very high. This is reflected in the variability of the composition that can be found in the bale for shredded cotton. This creates difficulties for the composition of the textile products, as illustrated in Table 3. In Brazil, the correct identification of the final composition of a textile product is mandatory and subject to fines if carried out incorrectly. The variability that characterizes the composition of shredded textile bale is addressed in the law. The Portaria Inmetro n° 118 of 10 March 2021, states that "UNDEFINED COMPOSITON" or "DIVERSIFIED FIBERS" can only be used for textile products for which the textile composition can be difficult to determine, such as in the case of shredded cotton fibres.

Table 3. Reliability index.

Material	UHML (in)	Micronaire	STR (gf/Tex)	UI (%)	Spinning Consistency Index (SCI)
Cotton lot	1.2	4.32	31.55	81.14	130
Shredded cotton lot	0.83	4.47	27.84	58.74	5

The contamination that exists in the shredded fibre bale, such as parts of buttons, zips, and needles is also an element of difficulty for the production since it reduces the lifetime of the shredding equipment as illustrated in Table 3. IT reduces the efficiency of the shredding and carding as well as the risk of accidents for the workers that handle the materials (including the risk of fire). The reduction in the efficiency of the shredding process is a very important factor to monitor. Also, the reduction in the lifetime of the equipment derived from the metal contamination increases the maintenance costs. Fibres from obtained from shredding processes have an average length and resistance that have a direct impact for the reliability index.

The reliability can be measured according to the formula where:

$$CI = -322.98 + (2.89 \times STR) - (9.02 \times MIC) + (43.53 \times (UHML)) + (4.29 \times UI)$$

SCI (spinning consistency index)—index of reliability for the fibre. The bigger the value of the index, the better is the quality of the material, and the better the manufacturing performance.

STR (strength)—fibre resistance. This corresponds to the strength measured in grams which is necessary to break a bundle of fibre of 1 tex, measured in gr/tex.

MIC (micronaire)—index for fineness. Measures the linear density (mass/length) through maturity.

UHML (upper half mean length)—average length for the upper half of the 50% of the longer fibres, measured in inches.

UI (uniformity index)—index for the uniformity of the length of the fibre, corresponds to the relationship between the medium length (ML) and the medium length for the upper half of the longer fibres (UHML), measured in %.

The formula for the reliability index is recommended by the manufacturers of equipment for HVI (high volume instrument), that performs the measurement of the information about the intrinsic quality of the cotton feather. There are two main manufacturers globally for this type of equipment: Uster Technologies [41] and Premier Evolvics [42].

A lower level for SCI created difficulties for the processing of the fibres to produce the thread, requiring lower speed from the equipment and resulting in losses in production. Depending on the spinning method that is used to make the thread, there will be a stronger demand for length and resistance of the fibres, notably for threads that are very thin for some fashion segments. One alternative to produce thinner shredded threads is to make a mixture adding some proportion of other fibres such as cotton, polyester, and fibres from recycled materials or from certified or organic materials. Such a mixture helps to increase reliability, easing the process of obtaining threads and reducing losses. In Table 3, a comparative view is offered with examples for the reliability index for a shredded fibre and a virgin fibre.

Likewise, the thread from shredded fibres has a higher number of irregularities, including the mass coefficient (%), thin spots (−50%), thick spots (+50%), hairiness (%) and tenacity (cN/tex), when compared with virgin fibre threads. The irregularities in threads that are composed by shredded fibres also contribute to difficulties in the production both in shredding and in weaving when producing fabric and knitted fabric. In Figure 5, it is possible to observe a comparison between the thread produced from spinning 100% cotton and shredded cotton.

3.2. Assessment of the Environmental Impact from the Implementation of the Process for Recovering Textile Fibres

The company addressed in the study uses 4,600,000 kg of cotton per year to produce fabric, as displayed in Table 4. Before the implementation of the circular economy principles, the company used virgin cotton, carrying a negative environmental impact of: 39,560,000 per kg in the abiotic levels, that stand for global warming, soil, humidity, vegetation etc.; 13,340,000 kg in the abiotic compartment related to life; 31,344,400,000 kg in the air, due to the generated emissions; 12,604,000 kg in water, due to the consumption in the irrigation process; and 23,046,000 kg for erosion due to the cultivation and crops in a linear cycle. Overall, it added up to a total negative environmental impact of 31,432,950,000 kg.

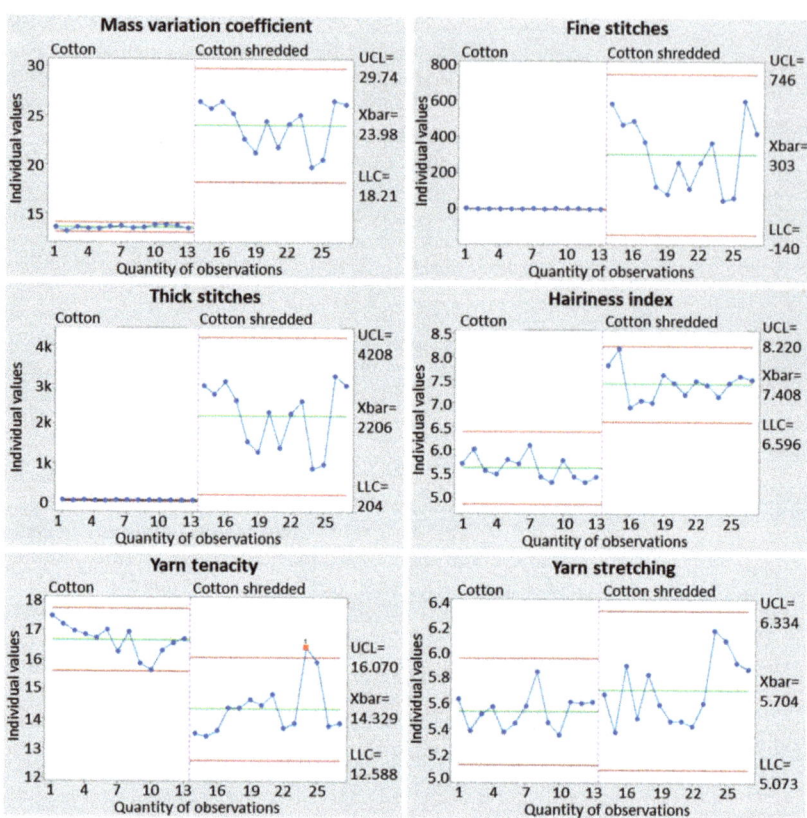

Figure 5. Comparative analysis for thread, produced from 100% cotton and from cotton shred.

The implementation of the principles of the circular economy led the company to focus on the process of recovering fibres by means of recycling and reuse of scrap from fabric in the textile industry, as well as for fabric that is discarded in the end of its lifetime. This led the company to engage in recycling processes with the goal of recovering 6,223,529.42 kg of fibres. However, 811,764.71 kg of this are non-recoverable and the waste is discarded in an adequate manner. In the process, 5,411,764.71 kg of the fibre is recovered with the original level of quality. The company implemented the process of mechanic recycling for textile waste that consumes 1,148,400 kW/h per year, with the use of perforating equipment, shredding equipment, baling machinery, compacting, and lighting equipment that generate a negative environmental impact of 97,181,901.82 kg when considering all of the compartments (i.e., abiotic, biotic, water, air, and erosion). The company also consumes 192 L of lubricants for the machines, adding to a total negative environmental impact of 1057.92 L.

This study offers a contribution for the advancement of current knowledge about circular economy by describing the evaluation of the environmental impacts for the various compartments (i.e., biotic, abiotic, water, air, and erosion). The values change from a level that was at 31,432,950,000 kg of negative impact before the implementation, and a reduction of the environmental impact amounting to 31,335,767,040.26 after the implementation of the circular economy, in the process of recovery of textile fibres. The existing studies on the topic by Subramanian et al. [16] and Fidan et al. [5] usually use only the mass balance for the process of analysing the lifecycle and do not offer an estimation for the environmental impact. In this context, the contribution of this study can also offer support to the managerial function about the opportunities for the recycling of textile waste by transforming them into fibres for replacing the consumption of cotton.

Table 4. Environmental assessment of the process for recovering textile fibers.

Components	Annual Consumption (kg/kWh)	Compartiments/Unit (kg/kWh)					MIT-Mass Intensity Total
		Abiotic	Biotic	Air	Water	Erosion	
"Before" the Adoption of Circular Economy							
Cotton fiber	4,600,000	8.6	2.9	6.814	2.74	5.01	31,432,950,000.00
		39,560,000.00	13,340,000.00	31,344,400,000.00	12,604,000.00	23,046,000.00	
MIC-Mass intensity per compartment-Cotton fiber		39,560,000.00	13,340,000.00	31,344,400,000.00	12,604,000.00	23,046,000.00	31,432,950,000.00
"After" the Implementation of Circular Economy							
Fabric Fiber Recovery							
Reduction from the reuse (cotton shredding)	5,411,764.71	8.6	2.9	6.814	2.74	5.01	36,979,941,204.61
		46,541,176.51	15,694,117.66	36,875,764,733.94	14,828,235.31	27,112,941.20	
Waste discarding (cotton shredding)	811,764.71	8.6	2.9	6.814	2.74	5.01	5,546,991,204.61
		6,981,176.51	2,354,117.66	5,531,364,733.94	2,224,235.31	4,066,941.20	
MIC-Mass intensity per compartment-fiber recovery		39,560,000.00	13,340,000.00	31,344,400,000.00	12,604,000.00	23,046,000.00	31,432,950,000.00
Energy Consumption							
Energy(perforating equipment)	156.975	2.67		37.92	0.64		10,802,234.63
		419,123.25		5,952,492.00	100,464.00		
Energy(shredding equipment)	376.740	2.67		37.92	0.64		25,925,363.10
		1,005,895.80		14,285,980.80	241,113.60		
Energy(shredding equipment)	112.185	2.67		37.92	0.64		7,719,997.01
		299,533.42		4,254,047.62	71,798.27		
Energy(Compacting)	418.600	2.67		37.92	0.64		43,945,255.90
		1,117,662.00		15,873,312.00	267,904.00		
Energy (Lighting)	83.720	2.67		37.92	0.64		8,789,051.18
		223,532.40		3,174,662.40	53,580.80		
MIC-Mass Intensity Per Compartment-Energy Consumption		3,065,746.87		43,540,494.82	734,860.67		97,181,901.82
Consuption of Lubricant for the Equipment							
Synthetic emulsifiable oil	192	1.22		4.28	0.01		1057.92
		234.24		821.76	1.92		
MIC-Mass Intensity per Compartment-Lubricants Consumption		234.24		821.76	1.92		1057.92
MIC-Mass Intensity per Compartment-Energy And Lubricants Consumption		3,065,981.11		43,541,316.58	734,862.59		97,182,959.74
MIT-Mass Intensity Total		36,494,018.89	13,340,000.00	31,300,858,683.42	11,869,137.41	23,046,000.00	31,335,767,040.26
Circularity Index			Before	31,432,950,000.00	After	31,335,767,040.00	99.69%

In the case addressed in the study, a total of 99.69% of circularity was obtained by dividing the 31,335,767,040.26 (corresponding to the reduction in the environmental impact) by 31,432,950,000 kg (corresponding to the negative environmental impact) in relative terms. The data suggest that despite using 100% of recycled fibre in the manufacturing process there was a loss of 97,182,959.74 due to the consumption of energy and lubricants. The valorisation of textile waste is an important point to produce clothing and the recycling alternatives have an important role for the creation of opportunities for circular economy. These findings support the work of Subramanian et al. [16] which states that even if mechanical recycling is the process that is most used for recycling waste from textiles, and which allows for mixing of recycled and virgin fibres, the process also has disadvantages, including the loss of quality in the fibre. As such, mechanical recycling does not generate 100% of circularity due to the productive resources that are consumed in the process (despite the fact that if offers a good contribution for the circular economy). Other technologies exist, such as the process of enzymatic hydrolysis for the treatments of fabric made of polyester and cotton [14] and for mixed fabrics of wool and polyester. These have proved to be an effective textile recycling process, relying on the shredding of fibres, and can bring the fashion industry to a new model of circular economy [15]. The bio-recycling approach also offers technologies that can contribute to the valorisation of the textiles while being environmentally friendly [16].

3.3. Economic Evaluation of the Implementation of the Process for Recovering Textile Fibre

Table 5 offers an overview of the cost analysis that allows for the identification of the reduction in the annual cost derived from the circular economy, considering the recycling of textile waste and the recovery of fibre for production. Considering the consumption of cotton for the annual production of 4,600,000 with a cost of US$2.90 per kg, it amounted to US$13,323,272.72. After the implementation of a circular economy, we observed a volume of direct costs of US$264,353.64 that resulted from hiring employees, including four for the operations of recycling and shredding and two for the maintenance of machinery and equipment (plus one manager and one administrative assistant). It is important to highlight the use of bags for conditioning the textile fibres after the recycling process and after being compacted, with the purpose of not using packaging that would generate waste in the textile chain. As such, they were considered an investment and not a direct cost. The indirect costs were also calculated at US$1,031,564.73, considering the costs of lubricants "synthetic emulsifiable oil" for the equipment, the consumption of electric energy, lab tests, food, storage costs, insurance, and indirect costs with the maintenance of machinery, equipment, and facilities. Also, and according to what was mentioned in the environmental assessment, the need to discard 811,764.71 kg of fibre added up to a loss of US$228,691.48 in the recycling due to the impurities that contaminate the cotton fibres. As such, the total cost amounted to US$1,524,609.75, considering the total of direct costs, the indirect costs, and the losses associated with the discarding of waste.

Table 5. Assessment of costs from the process of recovering textile fibres.

Costs (US$) before the Adoption of the Circular Economy	
Utilization of virgin cotton for yearly production	4,600,000
Average price for virgin cotton, per kg	2.90
Costs for the purchase of virgin cotton	13,323,272.72
Costs (US$) After the adoption of the Circular Economy	
Direct Costs (Year)	264,353.64
Fixed Human Resources	264,353.64
Indirect Costs (Year)	1,031,564.73
Lubricants	4809.09
Electric Energy	196,664.64
Lab tests	13,590.91
Food	117,090.91
Storage and Insurance	417,136.36
Indirect Costs for the recycling process	749,291.91
Maintenance of Machinery and Equipment	101,409.09
Professional Services for Machines and Equipment	87,818.18
Facilities Maintenance	93,045.45
Indirect Maintenance Costs	282,272.73
Losses from recycling, leading to waste	228,691.48
Total Costs (US$)	1,524,609.75
Reduction in annual costs (US$)	11,798,662.98

Replacing the purchase of virgin cotton with the utilization of recycled textile fibres led to a cost reduction of US$ 11,798,662.98, demonstrating a relevant pathway for textile companies to adopt the circular economy through the recycling and recovery of textile fibres. In this way, the company no longer purchases virgin cotton and contributes to the sustainability of the ecosystem while simultaneously reducing costs.

The company had to invest US$463,646 in infrastructures, including the acquisition of perforating equipment, shredding equipment, compacting machines, and 948 bags with a capacity of 1 m^3/1000 kg to protect the textile fibres from contamination and fungus. The reduction in annual costs amounted to US$11,798,662.98, as detailed in Table 5. As such leaving out the taxes and the values for depreciation, it is possible to observe a reduction in net cost of US$8,931,747 and a return on the capital invested of 30 days, suggesting a profitable opportunity for the textile industry, as displayed in Table 6. The study contributes to the advancement of theory by being the first work that evaluated costs and estimates the return on investment exploring real data from organizational experiences.

Table 6. Assessment of return on investment.

Investment in equipment	463,646					
Depreciation period (years)	10					
Annual Depreciation	46,365					
Annual Cost Reduction	11,798,663					
Annual Depreciation	−46,365					
Basis for Determining Income Tax (IR)	11,752,298					
IRPJ + CSLL (Social Contribution)	24.0%					
Value for IR + CSSL (Annual)	−2,820,552					
Net Cost Reduction (Annual)	8,931,747					
Net Cost Reduction (Annual)	8,931,747					
Depreciation (Annual)	46,365					
Income (Annual)	8,978,111					
Cash Flow	Year 0	Year 1	Year 2	Year 3	Year 4	Year 5
Investment	−463,646					
Income Flow (Annual)		8,978,111	8,978,111	8,978,111	8,978,111	8,978,111
Total Cash Flow	−463,646	8,978,111	8,978,111	8,978,111	8,978,111	8,978,111
ROI or TIR	1936.4%	per year				
Payback Discounted at 15% per year	0.08	years				

No other studies were identified offering an economic evaluation and using the estimation of the returns on investment. The existing studies are limited to the description of the economic gains in a descriptive and qualitative manner. Previously, only the work of Fidan et al. [5] used relative costs to offer an approximate description of the economic gains for the scenarios explored in the study.

4. Conclusions

The textile industry has an important share in the emission of CO_2 and is a big contributor for the environmental pollution. The textile and clothing sector generate tons of waste in the form of scrap that result from cutting of fabric, and most of the time these have as their final destination disposal in landfills. Moreover, in recent years the demand triggered by fast fashion trends increased the speed of production to meet the release of new collections and has been contributing to an excess in the volume of sold products, as well as unsold items that contribute to the discarding of textile materials in harmful ways for the environment. The textile chain carries social and environmental impacts. Despite contributing to the creation of jobs and income, it also contributes to environmental pollution, significant use of water resources, and the exploitation of human resources. Against this background, we are witnessing changes in consumer behaviour, as consumers are demanding products that are more sustainable and socially correct. There is a growing trend that has been acknowledged by fashion influencers for the demand of cyclical processes that make the reuse of cellulose from cotton fibres that are obtained from industrial waste in order to substitute the former linear processes of manufacturing, sale, use, and discarding.

According to the sustainable development goals set in the Agenda 2030 by the UN, the five guiding principles (people, planet, prosperity, peace, and partnership) will promote sustainability and innovation through the adoption of technologies that contribute for the implementation of industrial processes that are environmentally adequate. Accordingly, this study addresses the case of reuse of cellulose obtained from the waste generated in the textile manufacturing processes by producing shredded thread. The study supports the notion that the adoption of mechanical shredding for cotton cellulose allows for achieving economic gains of US$11,798,662.98 and a reduction in the environmental impacts of 31,335,767,040.26 kg when adding the several compartments (biotic, abiotic, water air, and erosion). These findings suggest opportunities for the circular economy in the textile sector around 99.69%, from the use of mechanical recycling, the consumption of energy, and lubricant in the recycling process, as an example of a way forward for the circular economy. Future research work should explore other recycling processes for textile fibres and explore both their impacts and circularity.

Author Contributions: Conceptualization, G.C.d.O.N. and M.M.T.; methodology, G.C.d.O.N., G.L.V.S. and V.D.A.; investigation, G.C.d.O.N. and M.M.T.; writing—original draft preparation, G.C.d.O.N., M.M.T., H.N.P.T. and M.A.; writing—review and editing, H.N.P.T. and M.A.; supervision, G.C.d.O.N.; funding acquisition, M.A. All authors have read and agreed to the published version of the manuscript.

Funding: This work was financially supported by the research unit on Governance, Competitiveness and Public Policy (UIDB/04058/2020)+(UIDP/04058/2020), funded by national funds through FCT—Fundação para a Ciência e a Tecnologia.

Institutional Review Board Statement: Not applicable.

Informed Consent Statement: Not applicable.

Conflicts of Interest: The authors declare no conflict of interest.

References

1. Ellen Macarthur Foundation. What Is a Circular Economy? Available online: https://ellenmacarthurfoundation.org/topics/circular-economy-introduction/overview (accessed on 8 January 2022).
2. Wysokińska, Z. Implementing the main circular economy principles within the concept of sustainable development in the global and European economy, with particular emphasis on Central and Eastern Europe—The case of Poland and the region of Lodz. *Comparat. Econ. Res. Central Eastern Eur.* **2018**, *21*, 75–93.
3. Esbeih, K.N.; Molina-Moreno, V.; Núñez-Cacho, P.; Silva-Santos, B. Transition to the Circular Economy in the Fashion Industry: The Case of the Inditex Family Business. *Sustainability* **2021**, *13*, 10202.
4. Ogunmakinde, O.E. A review of circular economy development models in China, Germany and Japan. *Recycling* **2019**, *4*, 27.
5. Fidan, F.Ş.; Aydoğan, E.K.; Uzal, N. An integrated life cycle assessment approach for denim fabric production using recycled cotton fibers and combined heat and power plant. *J. Clean. Prod.* **2021**, *287*, 125439.
6. Happonen, A.; Ghoreishi, M. *A Mapping Study of the Current Literature on Digitalization and Industry 4.0 Technologies Utilization for Sustainability and Circular Economy in Textile Industries*, Proceedings of Sixth International Congress on Information and Communication Technology, London, UK, 25–26 February 2021; Springer: Singapore, 2021; pp. 697–711.
7. Costa, C.; Azoia, N.G.; Coelho, L.; Freixo, R.; Batista, P.; Pintado, M. Proteins derived from the dairy losses and by-products as raw materials for non-food applications. *Foods* **2021**, *10*, 135.
8. Provin, A.P.; de Aguiar Dutra, A.R. Circular economy for fashion industry: Use of waste from the food industry for the production of biotextiles. *Technol. Forecast. Soc. Chang.* **2021**, *169*, 120858.
9. Stenton, M.; Kapsali, V.; Blackburn, R.S.; Houghton, J.A. From Clothing Rations to Fast Fashion: Utilising Regenerated Protein Fibres to Alleviate Pressures on Mass Production. *Energies* **2021**, *14*, 5654.
10. Ribul, M.; Lanot, A.; Pisapia, C.T.; Purnell, P.; McQueen-Mason, S.J.; Baurley, S. Mechanical, chemical, biological: Moving towards closed-loop bio-based recycling in a circular economy of sustainable textiles. *J. Clean. Prod.* **2021**, *326*, 129325.
11. Jiang, X.; Bai, Y.; Chen, X.; Liu, W. A review on raw materials, commercial production and properties of lyocell fiber. *J. Bioresour. Bioprod.* **2020**, *5*, 16–25.
12. Haslinger, S.; Hummel, M.; Anghelescu-Hakala, A.; Määttänen, M.; Sixta, H. Upcycling of cotton polyester blended textile waste to new man-made cellulose fibers. *Waste Manag.* **2019**, *97*, 88–96.

13. Majumdar, A.; Shukla, S.; Singh, A.A.; Arora, S. Circular fashion: Properties of fabrics made from mechanically recycled poly-ethylene terephthalate (PET) bottles. *Resour. Conserv. Recycl.* **2020**, *161*, 104915.
14. Piribauer, B.; Bartl, A.; Ipsmiller, W. Enzymatic textile recycling—Best practices and outlook. *Waste Manag. Res.* **2021**, *39*, 1277–1290. [PubMed]
15. Navone, L.; Moffitt, K.; Hansen, K.A.; Blinco, J.; Payne, A.; Speight, R. Closing the textile loop: Enzymatic fibre separation and recycling of wool/polyester fabric blends. *Waste Manag.* **2020**, *102*, 149–160. [PubMed]
16. Subramanian, K.; Chopra, S.S.; Cakin, E.; Li, X.; Lin, C.S.K. Environmental life cycle assessment of textile bio-recycling–valorizing cotton-polyester textile waste to pet fiber and glucose syrup. *Resour. Conserv. Recycl.* **2020**, *161*, 104989.
17. Piribauer, B.; Jenull-Halver, U.; Quartinello, F.; Ipsmiller, W.; Laminger, T.; Koch, D.; Bartl, A. Tex2mat–Next Level Textile Recycling with Biocatalysts. *Detritus Multidiscip. J. Waste Resour. Residues* **2020**, *13*, 78–86.
18. Serra, A.; Serra-Parareda, F.; Vilaseca, F.; Delgado-Aguilar, M.; Espinach, F.X.; Tarrés, Q. Exploring the Potential of Cotton Industry Byproducts in the Plastic Composite Sector: Macro and Micromechanics Study of the Flexural Modulus. *Materials* **2021**, *14*, 4787.
19. Patti, A.; Cicala, G.; Acierno, D. Eco-Sustainability of the Textile Production: Waste Recovery and Current Recycling in the Composites World. *Polymers* **2021**, *13*, 134.
20. Stanescu, M.D. State of the art of post-consumer textile waste upcycling to reach the zero waste milestone. *Environ. Sci. Pollut. Res.* **2021**, *28*, 14253–14270.
21. Riba, J.R.; Cantero, R.; Canals, T.; Puig, R. Circular economy of post-consumer textile waste: Classification through infrared spectroscopy. *J. Clean. Prod.* **2020**, *272*, 123011.
22. Sorrentino, G. Introduction to emerging industrial applications of cannabis (*Cannabis sativa* L.). *Rend. Lincei Sci. Fis. Nat.* **2021**, *32*, 233–243.
23. Ivanović, T.; Hischier, R.; Som, C. Bio-Based Polyester Fiber Substitutes: From GWP to a More Comprehensive Environmental Analysis. *Appl. Sci.* **2021**, *11*, 2993.
24. Ribul, M.; Goldsworthy, K.; Collet, C. Material-Driven Textile Design (MDTD): A methodology for designing circular material-driven fabrication and finishing processes in the materials science laboratory. *Sustainability* **2021**, *13*, 1268.
25. Subramanian, K.; Sarkar, M.K.; Wang, H.; Qin, Z.H.; Chopra, S.S.; Jin, M.; Lin, C.S.K. An overview of cotton and polyester, and their blended waste textile valorisation to value-added products: A circular economy approach–research trends, opportunities and challenges. *Crit. Rev. Environ. Sci. Technol.* **2021**, 1–22. [CrossRef]
26. Weber, H. Nazi German waste recovery and the vision of a circular economy: The case of waste paper and rags. *Bus. Hist.* **2021**, *63*, 1–22. [CrossRef]
27. Leon, A.L.; Potop, G.L.; Hristian, L.; Manea, L.R. Efficient technical solution for recycling textile materials by manufacturing nonwoven geotextiles. In *IOP Conference Series: Materials Science and Engineering*; IOP Publishing: Bristol, UK, 2016; p. 022022.
28. Stenton, M.; Houghton, J.A.; Kapsali, V.; Blackburn, R.S. The Potential for Regenerated Protein Fibres within a Circular Economy: Lessons from the Past Can Inform Sustainable Innovation in the Textiles Industry. *Sustainability* **2021**, *13*, 2328.
29. Piribauer, B.; Bartl, A. Textile recycling processes, state of the art and current developments: A mini review. *Waste Manag. Res.* **2019**, *37*, 112–119.
30. Angelova, R.A.; Velichkova, R.; Sofronova, D.; Ganev, I.; Stankov, P. Consumption of Electric Energy in the Production of Cotton Textiles and Garments. In *IOP Conference Series: Materials Science and Engineering*; IOP Publishing: Bristol, UK, 2021; p. 012030.
31. Komatsu, J.S.; Motta, B.M.; Tiburcio, R.S.; Pereira, A.M.; de Pape, P.W.; Mandelli, D.; Carvalho, W.A. Iron Nitrate Modified Cotton and Polyester Textile Fabric Applied for Reactive Dye Removal from Water Solution. *J. Braz. Chem. Soc.* **2021**, *32*, 964–977.
32. Yin, R.K. *Case Study Research and Applications: Design and Methods*; Sage Publications: Thousand Oaks, CA, USA, 2017.
33. Eisenhardt, K.M. Building theories from case study research. *Acad. Manag. Rev.* **1989**, *14*, 532–550.
34. Bryman, A. *Research Methods and Organization Studies*; Routledge: London, UK, 2003; Volume 20.
35. Collins, J.; Hussey, R. *Business Research Methods*; McGraw-Hill: New York, NY, USA, 2007.
36. De Oliveira Neto, G.C.; Tucci, H.N.P.; Pinto, L.F.R.; Costa, I.; Leite, R.R. Economic and environmental advantages of rubber recycling. In Proceedings of the IFIP International Conference on Advances in Production Management Systems, Iguassu Falls, Brazil, 3–7 September 2016; pp. 818–824.
37. Tucci, H.N.P.; de Oliveira Neto, G.C.; Rodrigues, F.L.; Giannetti, B.F.; de Almeida, C.M.V.B. Six sigma with the blue economy fundamentals to assess the economic and environmental performance in the aircraft refueling process. *Renew. Sustain. Energy Rev.* **2021**, *150*, 111424.
38. Wuppertal Institute. Calculating MIPs, Resources Productivity of Products and Services. Available online: https://wupperinst.org/uploads/tx_wupperinst/MIT_2014.pdf (accessed on 10 January 2021).
39. Bauer, J. Bases of Valuation in the Control of Return on Public Utility Investments. *Am. Econ. Rev.* **1916**, *6*, 568–588.
40. Digrius, B.; Keen, J. *Making Technology Investments Profitable: ROI Road Map to Better Business Cases*; Wiley: Hoboken, NJ, USA, 2002.
41. Uster Technologies. Uster HVI 1000 the Fiber Classification and Analysis System. Available online: https://www.uster.com/products/fiber-testing/uster-hvi/ (accessed on 21 January 2021).
42. Premier Evolvics. ART 2 Cotton Tester Brochure. Available online: https://premierevolvics.com/pdf/ART2.pdf (accessed on 21 January 2021).

Article

Extraction and Isolation of Cellulose Nanofibers from Carpet Wastes Using Supercritical Carbon Dioxide Approach

Halimatuddahliana Nasution [1], Esam Bashir Yahya [2,*], H. P. S. Abdul Khalil [2,3,*], Marwan Abdulhakim Shaah [2], A. B. Suriani [3], Azmi Mohamed [3], Tata Alfatah [2] and C. K. Abdullah [2]

[1] Department of Chemical Engineering, Faculty of Engineering, Universitas Sumatera Utara, Medan 20155, Indonesia; halimatuddahliana@usu.ac.id
[2] School of Industrial Technology, Universiti Sains Malaysia, Penang 11800, Malaysia; marwanshaah90@gmail.com (M.A.S.); tataalfatah83@gmail.com (T.A.); ck_abdullah@usm.my (C.K.A.)
[3] Nanotechnology Research Centre, Faculty of Science and Mathematics, Universiti Pendidikan Sultan Idris (UPSI), Tanjung Malim 35900, Perak, Malaysia; suriani@fsmt.upsi.edu.my (A.B.S.); azmi.mohamed@fsmt.upsi.edu.my (A.M.)
* Correspondence: essam912013@gmail.com (E.B.Y.); akhalilhps@gmail.com (H.P.S.A.K.)

Abstract: Cellulose nanofibers (CNFs) are the most advanced bio-nanomaterial utilized in various applications due to their unique physical and structural properties, renewability, biodegradability, and biocompatibility. It has been isolated from diverse sources including plants as well as textile wastes using different isolation techniques, such as acid hydrolysis, high-intensity ultrasonication, and steam explosion process. Here, we planned to extract and isolate CNFs from carpet wastes using a supercritical carbon dioxide (Sc.CO_2) treatment approach. The mechanism of defibrillation and defragmentation caused by Sc.CO_2 treatment was also explained. The morphological analysis of bleached fibers showed that Sc.CO_2 treatment induced several longitudinal fractions along with each fiber due to the supercritical condition of temperature and pressure. Such conditions removed th fiber's impurities and produced more fragile fibers compared to untreated samples. The particle size analysis and Transmission Electron Microscopes (TEM) confirm the effect of Sc.CO_2 treatment. The average fiber length and diameter of Sc.CO_2 treated CNFs were 53.72 and 7.14 nm, respectively. In comparison, untreated samples had longer fiber length and diameter (302.87 and 97.93 nm). The Sc.CO_2-treated CNFs also had significantly higher thermal stability by more than 27% and zeta potential value of -38.9 ± 5.1 mV, compared to untreated CNFs (-33.1 ± 3.0 mV). The vibrational band frequency and chemical composition analysis data confirm the presence of cellulose function groups without any contamination with lignin and hemicellulose. The Sc.CO_2 treatment method is a green approach for enhancing the isolation yield of CNFs from carpet wastes and produce better quality nanocellulose for advanced applications.

Keywords: cellulose nanofibers isolation; carpet wastes; supercritical carbon dioxide; enhanced properties

1. Introduction

The production of nanocellulose has attracted tremendous attention in the past few years due to its suitability for a wide range of applications in medical and other fields [1,2]. Cellulose nano fibers (CNFs) have excellent mechanical properties, thermal conductivity, and electrical conductivity, making these nanomaterials imparted to various matrices such as thermoplastics and elastomers, thermosets, ceramics, and even metals [3–6]. In addition, the significant biological properties of CNFs, such as biocompatibility, biodegradability, non-toxicity, and non-immunogenicity accompanied by eco-friendly nature, are added advantages and highly extend their application in the biomedical fields such as drug delivery, tissue engineering and biosensing [7–9]. The past few years witnessed utilization of different industrial and agricultural wastes including textile wastes as a source of predominant compounds they contain such as nanocellulose and recyclable nylon [10].

Natural fiber carpets and mats wastes are one of textile wastes that mostly end up in landfill due to the high costs of their recyclability compared to their original price [11,12]. With the increase in carpet production, carpet wastes are continuously increasing, giving the need for safer utilization of these wastes. Natural fiber carpets are mainly composed of natural fibers; in Malaysia, carpets and mats are highly produced from *Hibiscus cannabinus* L. natural fibers, which contain high cellulose content and can be an excellent source of nanocellulose instead of ending up in landfill [13].

New methods for producing high-quality cellulose nanofibers from natural fibers, agricultural and textile wastes are highly desired and some of them are very popular such as blending electrospinning [14], coaxial electrospinning [15], and tri-axial electrospinning [16–18]. These techniques have been reported to produce CNFs with slightly different properties depending on the extraction conditions as well as used chemical agents. High-intensity ultrasonication, cryo crushing, high-pressure homogenization, grinding, steam explosion process, electrospinning, high-speed blending, and 2,2,6,6-Tetramethylpiperidin-1-yl)oxyl oxidation method are the most used techniques for CNFs isolation [19–22]. However, many of these techniques either use toxic chemicals or expensive approaches, which eventually raise the production costs and/or reduce the quality of CNFs due to the possible chemical contamination. Supercritical carbon dioxide (Sc.CO_2) is an environmentally friendly, inexpensive, and essentially nontoxic technique that has attracted significant attention in the past few years in extraction and isolation applications, which use relatively moderate critical temperature (31.1 °C) and pressure (73.8 bar) [23,24].

The interactions of supercritical fluids carbon dioxide with lignocellulosic fibers have been extensively investigated by researchers. Atiqah et al. [25] used Sc.CO_2 explosions with a low temperature followed by mild oxalic acid hydrolysis to extract CNFs. Li and Kiran [26] studied the extent of dissolutions in wood species and reported that the isolation in CO_2, ethylene, n-butane, and nitrous oxide was not significant. Schmidt et al. [27] reported severe damage in cellulose fabrics (viscose and cotton) after Sc.CO_2 treatment. However, the exact role and the mechanism of Sc.CO_2 in CNFs isolation is not yet thoroughly investigated. Many researchers have reported using Sc.CO_2 approach with natural fibers without comparing the outcomes and highlighting the effect of Sc.CO_2 on the isolated fibers or trying such technique in textile wastes. This research aimed to isolate cellulose nano fibers from traditional natural fiber carpet wastes and investigate the role of Sc.CO_2 treatment in enhancing the properties of CNFs. Experiments were performed to compare the properties of CNFs isolated with and without Sc.CO_2 approach.

2. Materials and Methods

2.1. Materials

Natural fiber carpet wastes were obtained from NVD carpets Sdn. Bhd. (Penang, Malaysia). Carbon dioxide for the supercritical process was procured from ZARM Scientific & Supplies Sdn. Bhd. (Penang, Malaysia). All the chemicals in this study were used without further purification and were of analytical grade, purchased from Sigma–Aldrich (Schnelldorf, Germany); sodium hydroxide (99%), sodium chlorite (80%), Hydrogen peroxide (35%), oxalic acid (98%) and glacial acetic acid (99.5%).

2.2. Isolation of CNFs from Natural Fiber Carpet Wastes

Natural fiber carpet wastes (300 g) were cut into small pieces (2–3 cm long) and washed several times with hot water and then placed in a digester (Pulping) with 26% sodium hydroxide (NaOH). The ratio of cooking was 1:7 and cooked at temperature 170 °C for 90 min. The fiber pulp was then washed and purified through 3-stage bleaching described in Table 1.

Table 1. The stages of bleaching condition for CNFs isolation from pulped carpet wastes.

Bleaching Stage	Chemical Charge	Reaction Time (min)	Temperature (°C)	Consistency (%)
D_1	2% $NaClO_2$ + 3% CH_3COOH	120	70	10
E_p	1.5% NaOH + 1% H_2O_2	90	70	10
D_2	1% $NaClO_2$ + 3% CH_3COOH	90	60	10

Bleached pulp was then divided into two samples; one undergoes Sc.CO_2 treatment with 50 MPa at 60 °C for 2 h followed by mild hydrolysis, while the other one undergoes only mild hydrolysis without Sc.CO_2 treatment. The mild hydrolysis was done using 5% oxalic acid, and the washed fiber suspension was eventually homogenized for 6 h in an OV5 homogenizer (VELP SCENTIFICA, Usmate Velate MB, Italy) to get CNFs. Figure 1 presents the overall process of CNFs preparation from natural fiber carpet wastes.

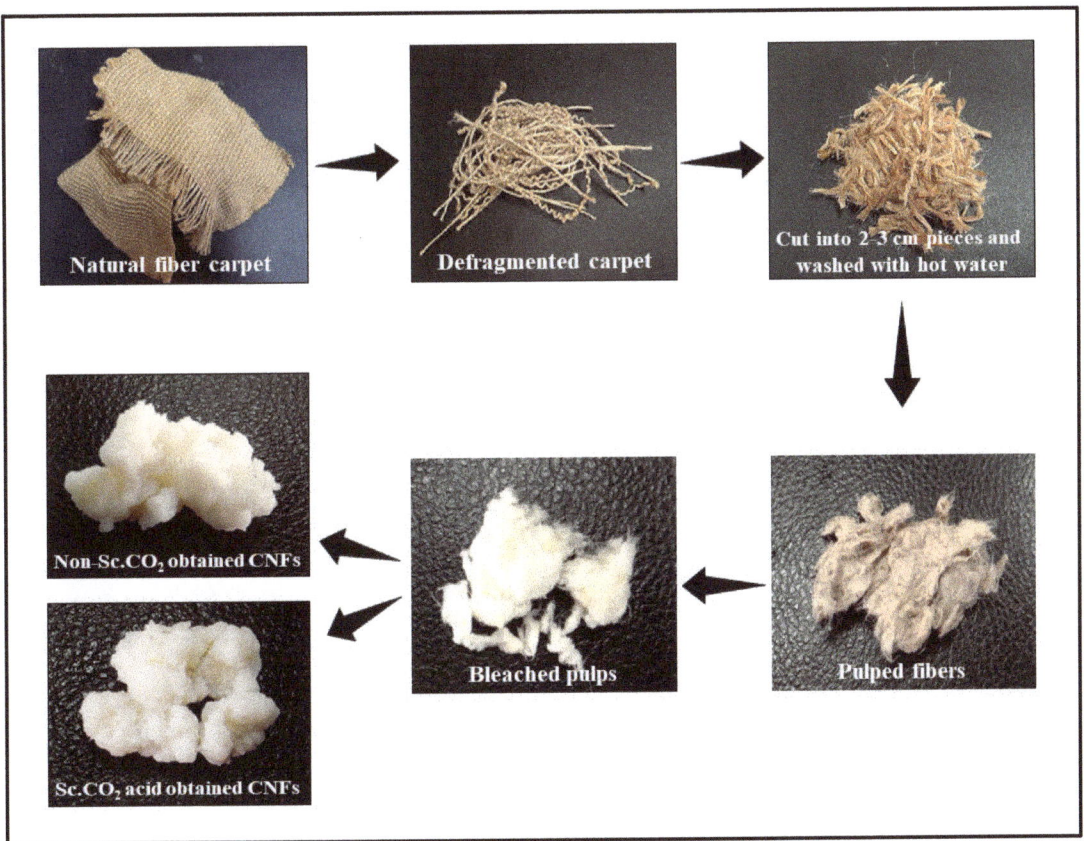

Figure 1. The overall process of preparing cellulose nano fibers (CNFs) from natural fiber carpet wastes.

2.3. Determination of the Fiber Yield and Chemical Analysis

The fibers were oven-dried and weighed before and after each stage of the isolation process. The calculation was repeated three times, and the average of three replications was considered as the result of the fiber yield. Standard TAPPI methods were used for determining the chemical composition of the samples. The cellulose content was assessed using TAPPI Test Method 203 om-93, while TAPPI Test Method T222 om-88 was used to evaluate hemicellulose and lignin percentage [28].

2.4. Microscopic Analysis

The morphology of bleached fibers before and after Sc.CO_2 treatment was characterized by using SEM model Leo Supra, 50 VP, Carl Zeiss, SMT (Carl Zeiss Group, Oberkochen, Germany) with high resolution. TEM model PHILIPS CM12 electron microscope equipped with Docu Verison 3.2 (Hamburg, Germany) was used to measure the dimension of the isolated CNFs.

2.5. Particle Size Distribution and Surface Charge

A laser diffraction analyzer was used to determine the size distribution and charge of all prepared CNFs using Nano-ZS90, Malvern, UK analysis machine. A 0.1 nm and 10 μm size range was used for analyzing the CNFs particles using a suspension of 0.01% CNFs consistency, which was dispersed with ultrasound for 15 min at 100% power. The results were the average of three replicated measurements.

2.6. Fourier Transform-Infrared (FTIR) Spectroscopy

FT-IR spectroscopy (Thermo Scientific model Nicolet I S10 spectrometer, Thermo Fisher Scientific, Waltham, MA, USA) was used to investigate the functional groups present in the isolated CNFs, using a Perkin Elmer spectrum 1000 for obtaining the spectrum. The CNFs suspension was freeze-dried before the analysis.

2.7. Crystallinity Analysis

Structural and crystallinity of the CNFs samples were determined using an X-ray diffractometer (XRD) test, Model Bruker D8 Advance with CuKα radiation. The Ni-filter was used to filter the CuKα radiation. A 2θ angle range from 0° to 50° in reflection mode was scanned at 2°/min. Crystallinity index for both samples was calculated using Origin software by using the following equation:

$$\text{Crystallinity \%} = \frac{\text{Area of crystalline peaks}}{\text{Area of all peaks}} \times 100$$

2.8. Thermo-Gravimetric Analysis (TGA)

A thermogravimetric analyzer (TGA/SDTA 851e, Brand Mettler Toledo, Mettler-Toledo International Inc., Columbus, OH, USA) was used to characterize all TGA curves for all the samples. The samples were heated from 25 °C to 600 °C at 10 °C/min^{-1}.

3. Results

3.1. Fiber Morphology

The microscopic analysis of the bleached fibers is presented in Figure 2, which can be observed the effect of Sc.CO_2 on the lignocellulosic fibers compared with untreated ones. It can be seen the differences of fractions among supercritical treated fibers (Figure 2a) as highlighted with the red arrows compared with untreated fibers (Figure 2b), which still seemed to be in contact without clear fractions. The combination of mild heat, supercritical carbon dioxide, and pressure could cause cellulosic fibers through acetal bonds [29]. The mechanism of Sc.CO_2 treatment in facilitating the acid hydrolysis process and generating nano cellulose fibers can be explained by the induction of several longitudinal fractions along with each fiber, which removed the impurities and resulted in cleaner and more fragile fibers (Figure 2c). The fragile fibers that contain the longitudinal fractions permit the acid to integrate within the fibers and thus facilitate their cleavage. The fibers undergo pressure and mild temperature, which increased their fragility and broke into smaller particle sizes, facilitating the acid hydrolysis process and generating smaller CNFs.

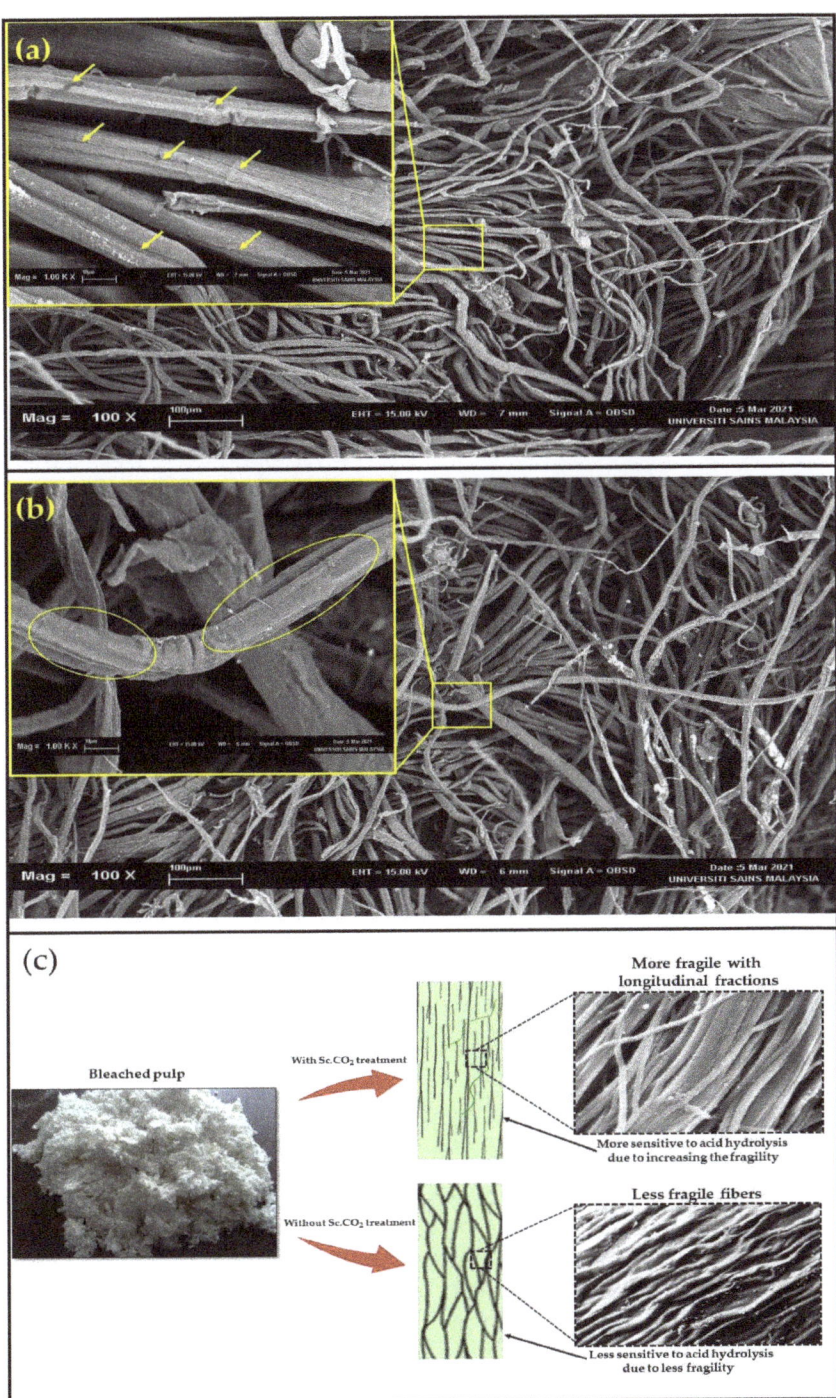

Figure 2. SEM images of bleached carpet pulp: (**a**) Sc.CO$_2$ treated fibers, (**b**) untreated bleached fibers, and (**c**) schematic drawing of the possible supercritical fraction mechanism on the carpet fibers.

The Sc.CO_2 destroyed and hydrolyzed the remaining lignin structure in the fibers, causing a plasticization effect in the supercritical state and higher crystallinity index. Multiple centrifugations were required to naturalize the acidic pH of CNF after the acid hydrolysis process, which was caused due to trapping the acid within the nano fraction of cellulosic fibers. In a previous study, it was reported a significant delignification effect of Sc.CO_2 on lignocellulosic rice husk without any reductions in enzymatic digestibility and crystallinity [18]. In a similar study, Atiqah et al. [25] used supercritical CO_2 in the acid hydrolysis process to produce CNFs from Kenaf bast and reported that supercritical facilitated the fibers fraction. In this study, results comparing the surface of Sc.CO_2 treated and untreated fibers can be observed with slight impurities and longitudinal fractions, confirm the finding. The ability of Sc.CO_2 treatment in endorsing such modifications in the carpet fibers, was further confirmed with chemical composition and crystallinity analysis.

3.2. Fibers Yield and Chemical Composition

Table 2 presents the yield, fiber length, and chemical composition of each stage of the CNFs isolation process. The weight reduction after each step of CNF isolation with increasing the cellulose content and elimination of other undesired materials, such as hemicellulose, lignin, and ash. This study achieved a CNF yield of 32% and 31.5% from the overall biomass fiber for Sc.CO_2 and non-Sc.CO_2. The cellulose content of raw fibers was found to be 63.2%, which ended up with 94.6 and 93.8% for Sc.CO_2 and non-Sc.CO_2 approaches, respectively. The cellulose content was higher than that obtained by Karimi et al. [30], who obtained CNFs from raw kenaf fiber with a cellulose content of 92.0 ± 0.5% and 91.8 ± 0.9%, respectively. Moreover, the CNFs yield obtained in this study was higher than the previous study, which used slightly different approaches. Lignin content slightly decreased after supercritical treatment from 0.8 in the non-treated approach to 0.6% in Sc.CO_2 samples. In this regard, François et al. [31] reported a similar finding, with a 24.5% decrease in lignin content after Sc.CO_2 treatment of hemp fibers. A slight difference in chemical composition and fiber yield was recorded among Sc.CO_2 and non-Sc.CO_2. However, the fiber length was significantly different, ranging from 100 nm to 120 nm in Sc.CO_2 approach compared with non-Sc.CO_2 approach, which was higher than 2000 nm.

Table 2. The fiber yield, fiber length, and chemical analysis after each stage of CNF preparation.

Biomass Stage	Fiber Yield	Fiber Length	Chemical Composition		
			Cellulose	Hemicellulose	Lignin
Raw carpet fibers	100%	2.0–3.0 cm	63.2 ± 0.7	18.3 ± 1.2	11.6 ± 0.5
Pulped fibers	91%	2.0–3.0 cm	65.4 ± 0.9	18.9 ± 1.4	12.1 ± 0.8
Bleached pulps	58%	0.1–1.1 cm	92.3 ± 0.6	3.8 ± 0.3	0.9 ± 0.2
Sc.CO_2 obtained CNFs	32%	97.0 nm	94.6 ± 0.4	3.1 ± 0.1	0.6 ± 0.2
Non-Sc.CO_2 obtained CNFs	31.5%	302 nm	93.8 ± 0.6	3.5 ± 0.3	0.8 ± 0.2

3.3. Surface Functional Group and Thermal Analysis

Figure 3 compares the results of surface functional groups of Sc.CO_2 and non-Sc.CO_2 obtained CNFs to assess the variations of any possible chemical structure changes. From FTIR spectra, it can be seen the little variation even in the dominant peaks between the two samples. However, the –OH stretches in both Sc.CO_2 and non-Sc.CO_2 obtained CNFs are observed in the 3800–3000 cm^{-1} range with a broad peak and great intensity. Non-Sc.CO_2 obtained CNFs showed some tiny shoulders in the 3700–3800 cm^{-1} range, which could be due to extractive or other impurities present in the fibers. Bigger CH-stretching can be observed in Sc.CO_2 obtained CNFs, and tiny shaft stretching vibrations could be due to the CO_2 supercritical treatment caused slight shifts for the characteristic bands [23].

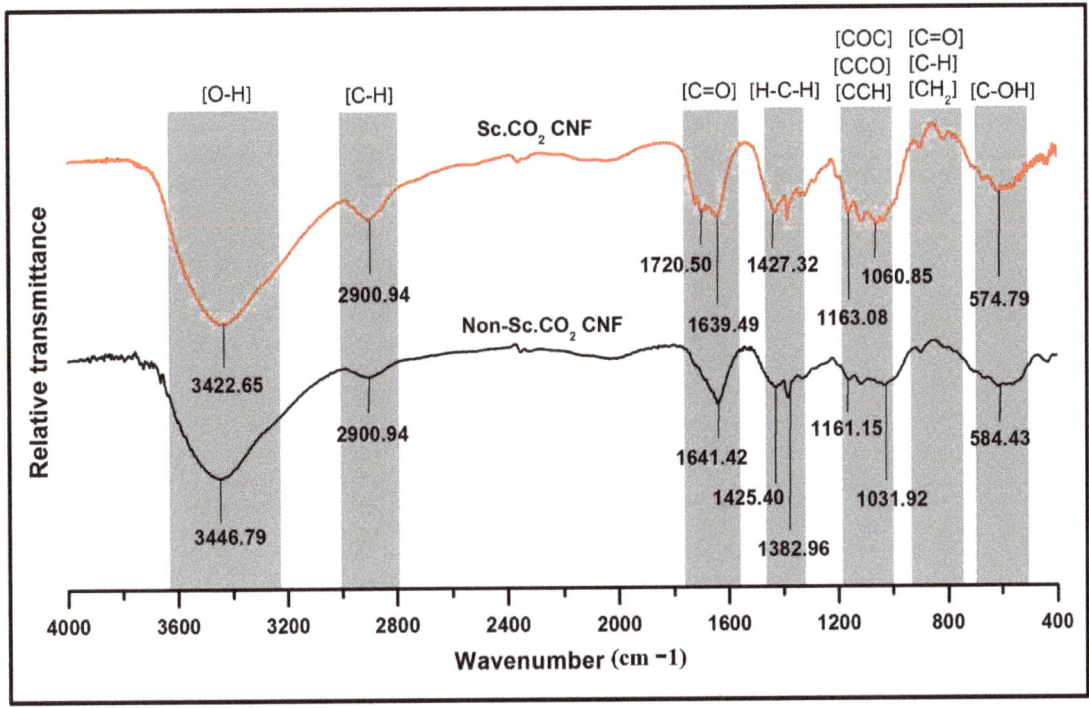

Figure 3. FT-IR spectra of Sc.CO$_2$ and non-Sc.CO$_2$ obtained CNFs.

The peak intensity in non-supercritical obtained CNFs more sharpens than supercritical ones, attributing to the high OH concentration, which could be produced from hydrogen bonds breakage in cellulose hydroxyl groups [32]—the peak at 1720 cm^{-1} in Sc.CO$_2$ obtained CNFs correspond to carbonyl shoulder (C–O), also found in lignin and hemicellulose. Sc.CO$_2$ treatment of bleached fibers could affect the functional groups at this spectrum, resulting in the removal of some carboxylic groups and thus resulting in higher intensity. This can also be observed among the peaks of each alkane (CH$_2$) ether and carbonyl(C–O) group; the Sc.CO$_2$ that obtained CNFs appeared in greater intensity. The decreasing intensity of the peaks at 1225–1250 cm^{-1} suggests the effective removal of lignin and hemicellulose, as this peak attributes to –CO stretching and syringyl ring [33]. Table 3 summarizes the peak location, shape, and size of all detected CNFs chemical functional groups.

Table 3. Summary of the peak location, shape, and size of the infrared bands of the main CNFs chemical functional groups obtained from FTIR analysis.

Wavenumber (cm^{-1})	Band Assignments	Peak Shape/Size	Remarks
3800–3000	Hydroxyl group (OH)	Very broad	Sc.CO$_2$ had greater intensity
2900–2700	Methyl group (CH)	Small	A larger peak in Sc.CO$_2$
2350	Carbon dioxide (COO)	Very small	Similar in both samples
1720	Carbonyl Shoulder (C–O)	Very small	Only in Sc.CO$_2$ obtained CNFs
1641–1639	Aldehyde group (C=O)	Broad and sharp	Sharper in non-Sc.CO$_2$
1427–1425	Alkane group (CH$_2$)	Tiny peak	The greater intensity in Sc.CO$_2$
1382	Alkane group (C–H)	Tiny peak	Only in non-Sc.CO$_2$ obtained CNFs
1163–1161	Ether group (C–O–C)	Small and sharp	The greater intensity in Sc.CO$_2$
1060–1031	Carbonyl group (C–O)	Medium and wide	Wider and greater in Sc.CO$_2$
891	Methyl group (C–H)	Very tiny shoulder	Similar in both samples
584–574	Carboxyl group (C–OH)	Small and wide	Wider and greater in non-Sc.CO$_2$

Cellulosic materials are known for their thermal sensitivity, which degrades at low to moderate temperatures [34]. Figure 4 presents the thermal degradation of Sc.CO$_2$ and non-Sc.CO$_2$ obtained CNFs. Generally, Sc.CO$_2$ obtained CNFs had better thermal stability than non-Sc.CO$_2$ obtained CNFs; it showed a T$_{onset}$ and T$_{max}$ of 312.68 and 343.33 °C, respectively, compared with non-Sc.CO$_2$ obtained CNFs (285.19 and 319.83 °C). At a low-temperature range of below 100 °C (phase I), evaporation of moisture, physisorbed water, and volatile compounds occurs, resulting in a slight weight loss of approximately 10% [35]. At this stage, the mass loss of non-Sc.CO$_2$ obtained CNFs was 7% higher than the Sc.CO$_2$ treated CNFs. However, with the increase in temperature (150–500 °C) (phase II) and the primary degradation of cellulosic fiber, major weight loss happened in both samples, with a significant difference; the mass loss of Sc.CO$_2$ obtained CNFs was lower by more than 27% than in non-treated fibers. The results suggest that supercritical could induce architecture changes in cellulose, making them more stable at such temperatures [36]. Phase III occurred at a higher temperature range (500 °C to 800 °C), decomposition of carbonaceous materials in the fiber sample occurs, leading to minor weight loss. The thermo-gravimetric analysis confirms the chemical composition results, as Sc.CO$_2$ obtained CNFs contain less lignin than non-Sc.CO$_2$ obtained CNFs. In general, lignin is characterized by slow weight loss over a broader temperature than cellulose and hemicellulose [37]. In derivative thermo-gravimetric curve of non-Sc.CO$_2$ obtained CNFs, a low temperature a shoulder can be seen, which could be attributed to higher hemicellulose content compared with Sc.CO$_2$ obtained CNFs [38,39].

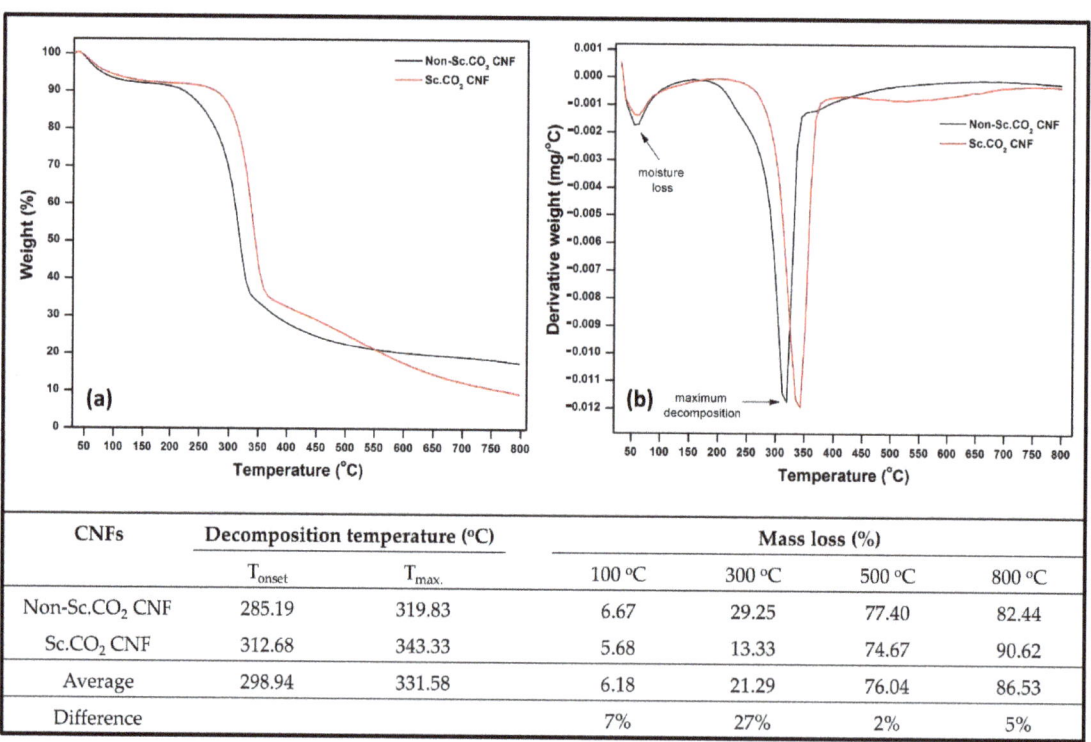

CNFs	Decomposition temperature (°C)		Mass loss (%)			
	T$_{onset}$	T$_{max.}$	100 °C	300 °C	500 °C	800 °C
Non-Sc.CO$_2$ CNF	285.19	319.83	6.67	29.25	77.40	82.44
Sc.CO$_2$ CNF	312.68	343.33	5.68	13.33	74.67	90.62
Average	298.94	331.58	6.18	21.29	76.04	86.53
Difference			7%	27%	2%	5%

Figure 4. Thermogravimetric analysis of obtained CNFs (**a**) TGA, (**b**) DTG, and decomposition temperature and mass loss data of the CNFs.

3.4. Particle Size Distribution and Surface Charge Analysis

The results of particle size distributions, surface charge analysis, and TEM images of Sc.CO_2 and non-Sc.CO_2 obtained CNFs are presented in Figure 5. Upon calculating the particle size, the Malvern Nano-ZS90 is known to assume spherical particles, which affected the general diameter of both samples. The average fiber length and diameter of Sc.CO_2 obtained CNFs was 53.72 and 7.14 nm, respectively, compared with non-Sc.CO_2 obtained CNFs, which had longer fiber length and diameter (302.87 and 97.93 nm). The uniform particle size of Sc.CO_2 obtained CNFs suggests the effectiveness of Sc.CO_2 treatment in defragmentation and breaking the fibers into smaller pieces. Furthermore, Ss.CO_2-assisted acid hydrolysis process to extract nano-size fibers from the bleached raw fibers. The lignocellulosic structure during the supercritical conditions may undergo chemical decomposition, as reported earlier [40]. Plant cellulose is often surrounded by a plethora of hemicellulose and lignin fibers, making them less susceptible to acid hydrolysis [30]. However, this study's morphological and particle size results confirm the effectiveness of supercritical in assisting the hydrolysis process.

Zeta potential test was used to estimate the surface charge for Sc.CO_2 and non-Sc.CO_2 obtained CNFs by tracking the rising rate of charged particles (positively or negatively charged) across an electric field. It can be seen in Figure 4g,h, that the high negative charge of Sc.CO_2 obtained CNFs -38.9 ± 5.11 mV compared with non-Sc.CO_2 obtained CNFs that had a negative value of -33.1 ± 2.99 mV. The higher negative value of supercritical treated fibers came from the oxalate groups formed during the acid hydrolysis. Due to the presence of hydroxyl and carboxyl functional groups, cellulosic surfaces possess a bipolar character with a predominant acidic contribution [41]. Hence, Sc.CO_2 assesses the hydrolysis by increasing the susceptibility of the fibers. Thus, treated fibers recorded higher negative zeta potential values. The absolute value of the zeta potential for both samples was higher than -25 mV, which refers to their stability, as reported by Abraham et al. [41].

3.5. Crystallinity

X-ray diffractometry (XRD) was used to determine the crystallinity of Sc.CO_2 and non-Sc.CO_2 obtained CNFs as presented in Figure 6. Generally, material crystallinity is an essential factor that determines its thermal and mechanical properties [42]. However, the intermolecular hydrogen bonding among the hydroxyl groups within the CNFs appears as a perfect crystalline packing. The excellent mechanical properties of CNF were due to hydrogen bonding within the CNFs [43]. From Figure 6, it can be seen that the crystallinity index between the Sc.CO_2 and non-Sc.CO_2 obtained CNFs was slightly different. However, the two samples showed similar diffraction peaks at $2\theta = 16.1°$ and $2\theta = 16.3°$ for Sc.CO_2 and non-Sc.CO_2 obtained CNFs, respectively. It represents the typical diffraction peaks of cellulose type I [40]. The crystallinity index of Sc.CO_2 obtained CNFs in the present study was also found to be 80.5%, which was higher than that of non-Sc.CO_2 obtained CNFs (75.5%). It is well known that CNFs typically consist of two regions: amorphous region and crystalline region. However, it has been reported that crystallinity of CNFs ranging from 40 to 80%, depending on the source of cellulose and the isolation approach [44]. The infiltration of Sc.CO_2 into the fibers caused swelling, leading to re-arrangements and re-crystallization of molecular chains and shifting the characteristic bands. The Sc.CO_2 breaks the remaining lignin structure in the fibers through hydrolysis, as confirmed by chemical composition and molecular bond vibration analysis. The breakdown of lignin caused a plasticization effect in the supercritical state of the CNFs. The increased crystallinity index of Sc.CO_2 obtained CNFs resulted from the increasing mobility of macro-molecular chains, which can induce their re-arrangement under supercritical conditions compared to non-supercritical conditions.

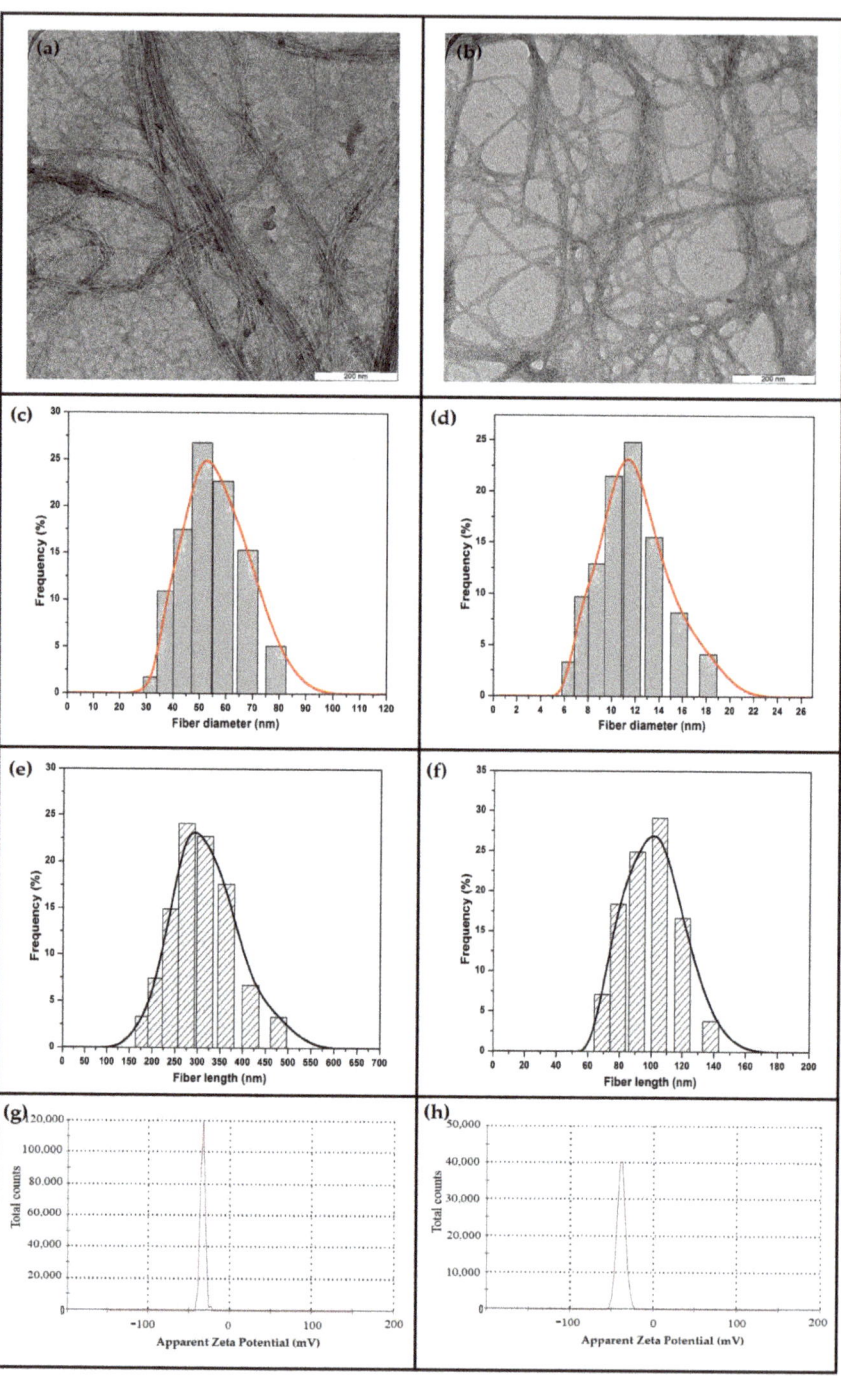

Figure 5. Particle size distribution and Surface charge analysis for non-Sc.CO$_2$ and Sc.CO$_2$ obtained CNFs respectively: (**a,b**) TEM micrograph, (**c,d**) fiber diameter, (**e,f**) fiber length, (**g,h**) zeta potential distribution.

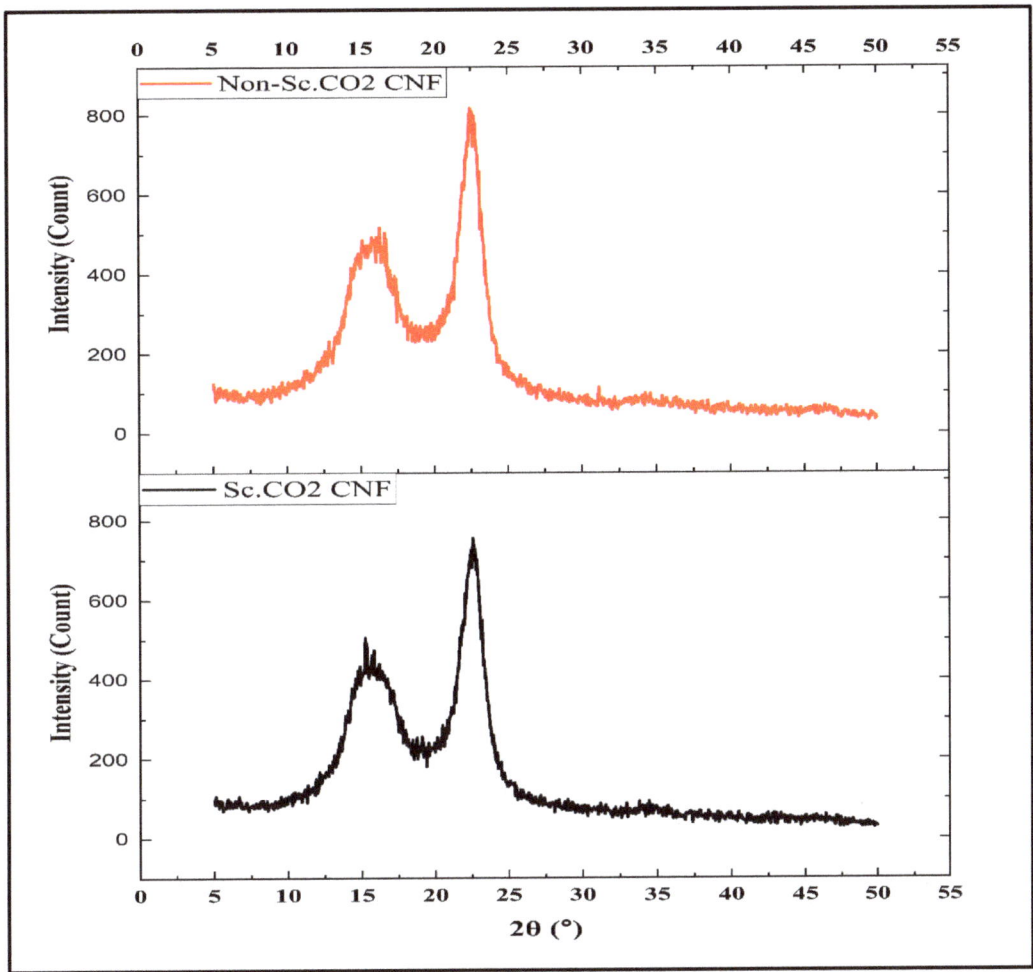

Figure 6. X-ray diffractograms of Sc.CO$_2$ and non-Sc.CO$_2$ obtained CNFs.

4. Conclusions

Cellulose nano fibers were successfully isolated from natural fiber carpet wastes using Sc.CO$_2$ and non-Sc.CO$_2$ approaches. Comparing the two approaches, it can be observed that the quality of CNFs isolated from the Sc.CO$_2$ approach was better than non-Sc.CO$_2$ approaches. The Sc.CO$_2$ treatment significantly enhanced their fragmentation and generated smaller particles than untreated fibers. The combination of mild heat, CO$_2$, and pressure caused depolymerization of cellulosic fibers due to cleavage of acetal bonds, which facilitated the acid hydrolysis process and generating smaller CNFs. The Sc.CO$_2$ helps in the cleavage of the remaining lignin structure in the fibers through hydrolysis, causing a plasticization effect in the supercritical state and a higher crystallinity index. The Sc.CO$_2$ obtained CNFs possess more thermal stability, better chemical composition and higher zeta potential value. The fiber diameter and length of Sc.CO$_2$ obtained CNFs was smaller than it in non-Sc.CO$_2$ obtained CNFs. The Sc.CO$_2$ has the potential to be used as a green approach for enhancing the isolation of CNFs and producing better quality nanocellulose for advanced applications that require high performance materials.

Author Contributions: Conceptualization, H.P.S.A.K., E.B.Y. and H.N.; Data curation, E.B.Y., T.A. and C.K.A.; Formal analysis, H.P.S.A.K., M.A.S. and A.M.; Funding acquisition, H.P.S.A.K. and H.N.; Investigation, E.B.Y.; Methodology, H.P.S.A.K. and E.B.Y.; Project administration, H.P.S.A.K. and C.K.A.; Resources, H.N. and H.P.S.A.K.; Software, A.B.S., M.A.S. and C.K.A.; Supervision, H.P.S.A.K.; Validation, M.A.S. and T.A.; Writing—original draft, E.B.Y.; Writing—review and editing, E.B.Y. and H.P.S.A.K. All authors have read and agreed to the published version of the manuscript.

Funding: This research was funded by the Ministry of Higher Education, Fundamental Research Grant Scheme—Malaysia's Research Star Award (FRGS-MRSA) with Project Code: FRGS/1/2019/TK05/USM/01/6.

Institutional Review Board Statement: Not applicable.

Informed Consent Statement: Not applicable.

Data Availability Statement: The data presented in this study are available on request from the corresponding author.

Acknowledgments: The authors would like to thank the collaboration between the Ministry of Higher Education, Fundamental Research Grant Scheme—Malaysia's Research Star Award (FRGS-MRSA) with Project Code: FRGS/1/2019/TK05/USM/01/6, Universitas Sumatera Utara, Medan, Indonesia, Universiti Pendidikan Sultan Idris, Perak, Malaysia, and Universiti Sains Malaysia, Penang, Malaysia that has made this work possible.

Conflicts of Interest: The authors declare no conflict of interest.

References

1. Abitbol, T.; Rivkin, A.; Cao, Y.; Nevo, Y.; Abraham, E.; Ben-Shalom, T.; Lapidot, S.; Shoseyov, O. Nanocellulose, a tiny fiber with huge applications. *Curr. Opin. Biotechnol.* **2016**, *39*, 76–88. [CrossRef] [PubMed]
2. Abdul Khalil, H.P.S.; Adnan, A.; Yahya, E.B.; Olaiya, N.; Safrida, S.; Hossain, M.; Balakrishnan, V.; Gopakumar, D.A.; Abdullah, C.; Oyekanmi, A. A Review on plant cellulose nanofibre-based aerogels for biomedical applications. *Polymers* **2020**, *12*, 1759. [CrossRef] [PubMed]
3. Yue, C.; Li, M.; Liu, Y.; Fang, Y.; Song, Y.; Xu, M.; Li, J. Three-dimensional Printing of Cellulose Nanofibers Reinforced PHB/PCL/Fe3O4 Magneto-responsive Shape Memory Polymer Composites with Excellent Mechanical Properties. *Addit. Manuf.* **2021**, *46*, 102146. [CrossRef]
4. Nishimura, T.; Shinonaga, Y.; Nagaishi, C.; Imataki, R.; Takemura, M.; Kagami, K.; Abe, Y.; Harada, K.; Arita, K. Effects of powdery cellulose nanofiber addition on the properties of glass ionomer cement. *Materials* **2019**, *12*, 3077. [CrossRef]
5. Katouah, H.A.; El-Sayed, R.; El-Metwaly, N.M. Solution blowing spinning technology and plasma-assisted oxidation-reduction process toward green development of electrically conductive cellulose nanofibers. *Environ. Sci. Pollut. Res.* **2021**, *28*, 56363–56375. [CrossRef]
6. Hassabo, A.G.; Mohamed, A.L.; Khattab, T.A. Preparation of cellulose-based electrospun fluorescent nanofibres doped with perylene encapsulated in silica nanoparticles for potential flexible electronics. *Luminescence* **2021**. [CrossRef]
7. Gupta, G.K.; Shukla, P. Lignocellulosic biomass for the synthesis of nanocellulose and its eco-friendly advanced applications. *Front. Chem.* **2020**, *8*, 1203. [CrossRef]
8. Yahya, E.B.; Amirul, A.; Abdul Khalil, H.P.S.; Olaiya, N.G.; Iqbal, M.O.; Jummaat, F.; AK, A.S.; Adnan, A. Insights into the Role of Biopolymer Aerogel Scaffolds in Tissue Engineering and Regenerative Medicine. *Polymers* **2021**, *13*, 1612. [CrossRef]
9. Song, J.; Chen, C.; Yang, Z.; Kuang, Y.; Li, T.; Li, Y.; Huang, H.; Kierzewski, I.; Liu, B.; He, S. Highly compressible, anisotropic aerogel with aligned cellulose nanofibers. *ACS Nano* **2018**, *12*, 140–147. [CrossRef]
10. Abdul Khalil, H.P.S.; Jummaat, F.; Yahya, E.B.; Olaiya, N.; Adnan, A.; Abdat, M.; NAM, N.; Halim, A.S.; Kumar, U.; Bairwan, R. A review on micro-to nanocellulose biopolymer scaffold forming for tissue engineering applications. *Polymers* **2020**, *12*, 2043. [CrossRef]
11. Miraftab, M.; Mirzababaei, M. Carpet Waste Utilisation, an Awakening realisation: A review. In *Conference Contribution*; CQUniversity: Melbourne, Australia, 2009.
12. Rizal, S.; Olaiya, F.G.; Saharudin, N.; Abdullah, C.; NG, O.; Mohamad Haafiz, M.; Yahya, E.B.; Sabaruddin, F.; Abdul Khalil, H.P.S. Isolation of textile waste cellulose nanofibrillated fibre reinforced in polylactic acid-chitin biodegradable composite for green packaging application. *Polymers* **2021**, *13*, 325. [CrossRef] [PubMed]
13. Juliana, A.; Aisyah, H.; Paridah, M.; Adrian, C.; Lee, S. Kenaf fiber: Structure and properties. In *Kenaf Fibers and Composites*; CRC Press: Boca Raton, FL, USA, 2018; pp. 23–36.
14. Kang, S.; Hou, S.; Chen, X.; Yu, D.-G.; Wang, L.; Li, X.; Williams, G.R. Energy-saving electrospinning with a concentric teflon-core rod spinneret to create medicated nanofibers. *Polymers* **2020**, *12*, 2421. [CrossRef] [PubMed]
15. Lv, H.; Guo, S.; Zhang, G.; He, W.; Wu, Y.; Yu, D.-G. Electrospun Structural Hybrids of Acyclovir-Polyacrylonitrile at Acyclovir for Modifying Drug Release. *Polymers* **2021**, *13*, 4286. [CrossRef] [PubMed]

16. Song, Y.; Jiang, W.; Zhang, Y.; Ben, H.; Han, G.; Ragauskas, A.J. Isolation and characterization of cellulosic fibers from kenaf bast using steam explosion and Fenton oxidation treatment. *Cellulose* **2018**, *25*, 4979–4992. [CrossRef]
17. Pennells, J.; Godwin, I.D.; Amiralian, N.; Martin, D.J. Trends in the production of cellulose nanofibers from non-wood sources. *Cellulose* **2020**, *27*, 575–593. [CrossRef]
18. Yu, D.G.; Wang, M.; Ge, R. Strategies for sustained drug release from electrospun multi-layer nanostructures. *Wiley Interdiscip. Rev. Nanomed. Nanobiotechnology* **2021**, e1772. [CrossRef]
19. Menon, M.P.; Selvakumar, R.; Ramakrishna, S. Extraction and modification of cellulose nanofibers derived from biomass for environmental application. *RSC Adv.* **2017**, *7*, 42750–42773. [CrossRef]
20. Yang, W.; Feng, Y.; He, H.; Yang, Z. Environmentally-friendly extraction of cellulose nanofibers from steam-explosion pretreated sugar beet pulp. *Materials* **2018**, *11*, 1160. [CrossRef]
21. Zhang, X.; Huang, H.; Qing, Y.; Wang, H.; Li, X. A comparison study on the characteristics of nanofibrils isolated from fibers and parenchyma cells in bamboo. *Materials* **2020**, *13*, 237. [CrossRef]
22. Rizal, S.; Yahya, E.B.; Abdul Khalil, H.P.S.; Abdullah, C.; Marwan, M.; Ikramullah, I.; Muksin, U. Preparation and Characterization of Nanocellulose/Chitosan Aerogel Scaffolds Using Chemical-Free Approach. *Gels* **2021**, *7*, 246. [CrossRef] [PubMed]
23. Seghini, M.C.; Touchard, F.; Chocinski-Arnault, L.; Placet, V.; François, C.; Plasseraud, L.; Bracciale, M.P.; Tirillò, J.; Sarasini, F. Environmentally friendly surface modification treatment of flax fibers by supercritical carbon dioxide. *Molecules* **2020**, *25*, 438. [CrossRef] [PubMed]
24. Baldino, L.; Cardea, S.; Reverchon, E. Supercritical Phase Inversion: A Powerful Tool for Generating Cellulose Acetate-AgNO$_3$ Antimicrobial Membranes. *Materials* **2020**, *13*, 1560. [CrossRef] [PubMed]
25. Atiqah, M.; Gopakumar, D.A.; FAT, O.; Pottathara, Y.B.; Rizal, S.; Aprilia, N.; Hermawan, D.; Paridah, M.; Thomas, S.; HPS, A.K. Extraction of cellulose nanofibers via eco-friendly supercritical carbon dioxide treatment followed by mild acid hydrolysis and the fabrication of cellulose nanopapers. *Polymers* **2019**, *11*, 1813. [CrossRef]
26. Li, L.; Kiran, E. Interaction of supercritical fluids with lignocellulosic materials. *Ind. Eng. Chem. Res.* **1988**, *27*, 1301–1312. [CrossRef]
27. Schmidt, A.; Bach, E.; Schollmeyer, E. Damage to natural and synthetic fibers treated in supercritical carbon dioxide at 300 bar and temperatures up to 160 °C. *Text. Res. J.* **2002**, *72*, 1023–1032. [CrossRef]
28. Wise, L.E. Chlorite holocellulose, its fractionation and bearing on summative wood analysis and on studies on the hemicelluloses. *Pap. Trade* **1946**, *122*, 35–43.
29. Serna, L.D.; Alzate, C.O.; Alzate, C.C. Supercritical fluids as a green technology for the pretreatment of lignocellulosic biomass. *Bioresour. Technol.* **2016**, *199*, 113–120. [CrossRef]
30. Karimi, S.; Tahir, P.M.; Karimi, A.; Dufresne, A.; Abdulkhani, A. Kenaf bast cellulosic fibers hierarchy: A comprehensive approach from micro to nano. *Carbohydr. Polym.* **2014**, *101*, 878–885. [CrossRef]
31. François, C.; Placet, V.; Beaugrand, J.; Pourchet, S.; Boni, G.; Champion, D.; Fontaine, S.; Plasseraud, L. Can supercritical carbon dioxide be suitable for the green pretreatment of plant fibres dedicated to composite applications? *J. Mater. Sci.* **2020**, *55*, 4671–4684. [CrossRef]
32. Chieng, B.W.; Lee, S.H.; Ibrahim, N.A.; Then, Y.Y.; Loo, Y.Y. Isolation and characterization of cellulose nanocrystals from oil palm mesocarp fiber. *Polymers* **2017**, *9*, 355. [CrossRef]
33. Zheng, D.; Zhang, Y.; Guo, Y.; Yue, J. Isolation and characterization of nanocellulose with a novel shape from walnut (*Juglans regia* L.) shell agricultural waste. *Polymers* **2019**, *11*, 1130. [CrossRef]
34. Hajaligol, M.; Waymack, B.; Kellogg, D. Low temperature formation of aromatic hydrocarbon from pyrolysis of cellulosic materials. *Fuel* **2001**, *80*, 1799–1807. [CrossRef]
35. Pasangulapati, V.; Ramachandriya, K.D.; Kumar, A.; Wilkins, M.R.; Jones, C.L.; Huhnke, R.L. Effects of cellulose, hemicellulose and lignin on thermochemical conversion characteristics of the selected biomass. *Bioresour. Technol.* **2012**, *114*, 663–669. [CrossRef] [PubMed]
36. Cherian, B.M.; Leão, A.L.; De Souza, S.F.; Thomas, S.; Pothan, L.A.; Kottaisamy, M. Isolation of nanocellulose from pineapple leaf fibres by steam explosion. *Carbohydr. Polym.* **2010**, *81*, 720–725. [CrossRef]
37. Spinace, M.A.; Lambert, C.S.; Fermoselli, K.K.; De Paoli, M.-A. Characterization of lignocellulosic curaua fibres. *Carbohydr. Polym.* **2009**, *77*, 47–53. [CrossRef]
38. Rosa, M.; Medeiros, E.; Malmonge, J.; Gregorski, K.; Wood, D.; Mattoso, L.; Glenn, G.; Orts, W.; Imam, S. Cellulose nanowhiskers from coconut husk fibers: Effect of preparation conditions on their thermal and morphological behavior. *Carbohydr. Polym.* **2010**, *81*, 83–92. [CrossRef]
39. Yi, T.; Zhao, H.; Mo, Q.; Pan, D.; Liu, Y.; Huang, L.; Xu, H.; Hu, B.; Song, H. From cellulose to cellulose nanofibrils—A comprehensive review of the preparation and modification of cellulose nanofibrils. *Materials* **2020**, *13*, 5062. [CrossRef] [PubMed]
40. Gopakumar, D.A.; Pasquini, D.; Henrique, M.A.; de Morais, L.C.; Grohens, Y.; Thomas, S. Meldrum's acid modified cellulose nanofiber-based polyvinylidene fluoride microfiltration membrane for dye water treatment and nanoparticle removal. *ACS Sustain. Chem. Eng.* **2017**, *5*, 2026–2033. [CrossRef]
41. Abraham, E.; Deepa, B.; Pothen, L.; Cintil, J.; Thomas, S.; John, M.J.; Anandjiwala, R.; Narine, S. Environmental friendly method for the extraction of coir fibre and isolation of nanofibre. *Carbohydr. Polym.* **2013**, *92*, 1477–1483. [CrossRef] [PubMed]

42. Mokhena, T.C.; Sadiku, E.R.; Mochane, M.J.; Ray, S.S.; John, M.J.; Mtibe, A. Mechanical properties of cellulose nanofibril papers and their bionanocomposites: A review. *Carbohydr. Polym.* **2021**, *273*, 118507. [CrossRef]
43. Iwamoto, S.; Isogai, A.; Iwata, T. Structure and mechanical properties of wet-spun fibers made from natural cellulose nanofibers. *Biomacromolecules* **2011**, *12*, 831–836. [CrossRef] [PubMed]
44. Usmani, M.; Khan, I.; Gazal, U.; Haafiz, M.M.; Bhat, A. Interplay of polymer bionanocomposites and significance of ionic liquids for heavy metal removal. In *Polymer-Based Nanocomposites for Energy and Environmental Applications*; Elsevier: Amsterdam, The Netherlands, 2018; pp. 441–463.

Article

Forensic Analysis of Polymeric Carpet Fibers Using Direct Analysis in Real Time Coupled to an AccuTOF™ Mass Spectrometer

Torki A. Zughaibi [1,2,*] and Robert R. Steiner [3]

1 Department of Medical Laboratory Technology, Faculty of Applied Medical Sciences, King Abdulaziz University, P.O. Box 80216, Jeddah 21589, Saudi Arabia
2 King Fahad Medical Research Center, King Abdulaziz University, P.O. Box 80216, Jeddah 21589, Saudi Arabia
3 Virginia Department of Forensic Science, Richmond, VA 23219-1416, USA; robert.steiner@dfs.virginia.gov
* Correspondence: taalzughaibi@kau.edu.sa

Abstract: Polymeric fibers are encountered in numerous forensic circumstances. This study focused on polymeric carpet fibers most encountered at a crime scene, which are nylons, polyesters and olefins. Analysis of the multiple polymer types was done using Direct Analysis in Real Time (DART™) coupled to an Accurate time-of-flight (AccuTOF™) mass spectrometer (MS). A DART gas temperature of 275 °C was determined as optimal. Twelve olefin, polyester, and nylon polymer standards were used for parameter optimization for the carpet fiber analysis. A successful identification and differentiation of all twelve polymer standards was completed using the DART-AccuTOF™. Thirty-two carpet samples of both known and unknown fiber composition were collected and subsequently analyzed. All samples with known fiber compositions were correctly identified by class. All of the remaining carpet samples with no known composition information were correctly identified by confirmation using Fourier-transform infrared spectroscopy (FTIR). The method was also capable of identifying sub-classes of nylon carpet fibers. The results exhibit the capability of DART-AccuTOF™ being applied as an addition to the sequence of tests conducted to analyze carpet fibers in a forensic laboratory.

Keywords: polymer; carpet fiber; direct analysis in real time; time of flight; mass spectrometry; function switching; oleamide

1. Introduction

Unlike DNA and fingerprints, fibers are associative evidence in that they establish a link between a person and a crime scene, object, or another person. Fibers are encountered as evidence in many different crimes such as hit and run, breaking, and entering, and rape. This exchange of material is a result of Locard's exchange principle, which states that when two objects are in contact with one another, a transfer of material will occur.

Nylons, also referred to as polyamides, are a group of synthetic polymers that are used to manufacture rope, fabric, carpet fibers, and more [1] Various kinds of nylons are differentiated based on their synthesis [2]. A few examples of the various nylons include nylons 6, 6/6, 11, etc. The number(s) following the nylon indicates the number of carbon atoms that originated from the dicarboxylic acid and diamine groups that form the different molecular structures [3].

Polyesters are also synthetically produced polymers used for apparel, textiles, and carpeting. In terms of amount produced and shipped to customers, polyesters rank second only to cotton. There are many different polyesters such as polyethylene terephthalate (PET), polybutylene terephthalate (PBT), and Poly (1,4,-cyclohexylenedimethylene terephthalate), (PCDT). PET is found more abundantly because it is the polymer of choice in over 95% of all polyester fibers manufactured [4,5]. Another kind of polyester, olefin fibers are known for to be tough fibers that have high resistance to abrasion, thus affording good application in carpet fibers [6]. Ethylene and propylene are the only two saturated hydrocarbons that serve to synthesize polyolefin for fibers [5].

The researchers have applied DART to a wide selection of problems of interest to forensic analytical chemistry, but the "maturity" of DART varies considerably across the field. DART-MS can analyze surfaces without extraction steps and without damaging the surface; it can be used to study substrates not ordinarily amenable to analytical methods, including plants and plant-derived materials. In 2011, Adams published a study using DART-MS to analyze paper, noting that the chemical composition of the pulp and additives present in paper can help to determine its age, aiding in determining its historical authenticity [7]. To analyze the chemical composition of fibers, Direct Analysis in Real Time (DART™) coupled to an accurate time-of-flight (AccuTOF™) mass spectrometer (MS) was utilized.

This involves an ambient ionization ion source which requires little-to-no sample preparation [8,9]. DART has been authorized for analyzing many types of samples encountered at a forensic lab ranging from accelerants and explosives, to inks and controlled substances [10–13]. The ability of DART to test solids, liquids, and gases (via headspace) with no prior sample preparation makes this technique ideal for the analysis of carpet fibers. Forensic analysis of carpet fibers involves using microscopic examinations, a sequence of microchemical tests, and either Fourier-transform infrared spectroscopy (FTIR) or pyrolysis-gas chromatography (pyrolysis-GC). While FTIR is capable of identifying polymer subtypes, it involves a more tedious sample preparation step when compared to this method. Whereas with the destructive techniques using pyrolysis-GC and microchemical tests, their main disadvantage compared to the method described in this study is they both take more time for analysis. With the DART, several samples, including a calibration standard, can be identified and analyzed within minutes. In this study, the capability of the DART-AccuTOF™ was tested to analyze carpet fiber samples by distinguishing the various polymer types and sub-types based on their chemical composition.

The novelty of the work lies in the DART-AccuTOF™ being capable of identifying the various polymer types in carpet fibers and could be implemented as one of the many series of tests conducted when analyzing fiber evidence to increase the power of discrimination. It could serve as a quick screening tool, especially in cases with many fibers.

2. Materials and Methods

2.1. Materials

The polyester standards used were PBT Poly (1, 4-butylene terephthalate) (Molecular weight of repeat unit: 220.23 g/mol.), PCDT Poly (1,4-cyclohexanedimethylene terephthalate) (Molecular weight of repeat unit: 226.32 g/mol.), and PET Poly (ethylene terephthalate) (Molecular weight of repeat unit: 192.2 g/mol) were obtained from Scientific Polymer Products, Inc. (Ontario, NY, USA) in a polymer sample kit. The olefin standards were isotactic polypropylene and polyethylene was purchased from Sigma-Aldrich (St. Louis, MO, USA). All standards were stored in glass vials and in pellet form (with the exception of polypropylene). Polypropylene was in a powder form, which was melted and allowed to solidify to facilitate sampling.

Characterization

Attenuated total reflection spectra in the range 4000–400 cm^{-1} of the polymers were measured with Fourier transform infrared (FTIR) spectroscopy (Nexus 8700, Thermo Nicolet, Madison, WI, USA) and Omnic software version number 9.1.26 (Thermo Fisher Scientific Inc., Waltham, MA, USA) and a uniform resolution of 2 cm^{-1} was maintained in case of polymers. Analysis was performed on a DART (IonSense, Inc. Saugus, MA, USA) ion source coupled to a JMS-T100LC AccuTOF™ (JEOL Inc., Peabody, MA, USA) mass spectrometer using modifications of a previously published method [14]. In brief, 2 mg/mL polyethylene glycol in methanol solution (PEG 600), was used for exact mass calibration. Calibration was done by dipping the sealed end of a capillary tube (Kimble Glass Company, Vineland, NJ, USA) in the PEG 600 calibration solution and "wanding" the capillary tube in the sample gap for a few seconds. JEOL MassCenter software version 1.3.4 m (JEOL Inc., Peabody, MA, USA) was used to gather and analyze the data. Each data file contained a calibration curve developed from the PEG 600 calibration standard and was subsequently applied to all polymer data collected.

2.2. Experimental

Thirty-two carpet samples of various blends and colors were collected from local hardware stores. Each sample was packaged separately in a plastic bag and labeled (Sample 1–32). Of the thirty-two samples, fifteen had their fiber compositions included on the label. The remaining seventeen samples were treated as unknowns. Of the fifteen known samples, only three specific fiber types were observed: nylon, polyester, and olefin. It is also common to encounter carpets that contain a blend of several of these fiber types.

To obtain optimum results, parameters were first established for the DART-AccuTOF for the different polymer standards. The major parameters of importance were (1) the temperature of the DART gas stream and (2) the orifice 1 voltage of the AccuTOF™. An initial range of 200–300 °C gas temperature was chosen because of the melting points of the polymers shown in Table 1.

Table 1. Manufacturer provided melting points of the various polymer standards used in this study (Scientific Polymer Products, Inc., Ontario, NY, USA).

Polymer	Melting Point °C
Nylon 6	220
Nylon 6/6	254
Nylon 6/9	210
Nylon 6/10	217
Nylon 6/12	250–260
Nylon 11	185
Nylon 12	178
PET	252
PBT	225
PCDT	218
Polypropylene	160
Polyethylene	121

PET: Polyethylene terephthalate, PBT: Polybutylene terephthalate, PCDT: Poly (1, 4-Cyclohexylene-Dimethylene terephthalate).

3. Results and Discussion

3.1. FTIR Spectroscopy

3.1.1. Nylon 6/6

IR spectroscopy has been usefully applied for identification of the basic structural units present in the chemical configuration of nylon-66. The FTIR spectra of the nylon-66 is presented in Figure 1. The complete vibrational band assignment is made available for the selected polymeric materials, thereby confirming their molecular structure. The vibrational frequencies of all the fundamental bands and probable assignments are given in Table 2. The primary motivation for determining the molecular structure of a polymer using FTIR spectroscopy is to confirm its presence in the carpet samples.

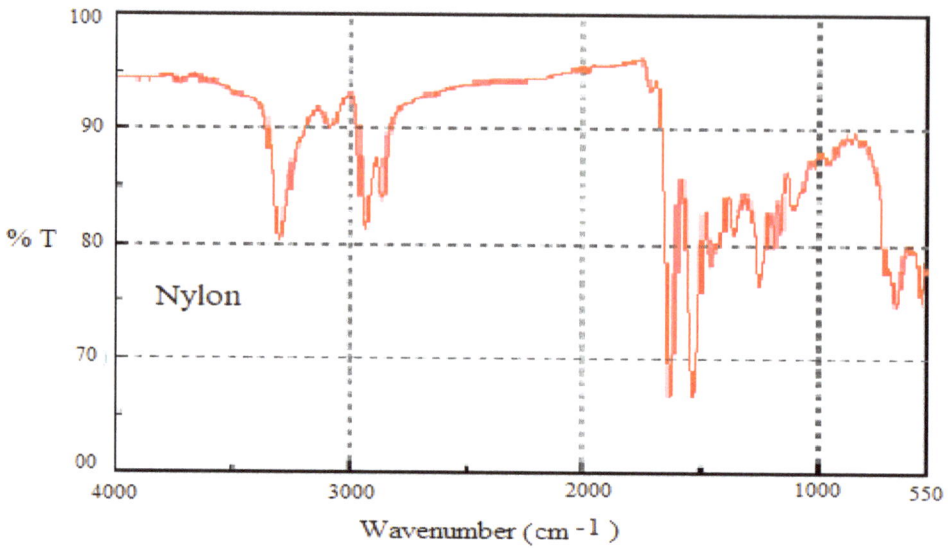

Figure 1. FTIR results for the Nylon 6/6.

Table 2. FTIR spectra and assignment of nylon-66.

Frequency, cm^{-1}	Assignment
3182	N-H stretching
3080	C-H asymmetric stretching
3020	C-H symmetric stretching
2841	CH$_2$ symmetric stretching
1745	C=O stretching
1660	Amide I band
1541	Amide II band
959	C-C stretching

3.1.2. Polyester

The peaks in the IR spectra of the polyester carpet fabric are shown in Figure 2 appeared in the range of 600–4000 cm^{-1}. The 1715 cm^{-1} shows C=O stretching vibration, 1409 cm^{-1} is attributed to aromatic ring, 1331 cm^{-1} and 1021 cm^{-1} shows carboxylic ester or anhydride, 1021 cm^{-1} indicates the presence of O=C–O–C or secondary alcohol, 967 cm^{-1} is of C=C stretching, and the 869 cm^{-1} peak shows five substituted H in benzene. The main structure of the polyester sample had ester, alcohol, anhydride, aromatic ring, and heterocyclic aromatic rings.

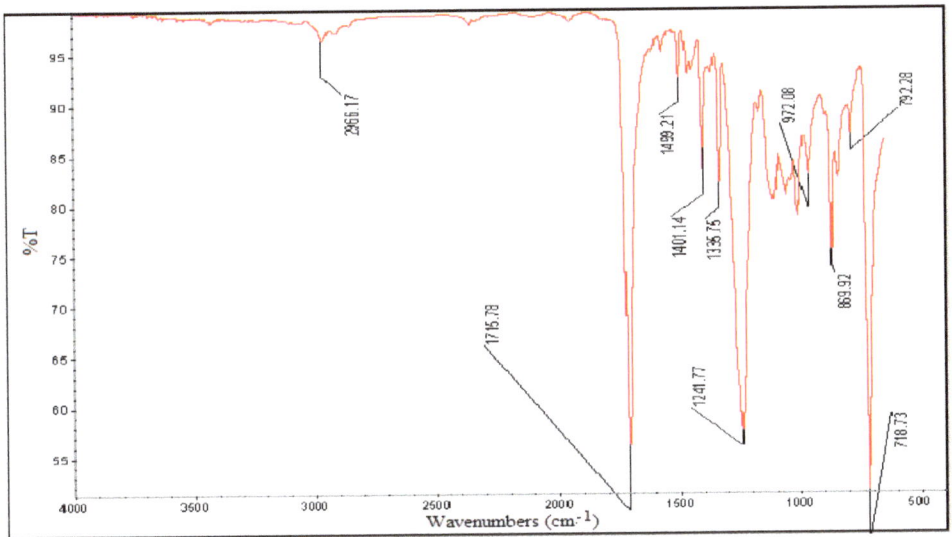

Figure 2. FTIR of Polyester.

3.1.3. Olefins

The major peaks for the olefins include C=C stretch from 1680–1640 cm^{-1}, =C–H stretch from 3100–3000 cm^{-1}, and =C–H bend from 1000–650 cm^{-1}.

The IR spectrum of olefins is shown in Figure 3. The band greater than 3000 cm^{-1} is attributed to the =C–H stretch and the several bands lower than 3000 cm^{-1} for –C–H stretch. The C = C stretch band is attributed to 1644 cm^{-1}. Bands for C–H scissoring (1465) and methyl rock (1378) are marked on this spectrum; in routine IR analysis, these bands are not specific to an alkene and are generally not noted because they are present in almost all organic molecules (and they are in the fingerprint region). The bands at 917 cm^{-1} and 1044 cm^{-1} are attributed to =C–H bends.

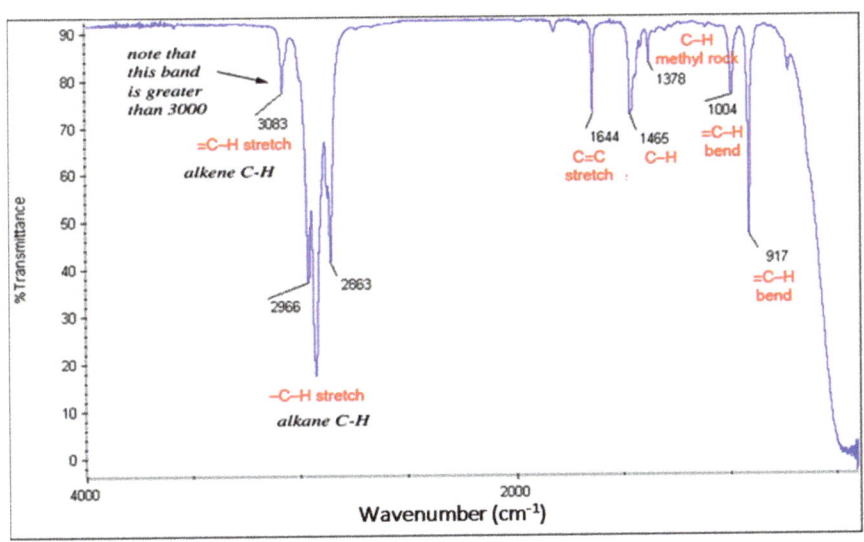

Figure 3. FTIR of Olefin.

3.2. Mass Spectra of Polymeric Samples

Parameters were selected to try and identify a temperature in which all polymer types could be identified. It was established that a helium gas temperature of 275 °C yielded the greatest compromise of response for the various polymer types. Figures 4 and 5 illustrate the total ion responses at temperatures ascending or descending in 25 °C intervals between 200–300 °C for PET and polypropylene, respectively.

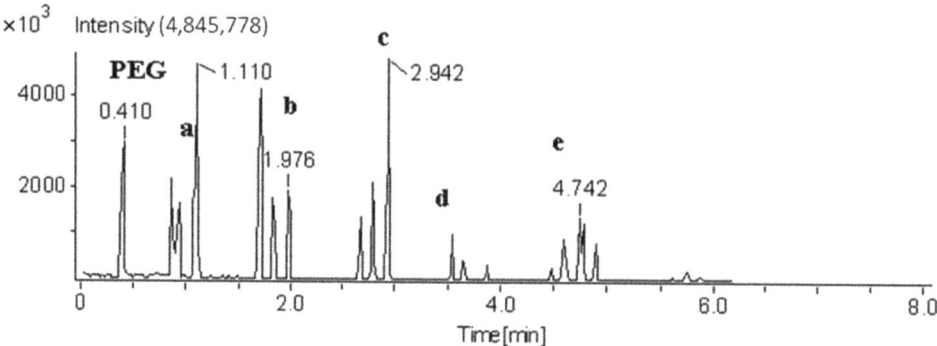

Figure 4. Total ion response of PET at temperatures: (**a**) 300 °C, (**b**) 275 °C, (**c**) 250 °C, (**d**) 225 °C, and (**e**) 200 °C.

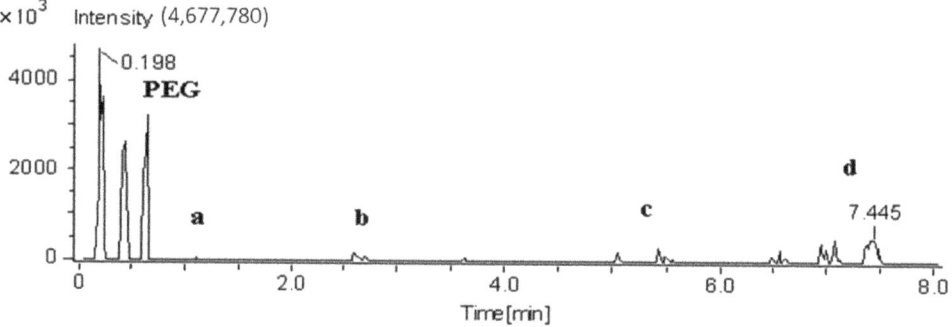

Figure 5. Total ion response of polypropylene at temperatures: (**a**) 200 °C, (**b**) 225 °C, (**c**) 250 °C, (**d**) 275 °C.

3.2.1. Nylon

When referring to Table 3, the protonated monomer [M+H] of nylon 6/6 is equal in weight to the nylon 6 protonated dimer [2M+H]. One potential cause for concern would have been the fragmentation of the nylon 6/6 monomer [M+H] into a 114.0919 m/z peak (equivalent to the nylon 6 monomer [M+H]). However, the protonated nylon 6/6 monomer [M+H] does not produce a fragment ion to a m/z equivalent to the nylon 6 protonated monomer [M+H], but rather into 100.1140 m/z that increases in intensity as the voltage of orifice 1 increases. In addition, as the orifice 1 voltage increases, the nylon 6 protonated dimer [2M+H] and trimer [3M+H] decreases in intensity, with the fragmentation resulting in a more abundant monomer [M+H]. While both the protonated monomer [M+H] and protonated dimer [2M+H] peaks of nylon 6/6 decrease in their abundances, a fragment ion is produced at m/z 100.1140, which in turn increases specificity among these similar yet different nylon types.

Table 3. Calculated exact masses of the monomers and dimers of nylons and polyesters, as well as their protonated exact masses.

Polymer	M	M+H	2M	2M+H
Nylon 6	113.0839	114.0919	226.1679	227.1759
Nylon 6/6	226.1679	227.1759	452.3361	453.3441
Nylon 6/9	268.2149	269.2229	536.43	537.4380
Nylon 6/10	282.2306	283.2386	564.4613	565.4693
Nylon 6/12	310.2618	311.2698	620.5239	621.5319
Nylon 11	183.1621	184.1701	366.3244	367.3324
Nylon 12	197.1778	198.1858	394.3558	395.3638
PET	192.0421	193.0501	384.0844	385.0924
PCDT	220.0734	221.0814	440.147	441.1550
PBT	274.1205	275.1285	548.2409	549.2489

Table 3 shows the differences in the exact masses for the different monomers of PBT, PCDT, and PET. The difference in mass spectra between the three polyester types is shown in Figure 6.

3.2.2. Olefin

The two olefins of interest proved to be particularly problematic in terms of identification. Because both polyethylene (C_2H_4) and polypropylene (C_3H_6) share the same empirical formula (C_nH_{2n}), many of their various units will have the same exact mass. Similar to the case with nylon 6 and 6/6, the protonated trimer of polyethylene has the same mass as the protonated dimer of polypropylene (m/z 85.1017). In the spectra of the two standards, an obvious difference can be seen (Figure 7).

In the polyethylene spectrum, peaks from the polymer backbone are observed ranging from approximately m/z 220–450. Each of those peaks was correctly identified as the different polyethylene monomers using the SearchFromList software. A difference of approximately 28 Da was observed between the repeat units of the polymer backbone, which is the mass of the polyethylene monomer [M] (C_2H_4 = 28 Da). However, with polypropylene, an expected difference of 42 Da (C_3H_6) between succeeding peaks was not observed. Although a noticeable difference in the spectra was observed, the majority of the peaks identified using SearchFromList were those that were common between both olefin types and thus could not be correctly identified as a polypropylene.

An interesting phenomenon occurred with the polypropylene standard. It is best explained when looking at the lower mass range in Figure 8. The differences observed between the "triplets" of peaks are a difference of 14 Da, which could theoretically be a methyl group. This is most likely due to the isotactic nature of the polypropylene standard. The m/z 85 peak is attributed to the protonated dimer [2M+H] of polypropylene [C_6H_{12} + H], and the m/z 99 peak is the protonated dimer of polypropylene plus a methyl group [C_6H_{12} + CH_2 + H]. Due to slightly different end group terminations, other protonated molecules are seen superimposed with the polypropylene dimer and other mers (Figure 8).

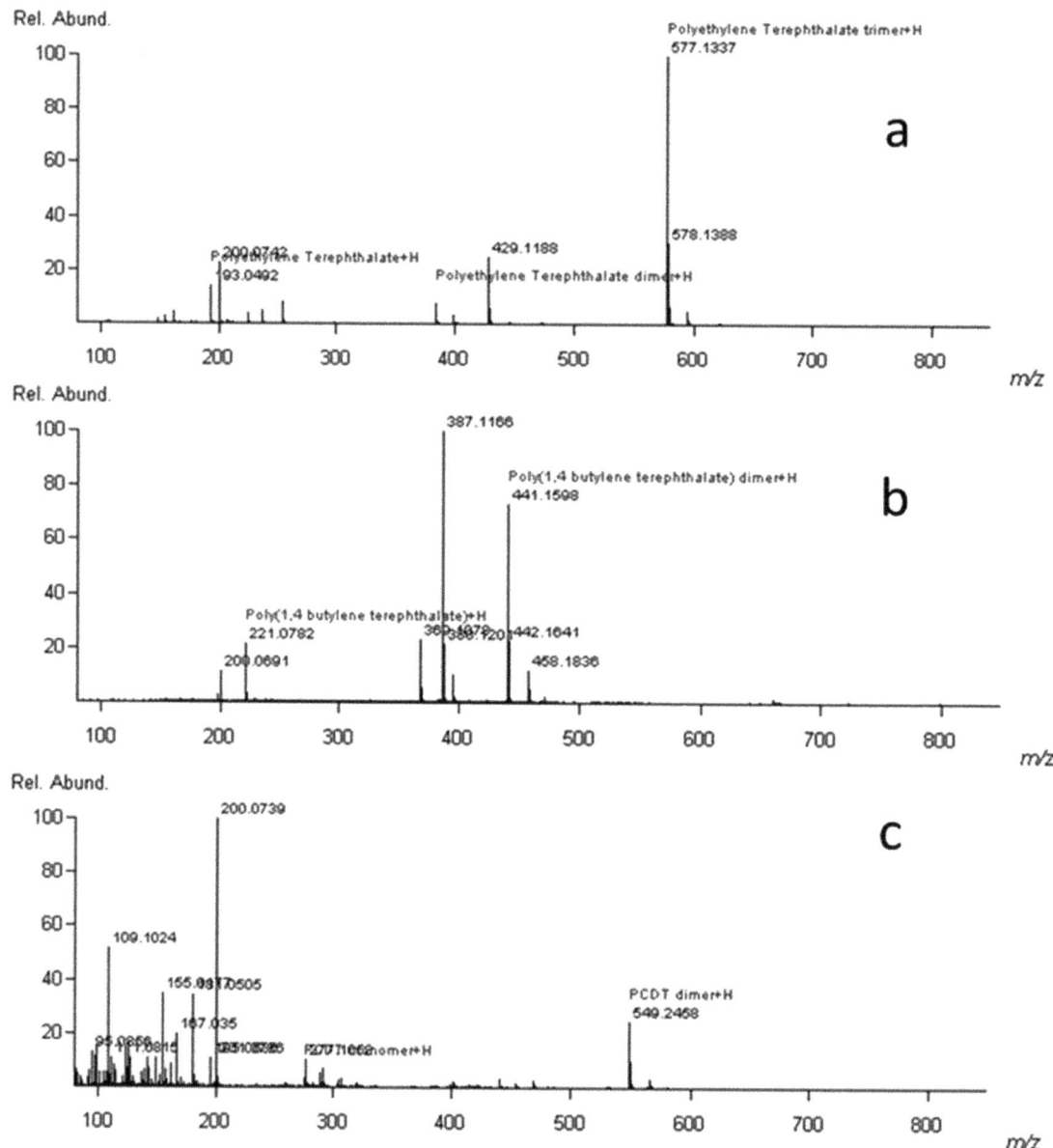

Figure 6. DART-AccuTOF™ mass spectra for standards (**a**) PET, (**b**) PBT, and (**c**) PCDT at 30 volts with correct peaks identified using SearchFromList software.

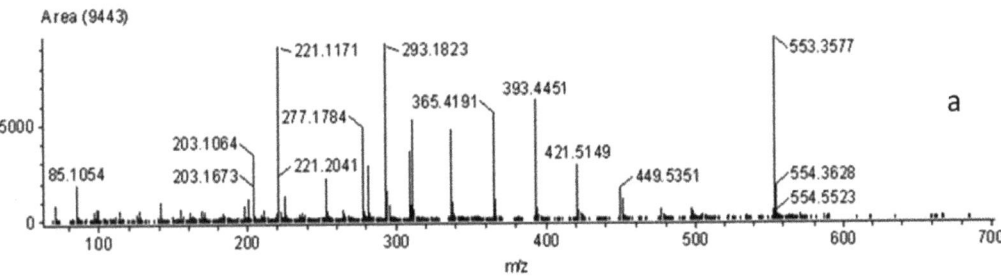

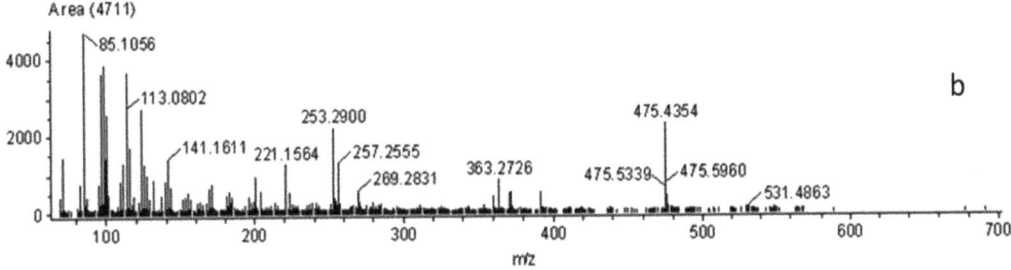

Figure 7. DART-AccuTOF™ mass spectra for standards (**a**) polyethylene and (**b**) polypropylene at 30 V.

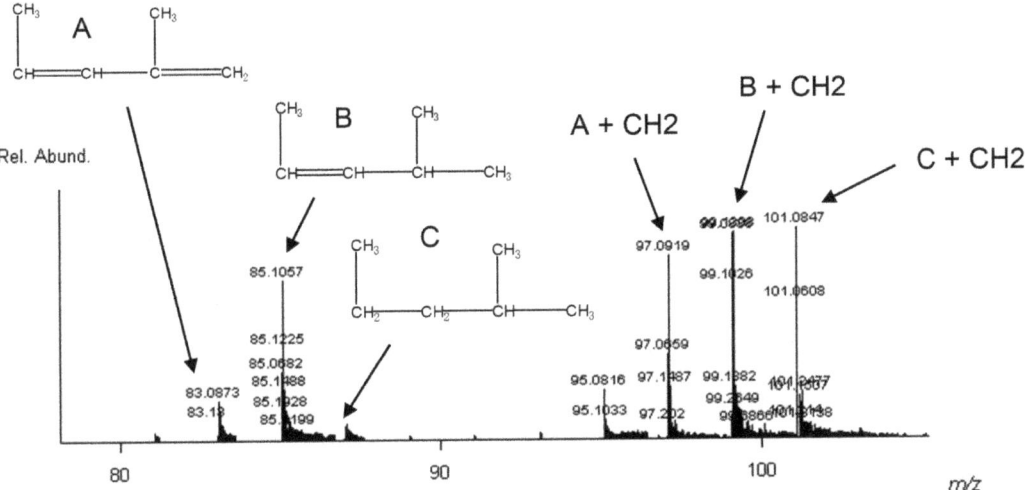

Figure 8. Close up of polypropylene standard mass spectrum (80–110 mass range). All peaks are protonated molecules of the indicated structures.

3.3. Carpet Samples

Following the successful differentiation of the various polymer standards, the carpet fibers were tested. Of the 15 known carpet samples, there were two labeled as 100% nylon carpet samples (#19,23), six labeled as 100% polyester carpet samples (#5,20,24,28–30), and four labeled as 100% olefin samples (#21,22,27,32). The remaining known samples (#25,26, and 31) were labeled as mixtures of 91% olefin and 9% nylon, 91% olefin and 9% nylon, and 80% polyester and 20% nylon, respectively.

All known nylon, polyester, and olefin carpet samples were correctly identified using the DART-AccuTOF™. Figure 9 shows representative spectra of the various fiber types.

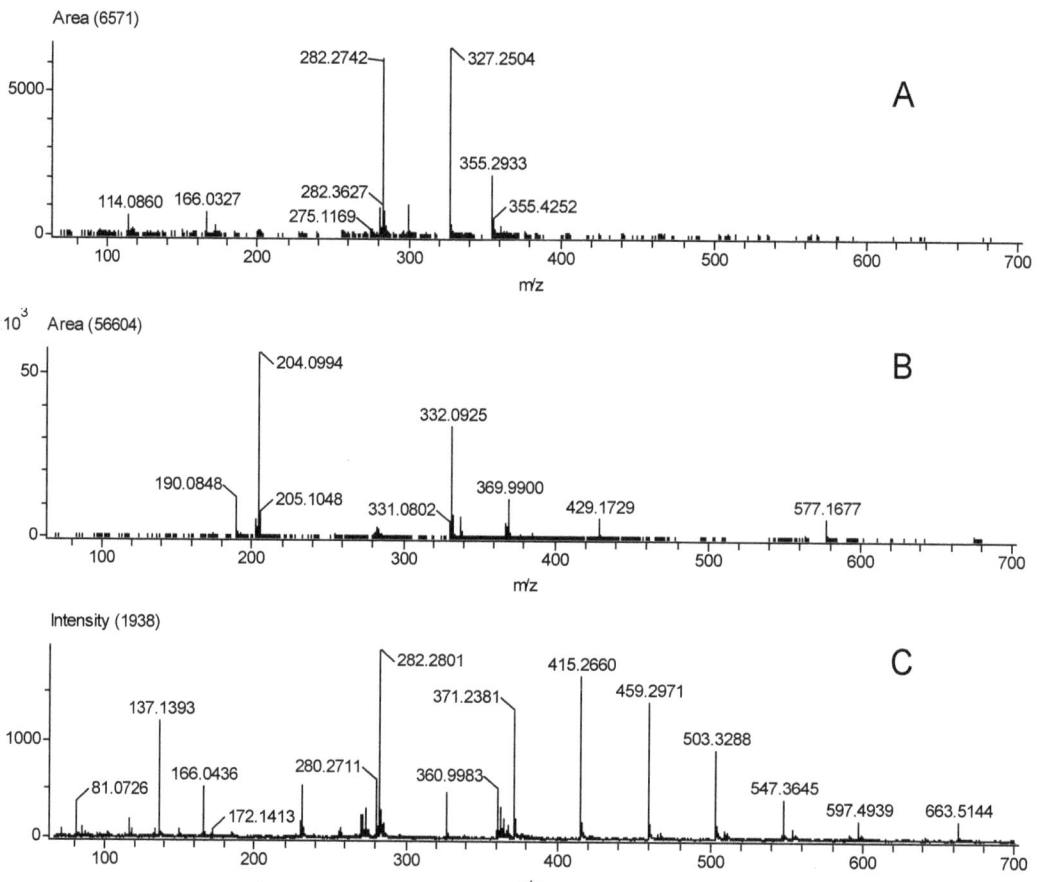

Figure 9. Mass spectra for carpet samples (**A**) 19 (nylon), (**B**) 20 (polyester) and (**C**) 21 (olefin) at 30 V.

Following the parameter optimization of the DART-AccuTOF™ and the sampling of the standards and known carpet fibers, it was determined whether this technique could compliment the FTIR results for unknown carpet fibers.

The DART-AccuTOF™ correctly identified all the remaining unknown samples as their corresponding polymer type. It was even able to distinguish the sub-type for two of the known nylon samples, where sample 3 was identified as nylon 6. Sample 4 was identified as nylon 6/6 as the latter did not produce a peak at m/z 114 (nylon 6 monomer [M+H]) [14]. The results were confirmed via a subsequent microchemical test using a few drops of 15.5% HCl, which instantaneously dissolves nylon 6 as opposed to other nylon sub-types as described in the author's previous publication [14]. The results of the micro solubility tests helped confirm that the DART-AccuTOF can differentiate the different polymer sub-types in addition to identifying the basic polymer types.

4. Conclusions

The carpet samples with no known composition information were correctly identified by using Fourier-transform infrared spectroscopy (FTIR). The method was also capable of identifying sub-classes of polyester, olefin, and nylon carpet fibers.

DART-AccuTOF effectively differentiated numerous polymer types and sub-types based on the polymer chemistry of their monomers [M+H], and their associated dimers [2M+H] and trimers [3M+H]. The DART gas stream temperature of 275 °C was used as a compromise due to the differences in melting points; however, data collected at that temperature for all polymer types produced acceptable results. This technique can be applied in the testing of other fabric fibers, as well as objects that are made using these polymers.

DART-AccuTOF™ successfully identified the carpet sample's polymer class (nylon vs. polyester vs. olefin) and successfully differentiated a nylon 6 carpet sample from a nylon 6/6. Due to the lack of other polyester carpet types, PET, which accounts for over 95% of the polyester production as previously stated, was identified in all the polyester carpet samples. In all the PET carpet samples, the m/z 204 ion peak was observed and therefore can serve as an aid in identifying the sample as a PET carpet sample. The lack of a PCDT or PBT carpet sample limits the conclusion that can be drawn with regard to the m/z 204 ion peak and polyester carpet samples. The same can be said about olefins, a differentiation of the olefin type could not be determined using DART-AccuTOF™. The presence of the m/z 85 ion peak and the polymer backbone will help assist with identifying the sample as an olefin. Additionally, due to the lack of known polypropylene or polyethylene carpet samples, a differentiating criterion could not be established.

In conclusion, DART-AccuTOF™ is capable of identifying the various polymer types in carpet fibers and could be implemented as one of the many series of tests conducted when analyzing fiber evidence to increase the power of discrimination. It could serve as a quick screening tool, especially in cases with many fibers.

Author Contributions: Conceptualization, T.A.Z. and R.R.S.; methodology, R.R.S.; software, R.R.S.; validation, T.A.Z. and R.R.S.; formal analysis, T.A.Z.; investigation, T.A.Z.; resources, T.A.Z. and R.R.S.; data curation, R.R.S.; writing—original draft preparation, T.A.Z.; writing—review and editing, R.R.S.; visualization, R.R.S.; supervision, R.R.S.; project administration, R.R.S. All authors have read and agreed to the published version of the manuscript.

Funding: This research received no external funding.

Institutional Review Board Statement: Not applicable.

Informed Consent Statement: Not applicable.

Data Availability Statement: Data is contained within the article.

Acknowledgments: The authors would like to thank Marilyn Miller from Virginia Commonwealth University, as well as Robyn Weimer and Joshua Kruger, Forensic Scientists in the Trace Evidence Section of the Virginia Department of Forensic Science in Richmond for their help and knowledge in polymer chemistry and their assistance throughout this project. In addition, the authors would also like to thank the entire Controlled Substances Section of the Virginia Department of Forensic Science in Richmond for their generosity, support, and contribution to this research.

Conflicts of Interest: The authors declare no conflict of interest.

References

1. Dasgupta, S.; Hammond, W.B.; Goddard, W.A. Crystal structures and properties of nylon polymers from theory. *J. Am. Chem. Soc.* **1996**, *118*, 12291–12301. [CrossRef]
2. Liang, J.; Frazier, J.; Benefield, V.; Chong, N.S.; Zhang, M. Forensic Fiber Analysis by Thermal Desorption/Pyrolysis-Direct Analysis in Real Time-Mass Spectrometry. *Anal. Chem.* **2020**, *92*, 1925–1933. [CrossRef] [PubMed]
3. Robertson, J.; Roux, C.; Wiggins, K.G. *Forensic Examination of Fibres*, 3rd ed.; Taylor and Francis: Milton Park, UK, 2017.
4. Pavlovich, M.J.; Musselman, B.; Hall, A.B. Direct analysis in real Time-Mass spectrometry (DART-MS) in forensic and security applications. *Mass Spectrom. Rev.* **2016**, *37*, 171–187. [CrossRef] [PubMed]

5. Hatch, K.L. *Textile Science*; West Pub: Saint Paul, MI, USA, 1993.
6. Needles, H.L. *Textile Fibers, Dyes, Finishes, and Processes: A Concise Guide*; Noyes Publications: Norwich, NY, USA, 1986.
7. Adams, J. Analysis of printing and writing papers by using direct analysis in real time mass spectrometry. *Int. J. Mass Spectrom.* **2011**, *301*, 109–126. [CrossRef]
8. Cody, R.B.; Laramée, J.A.; Durst, H.D. Versatile New Ion Source for the Analysis of Materials in Open Air under Ambient Conditions. *Anal. Chem.* **2005**, *77*, 2297–2302. [CrossRef] [PubMed]
9. Cody, R.B.; Laramée, J.A.; Nilles, J.M.; Durst, H.D. Direct Analysis in Real Time (DART) Mass Spectrometry 2005. *JEOL News* **2005**, *40*, 8–12.
10. Jones, R.W.; Cody, R.B.; McClelland, J.F. Differentiating writing inks using direct analysis in real time mass spectrometry. *J. Forensic Sci.* **2006**, *51*, 915–918. [CrossRef]
11. Howlett, S.E.; Steiner, R.R. Validation of Thin Layer Chromatography with AccuTOF-DARTTM Detection for Forensic Drug Analysis. *J. Forensic Sci.* **2011**, *56*, 1261–1267. [CrossRef] [PubMed]
12. Steiner, R.R.; Larson, R.L. Validation of the Direct Analysis in Real Time Source for Use in Forensic Drug Screening. *J. Forensic Sci.* **2009**, *54*, 617–622. [CrossRef] [PubMed]
13. Nilles, J.M.; Connell, T.R.; Stokes, S.T.; Dupont Durst, H. Explosives Detection Using Direct Analysis in Real Time (DART) Mass Spectrometry. *Propellants Explos. Pyrotech.* **2010**, *35*, 446–451. [CrossRef]
14. Zughaibi, T.; Steiner, R. Differentiating Nylons Using Direct Analysis in Real Time Coupled to an AccuTOF Time-of-Flight Mass Spectrometer. *J. Am. Soc. Mass Spectrom.* **2020**, *31*, 982–985. [CrossRef] [PubMed]

Article

Fabrication and Application of SERS-Active Cellulose Fibers Regenerated from Waste Resource

Shengjun Wang [1], Jiaqi Guo [2], Yibo Ma [3], Alan X. Wang [4], Xianming Kong [1,*] and Qian Yu [1,*]

1. School of Petrochemical Engineering, Liaoning Petrochemical University, Fushun 113001, China; wsj203203@sina.com
2. Jiangsu Co-Innovation Center for Efficient Processing and Utilization of Forest Resources and Joint International Research Lab of Lignocellulosic Functional Materials, Nanjing Forestry University, Nanjing 210037, China; jiaqi.guo@njfu.edu.cn
3. Department of Bioproducts and Biosystems, School of Chemical Engineering, Aalto University, FI-00076 Aalto, Finland; yibo.ma@aalto.fi
4. School of Electrical Engineering and Computer Science, Oregon State University, Corvallis, OR 97331, USA; alan.wang@oregonstate.edu
* Correspondence: xmkong@lnpu.edu.cn (X.K.); qyu@lnpu.edu.cn (Q.Y.)

Abstract: The flexible SERS substrate were prepared base on regenerated cellulose fibers, in which the Au nanoparticles were controllably assembled on fiber through electrostatic interaction. The cellulose fiber was regenerated from waste paper through the dry-jet wet spinning method, an eco-friendly and convenient approach by using ionic liquid. The Au NPs could be controllably distributed on the surface of fiber by adjusting the conditions during the process of assembling. Finite-difference time-domain theoretical simulations verified the intense local electromagnetic fields of plasmonic composites. The flexible SERS fibers show excellent SERS sensitivity and adsorption capability. A typical Raman probe molecule, 4-Mercaptobenzoicacid (4-MBA), was used to verify the SERS cellulose fibers, the sensitivity could achieve to 10^{-9} M. The flexible SERS fibers were successfully used for identifying dimetridazole (DMZ) from aqueous solution. Furthermore, the flexible SERS fibers were used for detecting DMZ from the surface of fish by simply swabbing process. It is clear that the fabricated plasmonic composite can be applied for the identifying toxins and chemicals.

Keywords: regenerated cellulose fiber; Au NP; controllably assembled; SERS; dimetridazole

1. Introduction

Cellulose is one of the most abundant biopolymers derived from biomass, which has been widely applied for synthesizing functional materials such as drug delivery, optical sensors, lithium-ion battery, textile, and biomedical engineering [1–6]. Cellulose is a cost-effective, eco-friendly, and biodegradable natural resource and the physical property and chemical reactivity of cellulose has attracted considerable research. Cellulose fibers exhibit many merits and advanced features such as the cheap price, abundant resource, light weight, biodegradability, and the capability for surface functionalization, which makes it a good matrix for incorporating various materials to construct composite with multiple advantages of cellulose fibers and guest materials. The β-cyclodextrins, polyacid, chitosan, and quinine were modified on the cellulose fibers to prepared cellulose fibers with special functionality [7–10].

With the development of nanotechnology, nanomaterials were incorporated with cellulose fibers and used in catalyst, supercapacitor electrodes, removal of metal ions, and biosensing [11–14]. Wang et al. deposited various TiO_2 nanobelts on the surface of cellulose fibers to construct a functional composite, which show excellent photocatalytic activity in degrading methylene blue and antibacterial ability to *E. col* [15]. Liu and coworkers decorated cellulose fabric with reduced graphene to fabricate multi-functional fabrics, the cellulose fabric/graphene was successfully used in pressure sensing and energy harvesting [16]. Compared to molecules, metal oxide, and graphene, plasmonic nanoparticles (NP) show excellent features in optical and thermal aspects due to their plasmonic properties. When the plasmonic nanoparticles are irradiated by light, the free electrons on the surface of NPs are driven by the electric field to collectively oscillate at a resonant frequency, the phenomenon is named the surface plasmon resonance (SPR) [17,18]. The plasmonic NPs have been numerously applied in biological microscopy, optical sensors, and catalyst. The plasmonic NPs such as Ag and Au were incorporated with cellulose fibers to enable new capabilities to the prepared composites. Tian et al. have fabricated plasmonic absorbent cotton by depositing Ag colloid on cotton fiber, the plasmonic absorbent cotton was successfully used for adsorption and detection of thiram from cucumber by Raman spectroscopy, and the limit of detection achieved 0.1 ppm [19]. Zheng et al. decorated cotton fabrics with Au nanorods. Au nanorods enable cotton fiber to present a broad range of colors varying from brownish red through green to purplish red, which is assigned to the SPR feature of Au nanorods [20].

Surface-enhanced Raman scattering (SERS) spectroscopy is an advanced spectral technology as the sensitivity and selectivity. Since the discovery in the 1970s by Fleischmann [21], SERS has attracted strong interests from many researchers. The performance of SERS is dependent on the enhanced substrate. The enhancement effect of SERS substrate is mainly attributed to the localized surface plasmon resonance (LSPR) of the plasmonic materials [22–25]. With the development of nanotechnology, various plasmonic nanostructures have been developed and used for SERS sensing [24,26,27]. In order to detect analytes from objectives with irregular surfaces, flexible SERS substrates have been proposed. The soft matrix, such as PDMS film, cotton fabrics, cotton gauze, and cotton fibers, were used for constructing flexible SERS substrates [28]. Qu and coworkers prepared plasmonic cotton swab by assembling plasmonic NPs on cotton swab, the flexible capability of such plasmonic substrate allows for contacting the surface of cucumber through the simple swabbing process [29]. Cai and calibrators developed a flexible SERS substrate by decorating Ag NPs on natural woven fabrics and the plasmonic cotton fabrics showed excellent sensitivity for detecting p-aminothiophenol (10^{-7} M) [30]. Cotton fiber is not enough to meet the rapidly growing demand for textile fibers. The regenerated cellulose fiber from a waste resource is an effective strategy to meet fiber consumption in the world, which also provides sustainable solution for the recycled resource and waste accumulation [31].

In this study, we decorated the recycled cellulose fibers with Au NPs. The fiber was recycled from cellulosic waste such as paper and cardboard. The regenerated cellulose fiber-Au composites are flexible, cheap, and effective SERS substrates. The regenerated cellulose fiber was firstly functionalized with an amino group to graft positive charges. After that, the Au NPs were self-assembled onto the surface of fiber. The distribution of Au NPs on the fibers were controlled by the assembling time. These plasmonic cellulose were used as SERS substrates to detect R6G at a concentration down to 1×10^{-9} M. The flexible SERS cellulose fiber are highly effective for capturing analytes from a target with an irregular surface. These plasmonic SERS cellulose fiber is cheap, eco-friendly, and disposable, which offers a good platform for SERS sensing.

2. Experiment
2.1. Materials

Sodium hydroxyl (NaOH), gold (III) chloride trihydrate (HAuCl$_4$ 3H$_2$O), (3-Aminopropyl) trimethoxysilane (APTMS), 4-Mercaptobenzoicacid (4-MBA), and trisodium citrate dehydrates (Na$_3$C$_6$H$_5$O$_7$) were obtained from Innochem Sci. & Tech. Co., Ltd. (Beijing, China). Dimetridazole (DMZ) was purchased from Aladdin (Shanghai, China). The regenerated cellulose fiber from waste resource was supplied by the Biorefinery Group in Aalto University.

2.2. Synthesis of Au Colloid

The Au NPs (40–50 nm) used in this study were prepared as the previous report with minor modification [32]. Briefly, 100 mL of HAuCl$_4$ 3H$_2$O (1 mM) aqueous solution was boiled to reflux. After that, 2.3 mL of Na$_3$C$_6$H$_5$O$_7$ solution (1%) were dropped into the boiling solution, and kept boiling for half an hour, then cooled down to room temperature.

2.3. Fabrication of Regenerated Cellulose Fiber-Au Composite

The regenerated cellulose fiber was firstly treated with NaOH, briefly, 10 mg of regenerated cellulose fibers were soaked in 0.1 M of aqueous solution of NaOH for 20 min. After that the fiber was washed thoroughly with water and ethanol and dried in oven, and then immersed the fiber in 40 mL of 1% ethanol solution of APTMS for 5 h. The amino group was modified on the surface of cellulose fiber for grafting positive charge. The citrate capped Au NPs exhibited negative charge. A total of 2 mg of regenerated cellulose fibers modified with APTMS was soaked into 4 mL of Au colloid at different times. After that, the composite was washed with water and used for further study.

2.4. Apparatus

The UV-vis absorption spectra of Au colloid were measured on UV2400 UV–Vis spectrophotometer (Sunny Hengping Instrument, Shanghai, China). The scanning electron microscope (SEM) images of regenerated cellulose fiber before and after Au NPs decorating were collected on a SU8010 field emission scanning electron microscope (Hitachi, Tokyo Japan). Fourier transform infrared (FTIR) spectrum of regenerated cellulose fiber-Au composites were acquired from Spectrum GX spectrometer (PerkinElmer, Wellesley, MA, USA). The SERS measurement were carried on the portable Raman spectrometer (BWS465 iRman; B&W Tek, Newark, NJ, USA), and the laser was 785 nm.

2.5. Detection of Dimetridazole Using the Regenerated Cellulose Fiber-Au

The powder (1 mg) of dimetridazole was dissolved in 1 mL of aqueous solution of HCl (30 mM) at an initial concentration of 1000 ppm. After that, the dimetridazole solutions at different concentrations were prepared by diluting the initial solution of dimetridazole, and the dry regenerated cellulose fiber-Au was dipped into a solution of dimetridazole with different concentrations. After 1 min, the SERS regenerated cellulose fiber were transferred for Raman measurements. In order to investigate the SERS performance of regenerated cellulose fiber-Au to detect illegal drugs, the dimetridazole was mixed with meat. The SERS regenerated cellulose fiber were dipping from the sample and the SERS measurement was carried using a portable Raman spectrometer.

3. Results and Discussion

3.1. Preparation of Regenerated Cellulose Fiber-Au NPs Composites

The Au NPs used in this study were prepared through the trisodium citrate reduction method, in which the trisodium citrate functioned as a reducing and stabling reagent. The UV-vis spectra of Au colloid NPs were shown in Figure S1, in which the characteristic band at 526 nm is assigned to the localized surface plasmon resonance (LSPR) of Au NPs. The surface morphology of Au NPs was determined through the TEM image as shown in Figure 1. The spherical Au NPs with an average size of 40 nm were observed. The TEM image is consistent with UV-vis results. In order to observe the structure of Au NPs in detail, high-resolution transmission electron microscopy (HRTEM) observations were developed. The HRTEM image of single Au NP was presented in Figure S2, in which the lattice planes of the Au NP are observed. The 0.2355 nm interplanar distance is corresponding to the (111) planes of face centered cubic (fcc) of Au.

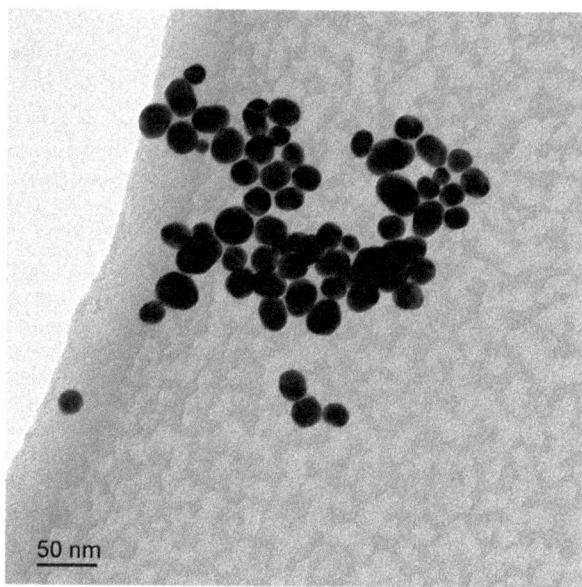

Figure 1. TEM image of colloidal gold nanoparticles.

The regenerated cellulose fiber showed a smooth surface as presented in the SEM image (Figure 2a,b). The diameter of the fiber was around 15 µm. After decorating Au NPs on the fiber, the rough surface fiber was observed as shown in Figure 2c. That result indicated that the Au NPs were assembled on the surface of the fiber successfully. SEM images with a high resolution were collected to observe the distribution of Au NPs on the surface of fiber. The different assemble time was used to control the density of Au NPs on the fiber during the self-assembly process. When the assembling time is 0.5 h, it could be observed that a small amount of Au NPs was decorated on the surface of the regenerated cellulose fiber (Figure 2d). As the assemble time increased to 2 h, the density of Au NPs distributed on the surface of cellulose fiber was increased significantly as present in Figure 2e. When the assembling time increases to 3 h, a dense layer of Au NPs was formed on the surface of fiber as shown in Figure 2f.

The property of the regenerated cellulose fiber before and after Au NPs decoration was also characterized by the FTIR spectrum as shown in Figure S3. The fiber composites were firstly cut to a small size and mixed with KBr for FTIR measurement. The feature IR peaks of cellulose were obtained. The peak at 3338 cm^{-1} is due to the stretching vibration of the OH group. The peaks at 2922 cm^{-1} and 2850 cm^{-1} were assigned to the anti-symmetrical vibration and symmetrical vibration of the methylene group. The prominent peak at 1060 cm^{-1} is attributed to the stretching vibration of the C–O–C group of pyranose ring in the cellulose [33]. The FTIR spectra of regenerated cellulose fiber after being modified with APTMS and Au NPs were similar with the blank fiber.

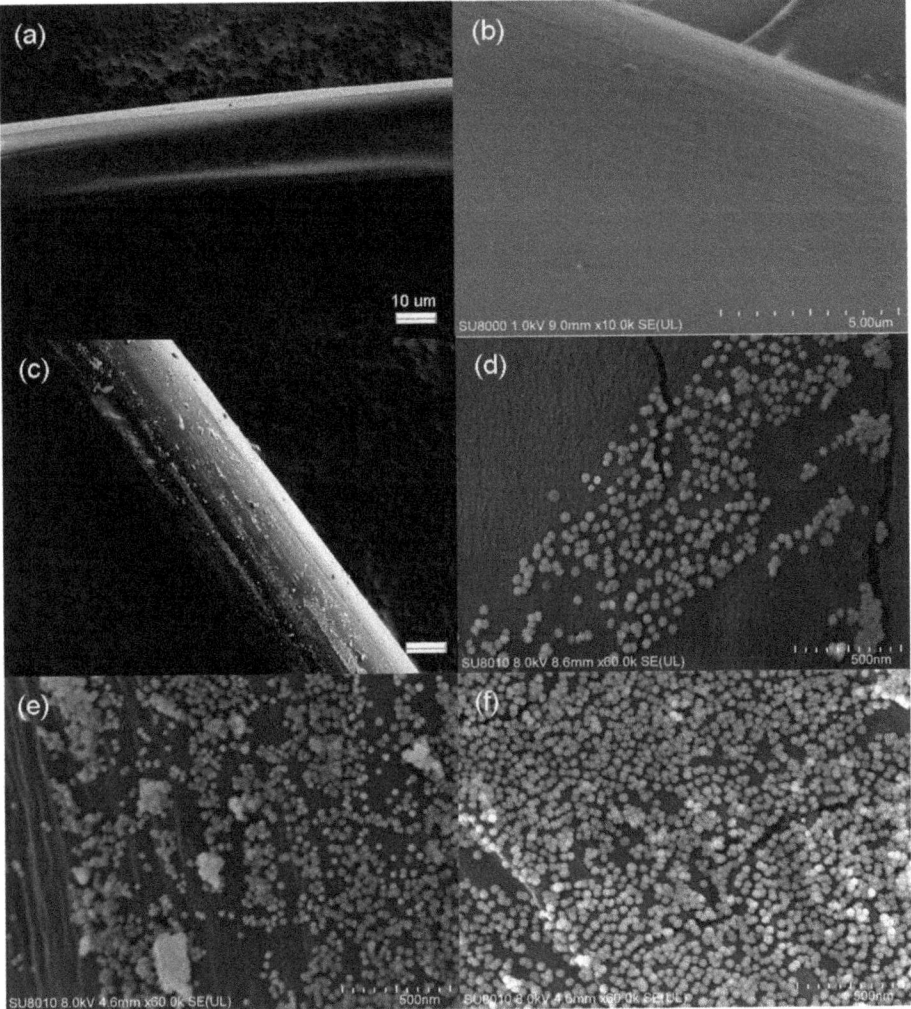

Figure 2. SEM images of regenerated cellulose fiber (**a**,**b**) and after decorating Au NPs (**c**–**f**).

In order to investigate the decorating of Au NPs on fiber quantitatively, the UV-vis spectra were used to characterize the Au colloid after the self-assembling process. The intensities of LSPR bands of the Au colloid were decreased after removing the regenerated cellulose fibers from the colloid as shown in Figure S4. The decrease in the intensity of UV-vis spectra indicates that the amount of Au NPs remaining in colloid was reduced. The intensity of LSPR peak of Au colloid decreased as the assemble time increased, which indicated that more Au NPs decorated on fiber through the electrostatic interaction. The amount of Au NPs deposited on the regenerated cellulose fiber was estimated base on the variation of the UV-vis spectra. The initial concentration of Au colloid is nearly at 0.92×10^{-10} M according to the Lambert–Beer's law, in which the molar extinction coefficient of colloid is 3.4×10^{10} M^{-1} cm^{-1} [34]. The 0.5-h assemble time corresponding to 1.48×10^{-13} M Au NPs decorated on the fiber. When the assembling time increased to 1 h, 2 h, 3 h, and 5 h, the amount of Au NPs on the cellulose fiber increased to 1.71×10^{-13} M, 2.31×10^{-13} M, 2.41×10^{-13} M, and 2.52×10^{-13} M, respectively.

The diameter of the fiber-Au composite was changed obviously under wet and dry conditions. As shown in Figure 3a, the diameter of the fiber-Au was nearly at 31 um under wet conditions. After dried in air conditions, the diameter of fiber-Au was decreased to 18 μm (Figure 3b). This variable feature can be explained as the cellulose of fiber were partially dissolved in the NaOH solutions, the free space in the fiber can adsorb more water during the swelling process. The shrinking process of the fiber-Au could decrease via swelling the distance of Au NPs, that could bring more "hot spots" for SERS measurement.

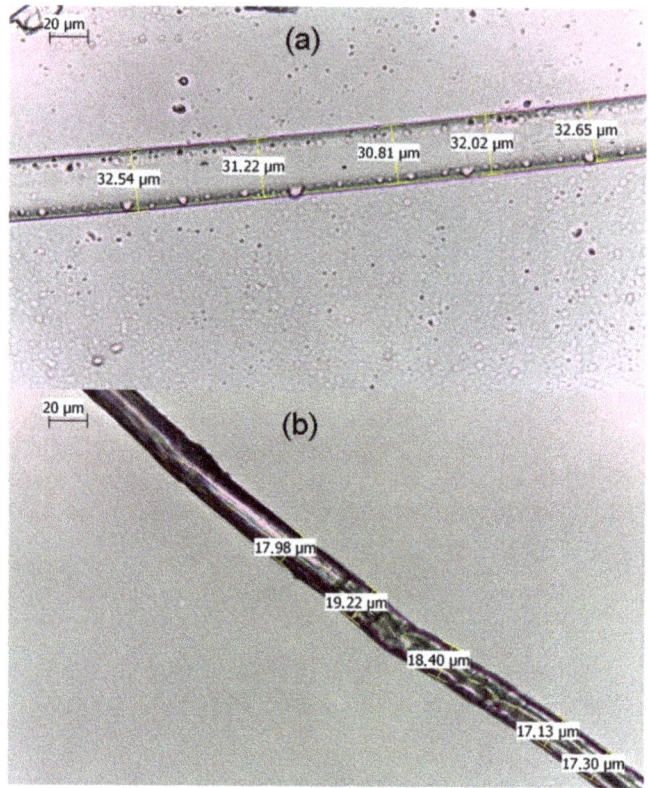

Figure 3. Microscope images of fiber-Au under wet (**a**) and dry (**b**) conditions.

3.2. SERS Application of Regenerated Cellulose Fiber-Au Composites

To evaluate the SERS performance of the plasmonic composite, MBA was chosen as the Raman probe molecule. Figure 4a presents the SERS spectra of MBA measured from different substrates. There are no Raman spectra of MBA observed from the regenerated cellulose fiber, which indicated no SERS enhancement effect on the regenerated cellulose fiber. After assembling Au NPs on the regenerated cellulose fiber, Raman spectra were obtained from the regenerated cellulose fiber-Au substrate. The SERS peaks mainly centered at 1070 and 1578 cm^{-1}. The intense peak at 1070 cm^{-1} is attributed to the in-plane ring breathing vibration combined with a_1 vibration mode of υ (C-S), the peak at 1578 cm^{-1} can be assigned to υ (CC), an a_1 vibration mode [35]. The intensities of SERS spectra were increased as the increase of the assemble times, the reason is that more Au NPs decorated on the regenerated cellulose fiber. The dense Au NPs produce high electromagnetic fields and bring more 'hot spots' in the SERS measurement. A total of 3 h of assemble time was chosen for the SERS measurement to provide enough SERS enhancement. Figure 4b presented the Raman spectra of MBA at different concentrations ranging from 10^{-5} to

10^{-9} M. The intensities of Raman spectra decreased as the concentration of MBA decreased. The SERS spectra of MBA was still observed as the concentration went down to 10^{-9} M, which suggested the regenerated cellulose fiber-Au has effective SERS enhancement.

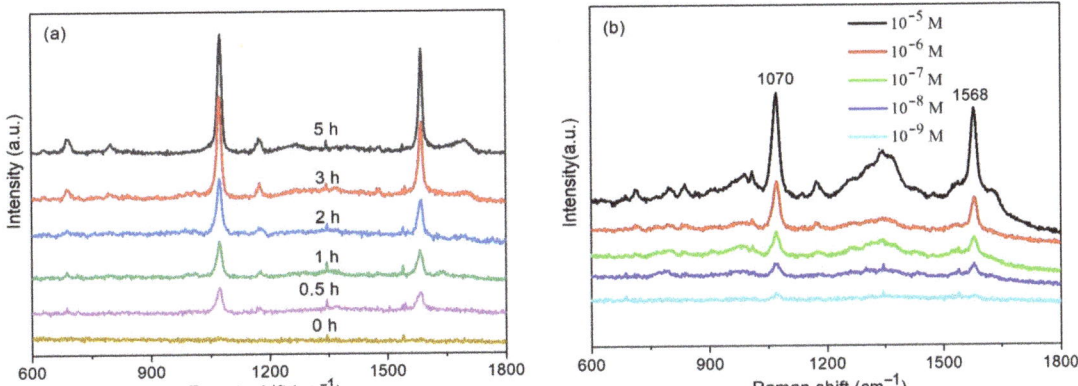

Figure 4. Raman spectra of MBA on cellulose fiber-Au with different assemble times (**a**), Raman spectra measured from the cellulose fiber-Au with a different concentration of MBA (**b**).

Dimetridazole (DMZ) is an effective veterinary drug that is usually used for controlling or treating bacterial and protozoal infections of animals. It has been proven that DMZ brings genotoxic and carcinogenic problems to public health. The European Union, United States, and China have forbidden the application of DMZ in food producing animals. The regenerated cellulose fiber-Au was used as the SERS substrate for detecting the DMZ molecule. The SERS spectra of DMZ with different concentrations were shown in Figure 5a. The intense Raman bands at 830, 1188, 1268, and 1360 cm^{-1} were presented. The prominent peak at 1188 cm^{-1} is due to the bending vibration of H–C–N, the most intense peak at 1360 cm^{-1} belongs to the deformational mode of H–C–N. The peak at 830 cm^{-1} results from the ring deformational mode combined with a wagging vibration of the NO$_2$ group, the peak at 1268 cm^{-1} is attributed to the ring deformational mode. The peak at 1188 cm^{-1} can be observed even as the concentration of DMZ is down to 10^{-9} M. The enhanced effect of the plasmonic composite is comparable to the current existing flexible SERS substrate [36]. The excellent SERS performance could be attributed to two aspects. First, the dense Au NPs on the cellulose fiber could provide more 'hot spots' for SERS. A FDTD theoretical calculation was used to verify the intensity of the electric field of Au NPs, the distance between Au NPs on the regenerated cellulose fiber were set as 10 nm and 4 nm corresponding to dimer and trimer as shown in Figure 5b,c. The electric field within the gaps of trimer is higher than that from dimer, due to the coupled LSPR at a 4 nm gap distance, which provides a higher electric field than a 10-nm gap. Second, the cellulose fiber has an absorption capability [19], which enable it to adsorb DMZ from the solution. The UV-Vis spectra were used to verify the absorption capability of the composite as presented in Figure 5d. The absorption peak of DMZ was decreased obviously after the regenerated cellulose fiber-Au immersed in the solution of DMZ. More DMZ molecules were captured on the plasmonic composite during the process of adsorption, which benefited the SERS measurement as the sample enrichment effect.

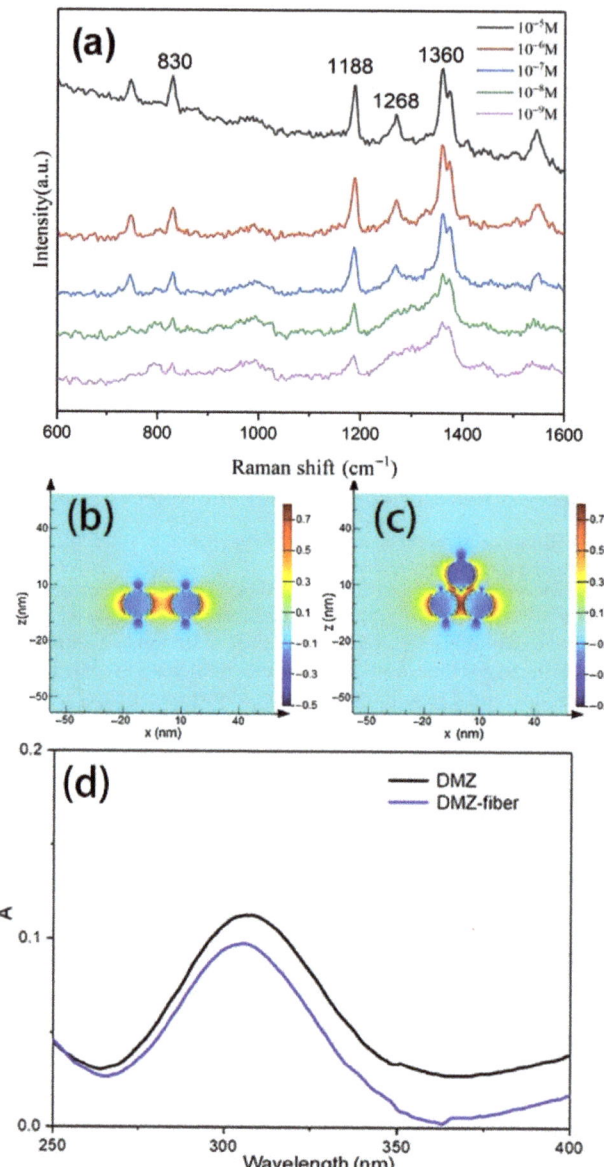

Figure 5. Raman spectra of DMZ with different concentrations: (**a**) FDTD simulations of SERS hot spots at Au NPs, dimer with a 10-nm gap (**b**) and trimer with a 4-nm gap (**c**), UV-vis spectra of DMZ (10^{-5} M), and after adsorbed by regenerated cellulose fiber-Au (**d**).

In real life, there is usually more than one kind of molecule that exists in the target system and multiple components in the mixture may bring interference for analysis. Sudan G is a kind of industrial dye with azo group. As it can provide a bright red color, Sudan G has been used as an illegal additive in food. The illegal additive (Sudan G) and drug (DMZ) were mixed and used as analytes. The regenerated cellulose fiber-Au was dipped into the solution of mixture and transferred for SERS measurement. The SERS spectra from regenerated cellulose fiber-Au was presented in Figure 6. The characteristic Raman peaks attributed to Sudan G and DMZ were observed simultaneously. The detection of DMZ from the target with multiple components achieved by the substrate proposed in this study.

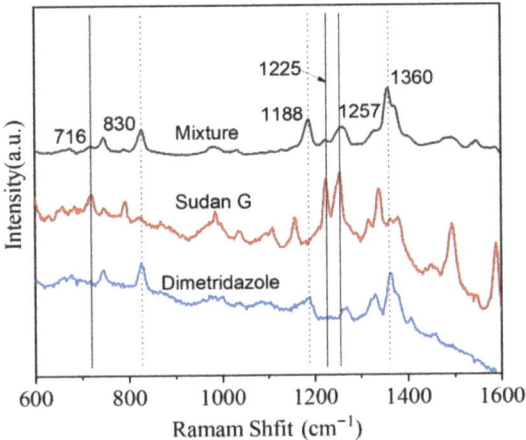

Figure 6. Raman spectra of DMZ, Sudan G, and their mixture from the regenerated cellulose fiber-Au.

In order to evaluate the SERS performance of fiber-Au in real food sample sensing, the fiber-Au composite was used as a flexible SERS substrate for DMZ detecting from fish. A total of 10 µL of DMZ with different concentrations were sprayed onto the surface of fish. The prepared fiber-Au was directly swabbed from the surface of a fish and for SERS spectra collecting. The SERS spectra are shown in Figure 7, the intensities of the SERS signal were decreased as the reduction of DMZ concentrations. The DMZ on the surface of fish could be easily adsorbed by fiber during the swabbing process.

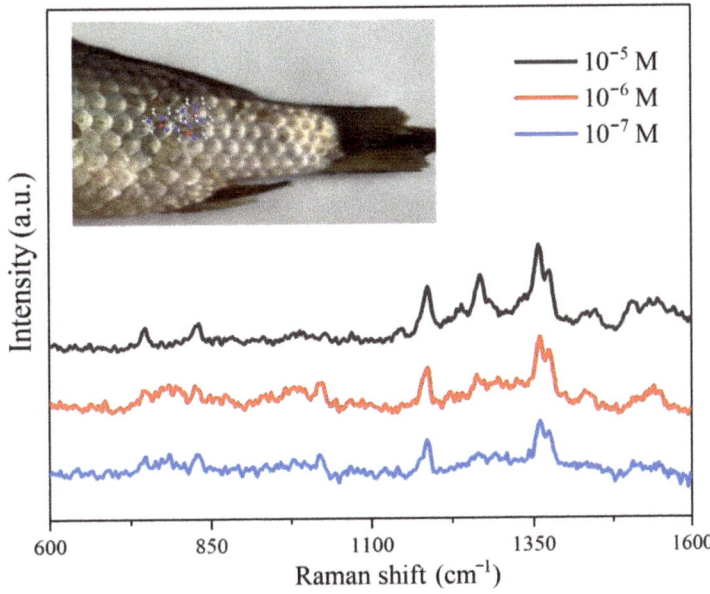

Figure 7. Swabbing detection of the DMZ from the surface of fish using fiber-Au as flexible SERS substrate.

4. Conclusions

The plasmonic composite was prepared based on regenerated cellulose fiber from waste paper. The regenerated cellulose fiber was firstly cationized with APTMS for grafting positive charges. The Au NPs were decorated on the surface of regenerated cellulose fiber by being deposited onto the surface of cotton gauze through electrostatic interaction. The distribution of Au NPs on the fiber could be controlled by adjusting the assembling time. The regenerated cellulose fiber-Au composite showed several advantages, such as eco-friendly, flexibility, adsorption ability, and active SERS enhancement. Proof-of-concept for the sensing application of the plasmonic composite was developed by identifying MBA and DMZ and the sensitivity could achieve 10^{-9} M. The fiber-Au also presented good selectivity for DMZ by SERS sensing. The plasmonic cellulose composite could be easily incorporated in paper or packages for in-situ monitoring of harmful ingredients in food and water.

Supplementary Materials: The following are available online at https://www.mdpi.com/article/10.3390/polym13132142/s1. Figure S1: UV-Vis spectra of colloidal gold nanoparticles, Figure S2: HRTEM image of Au NP, Figure S3: FTIR spectra of regenerated cellulose fiber and after modification, Figure S4: The UV-Vis spectra of Au colloid after exposure cellulose fibers with different assemble times.

Author Contributions: Conceptualization, X.K. and Q.Y.; methodology, S.W.; resources, Y.M.; writing—original draft preparation, S.W. and X.K.; writing—review and editing, A.X.W. and J.G. All authors have read and agreed to the published version of the manuscript.

Funding: This research was funded by Science Research Project of Education Department of Liaoning Province of China (No. L2019011) and the talent scientific research fund of LNPU (No. 2017XJJ-037).

Institutional Review Board Statement: Not applicable.

Informed Consent Statement: Not applicable.

Data Availability Statement: The data presented in this study are available on request from the corresponding author.

Acknowledgments: The authors would like to acknowledge the support from the Science Research Project of Education Department of Liaoning Province of China (no. L2019011) and the talent scientific research fund of LSHU (no. 2017XJJ-037).

Conflicts of Interest: The authors declare no conflict of interest.

References

1. Wang, S.; Zhang, L.; Zeng, Q.; Liu, X.; Lai, W.-Y.; Zhang, L. Cellulose microcrystals with brush-like architectures as flexible all-solid-state polymer electrolyte for lithium-ion battery. *ACS Sustain. Chem. Eng.* **2020**, *8*, 3200–3207. [CrossRef]
2. Roy, D.; Semsarilar, M.; Guthrie, J.T.; Perrier, S. Cellulose modification by polymer grafting: A review. *Chem. Soc. Rev.* **2009**, *38*, 2046–2064. [CrossRef]
3. Javanbakht, S.; Namazi, H. Doxorubicin loaded carboxymethyl cellulose/graphene quantum dot nanocomposite hydrogel films as a potential anticancer drug delivery system. *Mater. Sci. Eng. C* **2018**, *87*, 50–59. [CrossRef]
4. Ma, Y.; Asaadi, S.; Johansson, L.S.; Ahvenainen, P.; Reza, M.; Alekhina, M.; Rautkari, L.; Michud, A.; Hauru, L.; Hummel, M. High-strength composite fibers from cellulose–lignin blends regenerated from ionic liquid solution. *ChemSusChem* **2015**, *8*, 4030–4039. [CrossRef] [PubMed]
5. Moniri, M.; Moghaddam, A.B.; Azizi, S.; Rahim, R.A.; Ariff, A.B.; Saad, W.Z.; Navaderi, M.; Mohamad, R. Production and status of bacterial cellulose in biomedical engineering. *Nanomaterials* **2017**, *7*, 257. [CrossRef]
6. Jin, E.; Guo, J.; Yang, F.; Zhu, Y.; Song, J.; Jin, Y.; Rojas, O.J. On the polymorphic and morphological changes of cellulose nanocrystals (CNC-I) upon mercerization and conversion to CNC-II. *Carbohydr. Polym.* **2016**, *143*, 327–335. [CrossRef] [PubMed]
7. Lv, Y.; Ma, J.; Liu, K.; Jiang, Y.; Yang, G.; Liu, Y.; Lin, C.; Ye, X.; Shi, Y.; Liu, M.; et al. Rapid elimination of trace bisphenol pollutants with porous β-cyclodextrin modified cellulose nanofibrous membrane in water: Adsorption behavior and mechanism. *J. Hazard. Mater.* **2021**, *403*, 123666. [CrossRef]
8. Chitpong, N.; Husson, S.M. Polyacid functionalized cellulose nanofiber membranes for removal of heavy metals from impaired waters. *J. Membr. Sci.* **2017**, *523*, 418–429. [CrossRef]
9. Zhu, K.; Wang, Y.; Lu, A.; Fu, Q.; Hu, J.; Zhang, L. Cellulose/chitosan composite multifilament fibers with two-switch shape memory performance. *ACS Sustain. Chem. Eng.* **2019**, *7*, 6981–6990. [CrossRef]

10. Aka, E.C.; Nongbe, M.C.; Ekou, T.; Ekou, L.; Coeffard, V.; Felpin, F.-X. A fully bio-sourced adsorbent of heavy metals in water fabricated by immobilization of quinine on cellulose paper. *J. Environ. Sci.* **2019**, *84*, 174–183. [CrossRef]
11. Wan, C.; Jiao, Y.; Li, J. Flexible, highly conductive, and free-standing reduced graphene oxide/polypyrrole/cellulose hybrid papers for supercapacitor electrodes. *J. Mater. Chem. A* **2017**, *5*, 3819–3831. [CrossRef]
12. Shi, X.; Zhang, X.; Ma, L.; Xiang, C.; Li, L. TiO2-Doped chitosan microspheres supported on cellulose acetate fibers for adsorption and photocatalytic degradation of methyl orange. *Polymers* **2019**, *11*, 1293. [CrossRef]
13. Ali, A.; Mannan, A.; Hussain, I.; Hussain, I.; Zia, M. Effective removal of metal ions from aquous solution by silver and zinc nanoparticles functionalized cellulose: Isotherm, kinetics and statistical supposition of process. *Environ. Nanotechnol. Monit. Manag.* **2018**, *9*, 1–11. [CrossRef]
14. Liu, S.; Cui, R.; Ma, Y.; Yu, Q.; Kannegulla, A.; Wu, B.; Fan, H.; Wang, A.X.; Kong, X. Plasmonic cellulose textile fiber from waste paper for BPA sensing by SERS. *Spectrochim. Acta Part A Mol. Biomol. Spectrosc.* **2020**, *227*, 117664. [CrossRef]
15. Wang, J.; Liu, W.; Li, H.; Wang, H.; Wang, Z.; Zhou, W.; Liu, H. Preparation of cellulose fiber–TiO2 nanobelt–silver nanoparticle hierarchically structured hybrid paper and its photocatalytic and antibacterial properties. *Chem. Eng. J.* **2013**, *228*, 272–280. [CrossRef]
16. Wu, R.; Ma, L.; Patil, A.; Meng, Z.; Liu, S.; Hou, C.; Zhang, Y.; Yu, W.; Guo, W.; Liu, X.Y. Graphene decorated carbonized cellulose fabric for physiological signal monitoring and energy harvesting. *J. Mater. Chem. A* **2020**, *8*, 12665–12673. [CrossRef]
17. Lu, X.; Rycenga, M.; Skrabalak, S.E.; Wiley, B.; Xia, Y. Chemical synthesis of novel plasmonic nanoparticles. *Annu. Rev. Phys. Chem.* **2009**, *60*, 167–192. [CrossRef] [PubMed]
18. Sivashanmugan, K.; Squire, K.; Tan, A.; Zhao, Y.; Kraai, J.A.; Rorrer, G.L.; Wang, A.X. Trace detection of tetrahydrocannabinol in body fluid via surface-enhanced raman scattering and principal component analysis. *ACS Sens.* **2019**, *4*, 1109–1117. [CrossRef]
19. Tian, X.; Zhai, P.; Guo, J.; Yu, Q.; Xu, L.; Yu, X.; Wang, R.; Kong, X. Fabrication of plasmonic cotton gauze-Ag composite as versatile SERS substrate for detection of pesticides residue. *Spectrochim. Acta Part A Mol. Biomol. Spectrosc.* **2021**, *257*, 119766. [CrossRef] [PubMed]
20. Zheng, Y.; Xiao, M.; Jiang, S.; Ding, F.; Wang, J. Coating fabrics with gold nanorods for colouring, UV-protection, and antibacterial functions. *Nanoscale* **2013**, *5*, 788–795. [CrossRef] [PubMed]
21. Fleischmann, M.; Hendra, P.J.; McQuillan, A.J. Raman spectra of pyridine adsorbed at a silver electrode. *Chem. Phys. Lett.* **1974**, *26*, 163–166. [CrossRef]
22. Kong, X.; Chong, X.; Squire, K.; Wang, A.X. Microfluidic diatomite analytical devices for illicit drug sensing with ppb-Level sensitivity. *Sens. Actuators B Chem.* **2018**, *259*, 587–595. [CrossRef]
23. Zhang, C.; You, T.; Yang, N.; Gao, Y.; Jiang, L.; Yin, P. Hydrophobic paper-based SERS platform for direct-droplet quantitative determination of melamine. *Food Chem.* **2019**, *287*, 363–368. [CrossRef]
24. Sivashanmugan, K.; Squire, K.; Kraai, J.A.; Tan, A.; Zhao, Y.; Rorrer, G.L.; Wang, A.X. Biological photonic crystal-enhanced plasmonic mesocapsules: Approaching single-molecule optofluidic-SERS sensing. *Adv. Opt. Mater.* **2019**, *7*, 1900415. [CrossRef] [PubMed]
25. Huang, L.; Wu, C.; Xie, L.; Yuan, X.; Wei, X.; Huang, Q.; Chen, Y.; Lu, Y. Silver-Nanocellulose composite used as SERS substrate for detecting carbendazim. *Nanomaterials* **2019**, *9*, 355. [CrossRef] [PubMed]
26. Xu, T.; Wang, X.; Huang, Y.; Lai, K.; Fan, Y. Rapid detection of trace methylene blue and malachite green in four fish tissues by ultra-sensitive surface-enhanced Raman spectroscopy coated with gold nanorods. *Food Control* **2019**, *106*, 106720. [CrossRef]
27. Jiang, X.; Zhang, J.; Xu, L.; Wang, W.; Du, J.; Qu, M.; Han, X.; Yang, L.; Zhao, B. Ultrasensitive SERS detection of antitumor drug methotrexate based on modified Ag substrate. *Spectrochim. Acta Part A Mol. Biomol. Spectrosc.* **2020**, *240*, 118589. [CrossRef]
28. Qian, C.; Guo, Q.; Xu, M.; Yuan, Y.; Yao, J. Improving the SERS detection sensitivity of aromatic molecules by a PDMS-coated Au nanoparticle monolayer film. *RSC Adv.* **2015**, *5*, 53306–53312. [CrossRef]
29. Qu, L.-L.; Geng, Y.-Y.; Bao, Z.-N.; Riaz, S.; Li, H. Silver nanoparticles on cotton swabs for improved surface-enhanced Raman scattering, and its application to the detection of carbaryl. *Microchim. Acta* **2016**, *183*, 1307–1313. [CrossRef]
30. Chen, Y.; Ge, F.; Guang, S.; Cai, Z. Self-assembly of Ag nanoparticles on the woven cotton fabrics as mechanical flexible substrates for surface enhanced Raman scattering. *J. Alloy. Compd.* **2017**, *726*, 484–489. [CrossRef]
31. Ma, Y.; Hummel, M.; Määttänen, M.; Särkilahti, A.; Harlin, A.; Sixta, H. Upcycling of waste paper and cardboard to textiles. *Green Chem.* **2016**, *18*, 858–866. [CrossRef]
32. Grabar, K.C.; Freeman, R.G.; Hommer, M.B.; Natan, M.J. Preparation and characterization of Au colloid monolayers. *Anal. Chem.* **1995**, *67*, 735–743. [CrossRef]
33. Theivasanthi, T.; Christma, F.L.A.; Toyin, A.J.; Gopinath, S.C.B.; Ravichandran, R. Synthesis and characterization of cotton fiber-based nanocellulose. *Int. J. Biol. Macromol.* **2018**, *109*, 832–836. [CrossRef] [PubMed]
34. Navarro, J.R.; Werts, M.H. Resonant light scattering spectroscopy of gold, silver and gold–silver alloy nanoparticles and optical detection in microfluidic channels. *Analyst* **2013**, *138*, 583–592. [CrossRef]
35. Zhang, X.-Y.; Han, D.; Pang, Z.; Sun, Y.; Wang, Y.; Zhang, Y.; Yang, J.; Chen, L. Charge transfer in an ordered Ag/Cu2S/4-MBA system based on surface-enhanced raman scattering. *J. Phys. Chem. C* **2018**, *122*, 5599–5605. [CrossRef]
36. Xu, J.; Li, X.; Wang, Y.; Hua, T.; Guo, R.; Miao, D.; Jiang, S. Flexible, stable and sensitive surface-enhanced Raman scattering of graphite/titanium-cotton substrate for conformal rapid food safety detection. *Cellulose* **2020**, *27*, 941–954. [CrossRef]

Article

Green Synthesis of Thermo-Responsive Hydrogel from Oil Palm Empty Fruit Bunches Cellulose for Sustained Drug Delivery

Maha Mohammad Al-Rajabi [1] and Yeit Haan Teow [1,2,*]

[1] Department of Chemical and Process Engineering, Faculty of Engineering and Built Environment, Universiti Kebangsaan Malaysia, Bangi 43600, Selangor Darul Ehsan, Malaysia; p92216@siswa.ukm.edu.my
[2] Research Centre for Sustainable Process Technology (CESPRO), Faculty of Engineering and Built Environment, Universiti Kebangsaan Malaysia, Bangi 43600, Selangor Darul Ehsan, Malaysia
* Correspondence: yh_teow@ukm.edu.my; Tel.:+60-389-217-095

Abstract: Drug delivery is a difficult task in the field of dermal therapeutics, particularly in the treatment of burns, wounds, and skin diseases. Conventional drug delivery mediums have some limitations, including poor retention on skin/wound, inconvenience in administration, and uncontrolled drug release profile. Hydrogels able to absorb large amount of water and give a spontaneous response to stimuli imposed on them are an attractive solution to overcome the limitations of conventional drug delivery media. The objective of this study is to explore a green synthesis method for the development of thermo-responsive cellulose hydrogel using cellulose extracted from oil palm empty fruit bunches (OPEFB). A cold method was employed to prepare thermo-responsive cellulose hydrogels by incorporating OPEFB-extracted cellulose and Pluronic F127 (PF127) polymer. The performance of the synthesized thermo-responsive cellulose hydrogels were evaluated in terms of their swelling ratio, percentage of degradation, and in-vitro silver sulfadiazine (SSD) drug release. H8 thermo-responsive cellulose hydrogel with 20 w/v% PF127 and 3 w/v% OPEFB extracted cellulose content was the best formulation, given its high storage modulus and complex viscosity (81 kPa and 9.6 kPa.s, respectively), high swelling ratio (4.22 ± 0.70), and low degradation rate (31.3 ± 5.9%), in addition to high $t_{50\%}$ value of 24 h in SSD in-vitro drug release to accomplish sustained drug release. The exploration of thermo-responsive cellulose hydrogel from OPEFB would promote cost-effective and sustainable drug delivery system with using abundantly available agricultural biomass.

Keywords: cellulose hydrogel; thermo-responsive; sustained release; silver sulfadiazine; burn wound

1. Introduction

Drug delivery is a difficult task in the field of dermal therapeutics, particulalry for the treatment of burns, ulcers, and wounds [1]. Conventional drug delivery medium exists in many forms such as, topical liquid (solutions, suspensions, and emulsions), semi-solid formulations (ointments and creams), as well as dry traditional dressings including cotton wool, natural or synthetic bandages and gauzes [2]. However, topical liquid and semi-solid formulations have poor retention on the skin/wound surface where repeated application is required [3]. In addition, the conventional drug delivery media are inconvenient to administer. The dosage forms need to be rubbed into the skin for dispersing the formulations [4], causing pain, inflammation, and irritation [5]. Although dry traditional dressings are easier to use, they do not provide the moist environment that is required for wound healing [2]. On the top of that, conventional methods show uncontrolled drug release profiles, where the drug delivery is rapid and a high topical concentration might result in toxic effects [6]. To address these drawbacks, an alternative drug delivery method—hydrogels—was introduced [2,5].

Hydrogels are basically networks of three-dimensional synthetic or natural polymeric chains, cross-linked by physical or chemical bonds, which are capable of absorbing, swelling, and deswelling (releasing) extensive amount of trapped water, solvent or biological fluids without dissolving into it [7]. The ability of hydrogels to absorb large amounts of water (up to 99% of their weight) [8] and to give a spontaneous response to stimuli imposed on it such as temperature [9], pH [10,11], ionic strength [12], light [13], electric and magnetic fields [14], making them particularly useful for application in various biomedical field, especially as drug delivery media.

Thermo-responsive polymers have the ability to respond to changes in temperature, which make them useful as a dermal drug delivery medium. They exhibit a phase transition at a specific temperature, which causes a sudden change in the solubility. Polymers which become insoluble upon heating are called lower critical solution temperature (LCST) polymers. Whereas polymers which become soluble upon heating are named upper critical solution temperature (UCST) polymers [15]. Pluronic F127 (PF127), a kind of LCST polymer, presents good potential to be used as a hydrogel material as this polymer is in sol-phase below the phase transition temperature (room temperature) and changes to a gel-phase as the temperature is increased above the transition temperature (body temperature) [16]. Due to this sol-gel transition characteristic, it could be poured onto the skin to fill the wound/burn surfaces in sol-phase [17], and transform into a rigid hydrogel. In addition, PF127 was reported to be non-toxic [18] and has been approved by the Food and Drug Administration (FDA) for use as a pharmaceutical ingredient [19]. However, PF127 could not be used alone for hydrogel formation due to its inadequate mechanical strength and mechanical stability [20]. For this, another polymer (or polymers) has to be added to the hydrogel's network for mechanical strength enhancement. Cellulose is a sustainable, biodegradable, biocompatible, and low toxicity biopolymer. Multiple research works had been developed employing PF127 thermo-responsive polymer with cellulose/cellulose derivatives for hydrogel formation. Kim et al., (2012) synthesized methylcellulose-PF127 hydrogel for anti-cancer docetaxel (DTX) drug delivery. Sustained release pattern of methylcellulose-PF127 hydrogel was significantly enhanced anti-cancer effects of DTX [21]. On the other hand, carboxymethyl cellulose sodium (CMCs)-PF127 hydrogel was developed by Wang et al., (2016) to deliver cortex moutan drug for atopic dermatitis treatment. It was found that the presence of CMCs can appreciably improve the physical properties of PF127 hydrogel, which makes it more suitable for tailored drug loading [19]. However, efforts have been taken to produce more sustainable and green hydrogels, such as using cellulose extracted from agricultural biomass [22], engineered hydrogels based on sustainable cellulose acetate [23], as well as manufacturing of cellulose/alginate monolithic hydrogel for environmental applications [24].

Palm oil industry is considered the vital manufacturing industry in Malaysia [25]. However, this industry has long been associated with negative results such as tropical deforestation, biodiversity loss, and water pollution [26]. In addition, the production of palm oil faces numerous environmental challenges due to its waste generation during the production processes [27]. Oil palm empty fruit bunches (OPEFB) is the most abundant plantation biomass waste. Malaysia produces around 7.78 million tonnes of OPEFB annually [28] and these OPEFB are burned in boilers as a power source. However, this practice had resulted air pollution and raised awareness on the resource management issues [29]. OPEFB consists of 37.3–46.5% cellulose, 25.3–33.8% hemicellulose, and 27.6–32.5% lignin [30]. High cellulose content in OPEFB makes it a good option to be selected as the sustainable polymer to support the network structure of hydrogel for its mechanical strength enhancement.

The objective of this study is to explore a green synthesis method for the development of thermo-responsive cellulose hydrogels using cellulose extracted from OPEFB. The performance of the synthesized thermo-responsive cellulose hydrogels are evaluated in terms of their swelling ratio, percentage of degradation, and in-vitro drug (silver sulfadiazine) release. The exploration of thermo-responsive cellulose hydrogel with the use of cellulose extracted from OPEFB is a good innovation research in utilizing agricultural waste as a new source of cellulose in drug delivery medium. Long-term economic, social, and environmental sustainability could be ensured with the use of cost-effective and abundantly available raw materials.

2. Materials and Methods

2.1. Materials

Oil palm empty fruit bunches (OPEFB) was collected from the Tennamaram palm oil mill located at Bestari Jaya (Selangor Darul Ehsan, Malaysia). NaOH and hydrogen peroxide (30 w/w%) were obtained from Classic Chemicals Sdn. Bhd. (Selangor Darul Ehsan, Malaysia). Formic acid (98–100 w/w%) was purchased from Thermo Fisher Scientific (Waltham, MA, USA). Ethanol (99.5 v/v%) was supplied by Scienfield Expertise PLT (Selangor Darul Ehsan, Malaysia). Pluronic F127 (PF127) (molecular weight: 12,600 Da, 70 w/w% polyethylene oxide (PEO)) was obtained from Sigma-Aldrich (Hamburg, Germany). Silver sulfadiazine (SSD) (>98 w/w%) was supplied by Tokyo Chemical Industry (Tokyo, Japan). All chemicals were ACS grade and used as purchased.

2.2. Synthesis of Thermo-Responsive Cellulose Hydrogel

The cellulose extraction process was adapted and modified from the study conducted by Nazir et al. (2013) [31]. Firstly, OPEFB was washed with 1 w/v% detergent until the rinse water turned colorless. Next, washed OPEFB was dried at 100 ± 2 °C until a constant weight was obtained. Following, dried OPEFB was cut, and sieved using 1.18 mm opening mesh sieve. Consequently, dry OPEFB was de-waxed using 70 v/v% ethanol in soxhlet extraction apparatus for 6 h at 78 ± 2 °C. The OPEFB fibers were then washed with deionized (DI) water and dried at 100 ± 2 °C. 3 w/v% NaOH solution was added to de-waxed OPEFB fibers and heated to 121 °C for 1 h for delignification. Acid treatment was then started by soaking 10 g of delignified OPEFB fibers in 200 mL of 20 v/v% formic acid and 10 v/v% hydrogen peroxide mixture. The mixture was heated to 85 °C for 2 h. Finally, the light yellow OPEFB biocellulose fibers were bleached by suspending in 10 v/v% hydrogen peroxide with pH 9 at 60 °C for 90 min. Next, OPEFB biocellulose fibers were rinsed with DI water until neutral pH was obtained and dried in an oven at 60 ± 2 °C. Cold method was employed for the synthesis of thermo-responsive cellulose hydrogel [32]. Typically, 15–35 w/v% of PF127 was dissolved in DI water and kept in refrigerator (2–8 °C) for 20 h until complete dissolution was achieved. Next, 0.0–3.0 w/v% of extracted cellulose fibers were added into PF127 aqueous solution. The mixed solution was continuously stirred at 200 rpm for 7 days at low temperature (2–8 °C) to obtain a homogeneous thermo-responsive cellulose hydrogel. The formula of thermo-responsive cellulose hydrogel developed in this study was summarized in Table 1.

Table 1. Formula of thermo-responsive cellulose hydrogel.

Sample	PF127 Polymer (w/v%)	Cellulose Fibers (w/v%)	DI Water (w/v%)
H1		0.0	85
H2	15	1.0	84
H3		2.0	83
H4		3.0	82
H5		0.0	80
H6	20	1.0	79
H7		2.0	78
H8		3.0	77
H9		0.0	75
H10	25	1.0	74
H11		2.0	73
H12		3.0	72

Table 1. Cont.

Sample	PF127 Polymer (w/v%)	Cellulose Fibers (w/v%)	DI Water (w/v%)
H13		0.0	70
H14		1.0	69
H15	30	2.0	68
H16		3.0	67
H17		0.0	65
H18		1.0	64
H19	35	2.0	63
H20		3.0	62

2.3. Characterization of Thermo-Responsive Cellulose Hydrogel

2.3.1. Sol-Gel Transition Temperature (SGTT)

Tube inversion method was applied to determine the sol-gel transition temperature of the synthesized thermo-responsive cellulose hydrogel. 1 mL of thermo-responsive cellulose hydrogel was added into a glass vial and heated from 15 °C to 90 °C at heating rate of 1 °C/min for 15–25 °C and at heating rate of 5 °C/min for 25–90 °C in a water bath [33]. The flow behaviour of the thermo-responsive cellulose hydrogel at 4 °C, 20 °C, and 37 °C was observed by tilting the vial at different temperature: 4 °C (storing temperature in fridge), 20 °C (ambient temperature), and 37 °C (human body temperature) [34]. The flow behaviour of the thermo-responsive cellulose hydrogel was described in four categories: liquid free flow, slow flow (high viscosity), hard to flow (weak gel), solid-like behavior, and non-free flow. The temperature at which the thermo-responsive cellulose hydrogel was immobile was recorded as the gelation transition temperature [19]. Triplicate SGTT test was conducted for each thermo-responsive cellulose hydrogel and the average SGTT value was reported with standard deviation. Thermo-responsive cellulose hydrogels with selective flow behaviour were chosen as the medium for drug delivery. It was further characterized on its functional group, surface morphology and structure, and rheological property.

2.3.2. Functional Group

The functional groups presence on the synthesized thermo-responsive cellulose hydrogel were ascertained using fourier transform infrared spectroscopy (FTIR), Nicolet 6700 (Thermo Scientific, Waltham, MA, USA) at attenuated total reflectance (ATR) mode for wavenumber ranging from 500 to 4000 cm^{-1} and under 32 scans.

2.3.3. Surface Morphology and Structure

Top surface morphology and cross-section structure of the synthesized thermo-responsive cellulose hydrogel were observed using field emission scanning electron microscope (FESEM), Merlin Compact (Carl Zeiss, Jena, Germany). Firstly, the synthesized thermo-responsive cellulose hydrogel was soaking in liquid nitrogen, fractured, and mounted vertically onto the sample holder. It was then coated with a thin layer of iridium using the vacuum sputter coater, Q150T S (Quorum Technologies, Lewes, East Sussex, UK). The cross-section structure of thermo-responsive cellulose hydrogel was observed under 100× magnification. Whereas, the top surface morphology of thermo-responsive cellulose hydrogel was observed under 2.50k× magnification.

2.3.4. Rheological Property

Rheological property of the synthesized thermo-responsive cellulose hydrogel was measured using Physica MCR301 Rheometer (Anton Paar, Graz, Austria). The heating rate of rheometer was set constant at 1 °C/min with temperature range between 15 °C (ambient temperature under non-physiological condition) and 37 °C (human body temperature under physioligical condition). Storage modulus (G') and complex viscosity (η^*) values were recorded as the function of temperature at fixed angular frequency of 10 1/s. G' is a measure of energy stored and recovered per deformation cycle. High G' value reflects elastic or highly structured thermo-responsive cellulose hydrogel [35]. On the other hand, η* is an overall resistance to deformation, and it is a function of complex shear modulus (G*) [36]. G* is sum of the elastic and viscous components of the thermo-responsive cellulose hydrogel, represented by G' and loss modulus (G''), respectively. η* and G* are described in Equations (1) and (2), respectively.

$$\eta^* = \frac{G^*}{\omega} \quad (1)$$

where ω is the angular frequency (rad/s):

$$G^* = G' + iG'' \quad (2)$$

where i is the complex number ($\sqrt{-1}$).

2.4. Performance of Thermo-Responsive Cellulose Hydrogel

2.4.1. Swelling and Degradation

Dried thermo-responsive cellulose hydrogel (0.2 g) was weighed and allowed to swell in 2 mL DI water at room temperature [37]. The thermo-responsive cellulose hydrogel after 15 min, 30 min, 60 min, 120 min, 180 min, and 240 min of swelling was superficially dried with filter paper and weighed. DI water was added after each sampling to maintain the DI water volume throughout the swelling and degradation test. Thermo-responsive cellulose hydrogel swelling ratio was calculated using Equation (3), whereas the percentage of thermo-responsive cellulose hydrogel degradation was calculated using Equation (4) [37]:

$$\text{Swelling ratio} = \left[\frac{W_t - W_0}{W_0}\right] \quad (3)$$

where W_t is the weight of thermo-responsive cellulose hydrogel after swelling at time t (g) and W_0 is the initial weight of dry thermo-responsive cellulose hydrogel (g):

$$\text{Percentage of degradation (\%)} = \left[\frac{W_{s0} - W_d}{W_{s0}}\right] \times 100\% \quad (4)$$

where W_d is the weight of thermo-responsive cellulose hydrogel after saturation (maximum swelling) at time t (g) and W_{s0} is the maximum weight of thermo-responsive cellulose hydrogel in the swelling test (g).

2.4.2. In-Vitro Drug Delivery Study

SSD (100 mg) was loaded into thermo-responsive cellulose hydrogel (10 g) in sol-phase to simulate the commercial SSD formulation [38]. The thermo-responsive cellulose hydrogel loaded with SSD was then stirred at 200 rpm for 1 h under low temperature (2–8 °C) to obtain a homogeneous drug-loaded thermo-responsive cellulose hydrogel [34].

Vertical diffusion cell, Copley HDT 1000 (Copley, County Durham, UK) was used for SSD release from the thermo-responsive cellulose hydrogel. 0.25 v/v% ammonia in phosphate buffer solution at pH 7.4 was the receptor medium [39,40]. On the other hand, 0.45 µm regenerated cellulose filter pre-soaked in receptor medium for 30 min was used as membrane. Firstly, vertical diffusion cell was filled with 11 mL of receptor medium. It was then heated to 32 °C under continuous stirring at 600 rpm [41]. Thereafter, 0.2 g of drug-loaded thermo-responsive cellulose hydrogel was placed at the drug donor chamber and heated in an oven at 32 °C for 10 min to ensure complete solidification. Following, the pre-soaked regenerated cellulose filter was then placed between the drug-loaded thermo-responsive cellulose hydrogel and heated receptor medium in vertical diffusion cell to maintain a close contact between thermo-responsive cellulose hydrogel and receptor medium, allowing SSD releases from thermo-responsive cellulose hydrogel to receptor medium [42].

Receptor medium (1 mL) was withdrawn from the vertical diffusion cell at 15 min intervals for the first hour, followed by 1 h intervals for the following 25 h. Once the receptor medium was withdrawn, an equal volume of fresh receptor medium was added into the vertical diffusion cell. SSD concentration in sampled receptor medium was analyzed by UV-visible spectrophotometer, Genesys 10S UV-VIS (Thermo Scientific, Waltham, MA, USA) at the wavelength of 260 nm. Cumulative percentage of SSD release from thermo-responsive cellulose hydrogel was calculated using Equation (5) [43]:

$$\text{Cumulative percentage of SSD release (\%)} = \left[\frac{C_t V + v \sum_1^{t-1} C_t}{W_D}\right] \times 100\% \quad (5)$$

where C_t is the concentration of SSD released at time t (mg/mL), V is the volume of receptor medium (mL), v is volume of receptor medium being withdrawn (mL), and W_D is the initial amount of SSD in thermo-responsive cellulose hydrogel (mg).

2.4.3. Kinetic Study of Drug Release

Several kinetic models were applied to describe the kinetic governing SSD release from thermo-responsive cellulose hydrogel.

(i) Zero-Order Model

Zero-order model assumes the area of dosage form (thermo-responsive cellulose hydrogel) does not change significantly with time. This model supports slow release of drug as the dug loaded into dosage form is not disaggregate. The zero-order model is depicted by Equation (6) [44]:

$$Q_t = Q_0 + k_0 t \quad (6)$$

where Q_t is the percentage of drug released at time t (%), Q_0 is the initial percentage of drug in receptor medium (%), k_0 corresponds to Zero-order model constant (1/h), and t is the sampling time of receptor medium (h).

(ii) First-Order Model

First-order model often use to describe the absorption and release of water soluble drug. The rate of drug release is depends on its initial concentration. First-order model is depicted by Equation (7) [45]:

$$\text{Log} Q_t = \text{Log} Q_0 - \frac{k_1}{r} t \quad (7)$$

where k_1 corresponds to First-order model constant (1/h) and r is the conversion factor (2.303).

(iii) Higuchi Model

The Higuchi model describes the release of soluble and sparingly soluble drug. Higuchi model assumes that: (i) the initial drug concentration in dosage form is higher than drug solubility, (ii) drug only spread in one dimension, (iii) drug diffusivity does not change, (iv) sink condition is achieved where the receptor medium has high capacity to dissolve the drug [41], Higuchi model is depicted by Equation (8) [46]:

$$Q_t = k_H t^{1/2} \quad (8)$$

where k_H is Higuchi model constant ($1/h^{1/2}$).

2.4.4. Mechanism of Drug Release

Korsmeyer-Peppas model is a generalized model of Higuchi model to describe drug release mechanism from polymeric dosage form where erosion and/or dissolution of the dosage form occurs. Korsmeyer-Peppas model is depicted by Equation (9) [47]:

$$Q_t = k_r t^n \tag{9}$$

where k_r is the Korsmeyer-Peppas model constant $(1/(h)^n)$, n is the diffusion exponent indicates the mechanism of drug molecules transport from the dosage form.

2.5. Statistical Analysis

Each experiment was independently repeated in triplicate, and the results were presented as mean ± standard deviation (SD). SD was calculated using the STDEV formula in Excel.

3. Results and Discussion

3.1. Characterization of Thermo-Responsive Cellulose Hydrogel

3.1.1. Sol-Gel Transition Temperature (SGTT)

The SGTT and flow behavior of the thermo-responsive cellulose hydrogels are summarized in Table 2, while Figure 1 shows the phase diagram of the synthesized thermo-responsive cellulose hydrogels. Generally, SGTT consists of a lower critical solution temperature (LCST) and upper critical solution temperature (UCST). LCST is the temperature at which the thermo-responsive cellulose hydrogel in sol form transforms into gel form due to an increase of temperature. On the other hand, UCST is the temperature at which the thermo-responsive cellulose hydrogel in gel form transform back into gel form by further increase in temperature [48].

Table 2. SGTT and flow behavior of the thermo-responsive cellulose hydrogels.

Sample	LCST (°C)	UCST (°C)	Status at (4 °C)	Status at (20 °C)	Status at (37 °C)
H1	N/D	N/D	-	-	-
H2	N/D	N/D	-	+	+
H3	N/D	N/D	+	+	++
H4	N/D	N/D	+	++	++
H5	24.0 ± 1.0	58.3 ± 2.9	-	+	+++
H6	23.7 ± 0.6	61.7 ± 2.9	+	+	+++
H7	22.3 ± 1.2	68.3 ± 2.9	+	++	+++
H8	21.0 ± 1.0	78.3 ± 2.9	+	++	+++
H9	20.0 ± 1.0	73.3 ± 2.9	-	+++	+++
H10	17.0 ± 1.0	78.3 ± 2.9	+	+++	+++
H11	15 ± 0.0	86.7 ± 2.9	+	+++	+++
H12	<15	N/D	++	+++	+++
H13	17.7 ± 0.6	N/D	-	+++	+++
H14	15 ± 0.0	N/D	+	+++	+++
H15	<15	N/D	++	+++	+++
H16	<15	N/D	++	+++	+++
H17	16.7 ± 0.6	N/D	-	+++	+++
H18	<15	N/D	+	+++	+++
H19	<15	N/D	++	+++	+++
H20	<15	N/D	++	+++	+++

N/D: not defined; - Liquid free flowing; + High viscosity, slow flow; ++ Weak gel, hard to flow; +++ Solid-like behavior, non-free flow.

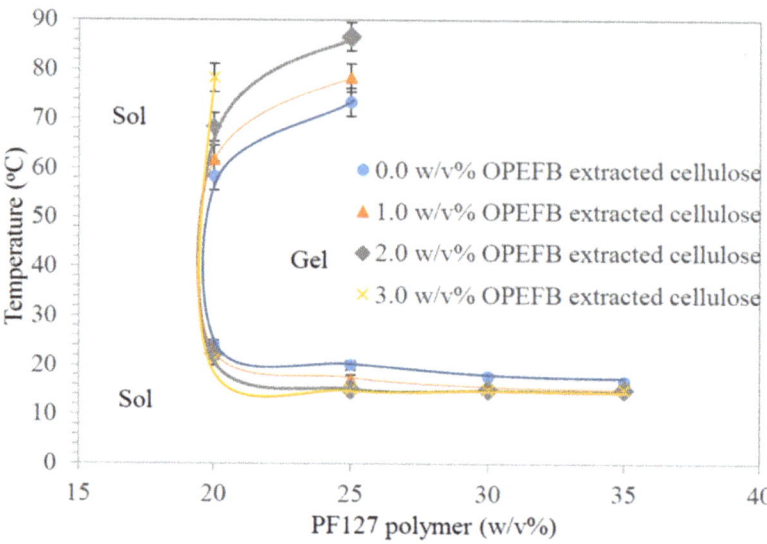

Figure 1. Phase diagram of the thermo-responsive cellulose hydrogels.

As presented in Table 2, SGTT is not measurable for H1-H4 thermo-responsive cellulose hydrogels. PF127 consists of hydrophilic PEO and hydrophobic polypropylene oxide (PPO) blocks (unimers) arranged in (PEOx-PPOy-PEOx) tri-block structure [49]. Unimers are water soluble, hence it prevents micelle formation, packing and entanglement for solid-like hydrogel formation at low PF127 percentage (15 w/v%) [50]. Whereas, LCST of H5-H20 thermo-responsive cellulose hydrogels is decreased with the increasing of PF127 percentage at constant OPEFB extracted cellulose percentage; while UCST of H5-H20 thermo-responsive cellulose hydrogels shows an opposite trend. This could be explained with the aid of Figure S1. When the PF127 percentage is exceeding critical micelle concentration (CMC) and/or critical micelle temperature (CMT), unimers are aggregate to form spherical micelles with hydrophobic PPO blocks as central core surrounded by hydrophilic PEO chains. This process is called micellization. The PF127 micelles are stable in water with PEO hydrophilic chains in-contact with water molecules through the formation of hydrogen bonds [17,51]. More PF127 micelles in the thermo-responsive cellulose hydrogel at higher PF127 percentage will drive the aggregation of unimers and promote for the formation solid-like hydrogel at lower temperature [52]. This explains the decrease of LCST with the increase of PF127 percentage at constant OPEFB extracted cellulose percentage.

When the temperature is raised above the LCST, hydrogen bonds between water molecules and PEO hydrophilic chains of PF127 are broken due to dehydration of hydrophilic PEO chains [53]. The occurrence of segregation between PEO chains and PPO blocks is attributed by shrinking of the gel phase structure. As hydrophobic PPO associations are dominany, this leads to solid-like hydrogel formation [52]. An increasing PF127 percentage in a thermo-responsive cellulose hydrogel formulation results in greater aggregation in the form of solid-like hydrogel. As such, a higher UCST is required to loosen the PF127 micelle interaction for transforming it from a gel state to a sol state [54]. A similar trend was observed for thermo-responsive cellulose hydrogels synthesized with different OPEFB -extracted cellulose percentages at constant PF127 percentage. The LCST of thermo-responsive cellulose hydrogel is decreased while the UCST of thermo-responsive cellulose hydrogel is increased with the increasing of OPEFB extracted cellulose percentage in formulation. This is because OPEFB-extracted cellulose could bind with hydrophilic PEO chains through intermolecular hydrogen bonding [55]. This would promote dehydration of hydrophilic PEO chains as the bonding between OPEFB-extracted cellulose and hydrophilic PEO chains decreases the hydrogen bonding between hydrophilic PEO chains and water molecules. This will cause an increase in entanglement of adjacent P127 micelles, thus promoting the occurrence of gelation at a lower temperature [55]. However, the UCST of H12-H20 thermo-responsive cellulose hydrogels is not detected. This is estimated to happen at temperatures higher than 90 °C.

For the aspect of flow behaviour, pristine hydrogel (H1) does not show changes in flow behavior at different temperatures but thermo-responsive cellulose hydrogels did. Generally, the flow behaviour of thermo-responsive cellulose hydrogels was changed from liquid free flowing to slow flow, hard flowing, and finally non-free flow with the increase of both OPEFB-extracted cellulose percentage and/or PF127 percentage. The change of flow behavior is attributed by higher viscosity in thermo-responsive cellulose hydrogel formulations [56]. The ideal thermo-responsive cellulose hydrogel should be free flowing under the preparation conditions (4 °C) to ease the drug loading process [57], easy flow at room temperature (20 °C) for applying onto the skin [17] and form a non-flowing solid hydrogel at body temperature (37 °C) for its application as a drug delivery medium [17]. As presented in Table 2, H5-H8 thermo-responsive cellulose hydrogels with 20 w/v% PF127 fulfilled the prescribed properties and thus are suitable for use as a drug delivery medium. With this, H5-H8 thermo-responsive cellulose hydrogels were further characterized regarding their functional groups, surface morphology and structure, and rheological properties.

3.1.2. Functional Groups

The FTIR spectrum of OPEFB-extracted cellulose, PF127, and H6-H8 thermo-responsive cellulose hydrogels are shown in Figure 2. As illustrated in Figure 2a, a broad peak was observed for H6 thermo-responsive cellulose hydrogel at 3529 cm^{-1}, confirming the presence of O–H stretching vibrations [58]. On the other hand, the peak at 1647 cm^{-1} was attributed by O–H bending [59]. The intensity of the O–H stretching vibration peak and O–H bending peak for H6 thermo-responsive cellulose hydrogel was the highest compared to PF127 and OPEFB-extracted cellulose. This indicates that H6 thermo-responsive cellulose hydrogel has higher hydrophilicity than its base material (PF127) due to the alteration of its properties by hydrophilic OPEFB-extracted cellulose. The influence of hydrophilic OPEFB-extracted cellulose could also reflected by thermo-responsive cellulose hydrogels at different OPEFB-extracted cellulose percentages. By varying the OPEFB extracted cellulose percentage from 1 to 3 w/v% for H6-H8 thermo-responsive cellulose hydrogels, the intensity of the O–H stretching vibration peak and O–H bending peak was increased.

On the other hand, there was a slight shift of the O–H stretching vibration peak in thermo-responsive cellulose hydrogels compared to both PF127 and OPEFB-extracted cellulose. The shift of the O–H stretching vibration peak to a higher wavenumber was probably due to the increase of O–H bond strength [60]. Strong O–H bond strength in thermo-responsive cellulose hydrogels indicates strong intermolecular hydrogen bonding between PF127 and OPEFB-extracted cellulose.

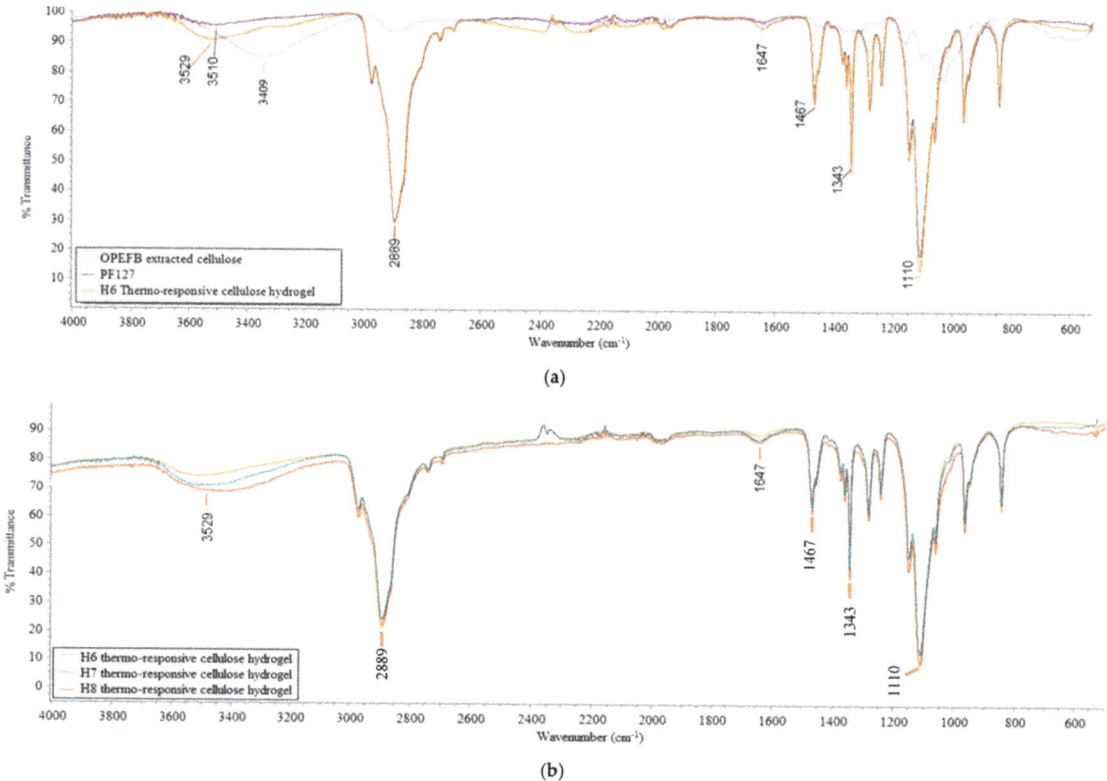

Figure 2. FTIR spectrum of (**a**) OPEFB extracted cellulose, PF127, H6 thermo-responsive cellulose hydrogel (**b**) H6-H8 thermo-responsive cellulose hydrogels.

3.1.3. Surface Morphology and Cross-Section View

Figure 3 presents the surface morphology and cross-sectional view of H5-H8 thermo-responsive cellulose hydrogels. Generally, the surface of thermo-responsive cellulose hydrogels was not smooth, with fibers deposited on the thermo-responsive cellulose hydrogels' surface. H5 and H6 thermo-responsive cellulose hydrogels were associated with some voids at a diameter range between 0.22–0.76 μm and 0.22–0.66 μm, respectively. In comparison, the surfaces of H7 and H8 thermo-responsive cellulose hydrogels were smoother with no voids on the surface. The smoother H7 and H8 thermo-responsive cellulose hydrogels' surface was possibly due to a stronger cross-linking effect between OPEFB-extracted cellulose and PF127 at higher OPEFB-extracted cellulose percent. High cross-linking density was proven to be able to produce a dense and smooth surface [61]. A smoother hydrogel surface was observed by Rasoulzadeh and Namazi (2017) with the increase of graphene oxide content in carboxymethyl cellulose hydrogel and attributed to stronger hydrogen bonding interactions between graphene oxide and carboxymethyl cellulose [62].

On the other hand, the cross-section morphology of H6-H8 thermo-responsive cellulose hydrogels were rougher than that of pristine thermo-responsive hydrogel, H5. Similar to surface morphology, it was probably due to the presence of OPEFB extracted cellulose in thermo-responsive hydrogel's matrix. However, there was not significant aggregation of OPEFB-extracted cellulose fibers in thermo-responsive hydrogel matrix. Hydrophilic-hydrophilic interaction between OPEFB-extracted cellulose and PF127 had resulted well distribution of OPEFB-extracted cellulose fiber in the thermo-responsive cellulose hydrogels' matrix.

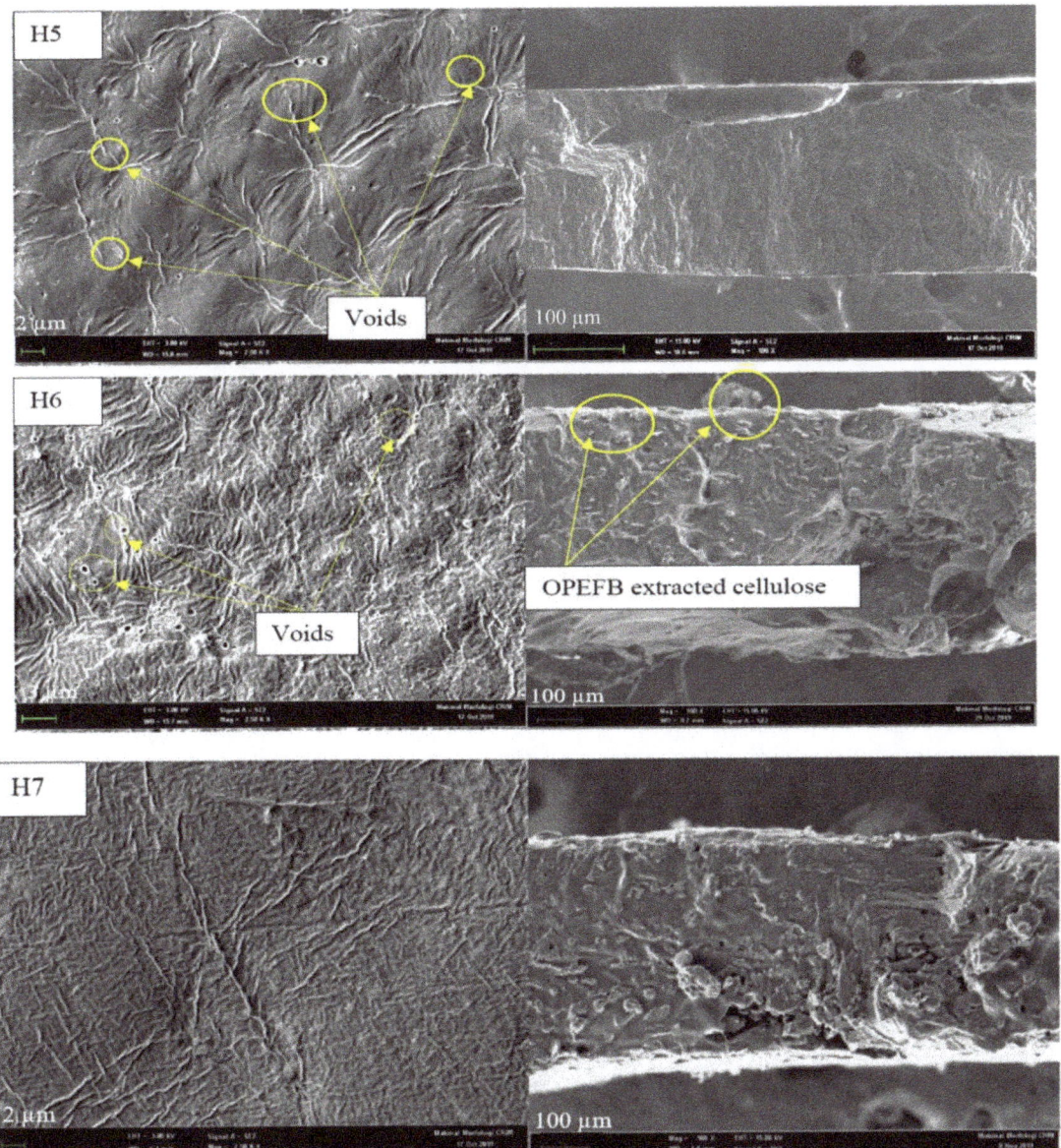

Figure 3. *Cont.*

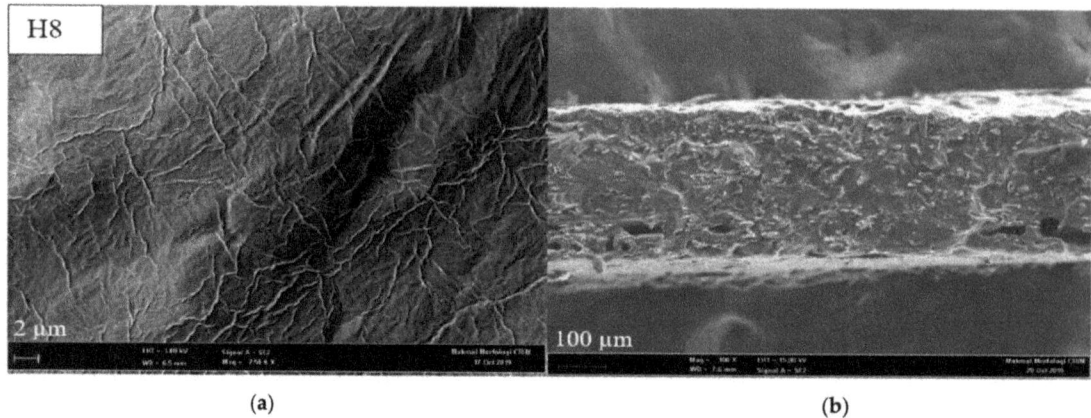

Figure 3. (a) Surface morphology and (b) cross-sectional view of thermo-responsive cellulose hydrogels at the magnification of 2.50k× and 100×, respectively.

3.1.4. Rheological Property

Figure 4 presents the rheological behavior (storage modulus (G′) and complex viscosity (η*)) of thermo-responsive cellulose hydrogels as a function of temperature. As shown in Figure 4, both the G′ and η* values of thermo-responsive cellulose hydrogels are temperature-dependent, confirming the thermo-responsive behaviour of the hydrogels. At phase I where the temperature is between 15 °C and LCST, the pristine thermo-responsive hydrogel H5 recorded a zero value for both G′ and η*. Whereas, the G′ and η* values were increased with the increase of OPEFB-extracted cellulose percent in thermo-responsive hydrogel formulations. This indicates an increase of the elasticity and viscosity of thermo-responsive cellulose hydrogels with the increase of OPEFB-extracted cellulose percent in the thermo-responsive hydrogels' formulation. The results agree well with the flow behavior of thermo-responsive cellulose hydrogels, where it changed from liquid free flow to slow flow, hard flowing, and finally non-free flow with the increase of OPEFB-extracted cellulose percent due to higher elasticity and viscosity of the formulation. A viscous thermo-responsive cellulose hydrogel at room temperature is important for a drug delivery medium in topical applications. A viscous thermo-responsive cellulose hydrogel is needed for it to stay on the skin and not flow off after application [63]. On the other hand, both G′ and η* values of thermo-responsive cellulose hydrogels were decreased with the increase of temperature in phase I. As the temperature increased, the thermo-responsive cellulose hydrogel structure was deformed and transformed into a sol-phase. The decrease of viscosity therefore led to the reduction of both G′ and η* values.

In phase II, where the temperature is between LCST and 30 °C, G′ and η* values of thermo-responsive cellulose hydrogels were increased with the increase of temperature, revealing the transition of thermo-responsive cellulose hydrogels from sol-phase to gel-phase. As the temperature increased in this transition phase (phase II), PF127 micellization and its interaction with OPEFB-extracted cellulose occurred, leading to the formation of dense, solid-like network with predominant elastic properties [52,64]. It is therefore gives larger G′ and η* values at higher temperature. However, in phase III where the temperature is between 30 °C and 37 °C, G′ and η* values reached a plateau. This signified that the thermo-responsive cellulose hydrogels' network does not undergo further structural transition at temperatures above 30 °C. The stability of thermo-responsive cellulose hydrogel with gel-like behaviour [65] during phase III is desired for thermo-responsive cellulose hydrogel application on human skin. On top of that, the plateau values of G′ and η* at phase III were used to indicate the strength of thermo-responsive cellulose hydrogels and their cross-linking density [66]. The G′ and η* values of thermo-responsive cellulose hydrogels ranged between 48–81 kPa and 5.6–9.6 kPa.s, respectively. Thus was higher than that of a thermo-responsive nanocrystal cellulose hydrogel (G′ value of 15–40 kPa) [56] and a thermo-responsive pentablock (polyethylene glycol-polycaprolactone polylactide-polycaprolactone-polyethylene glycol) one (η^* value of 1.1–2.9 kPa.s) [67], validating the favorable use of thermo-responsive cellulose hydrogel as a drug delivery medium.

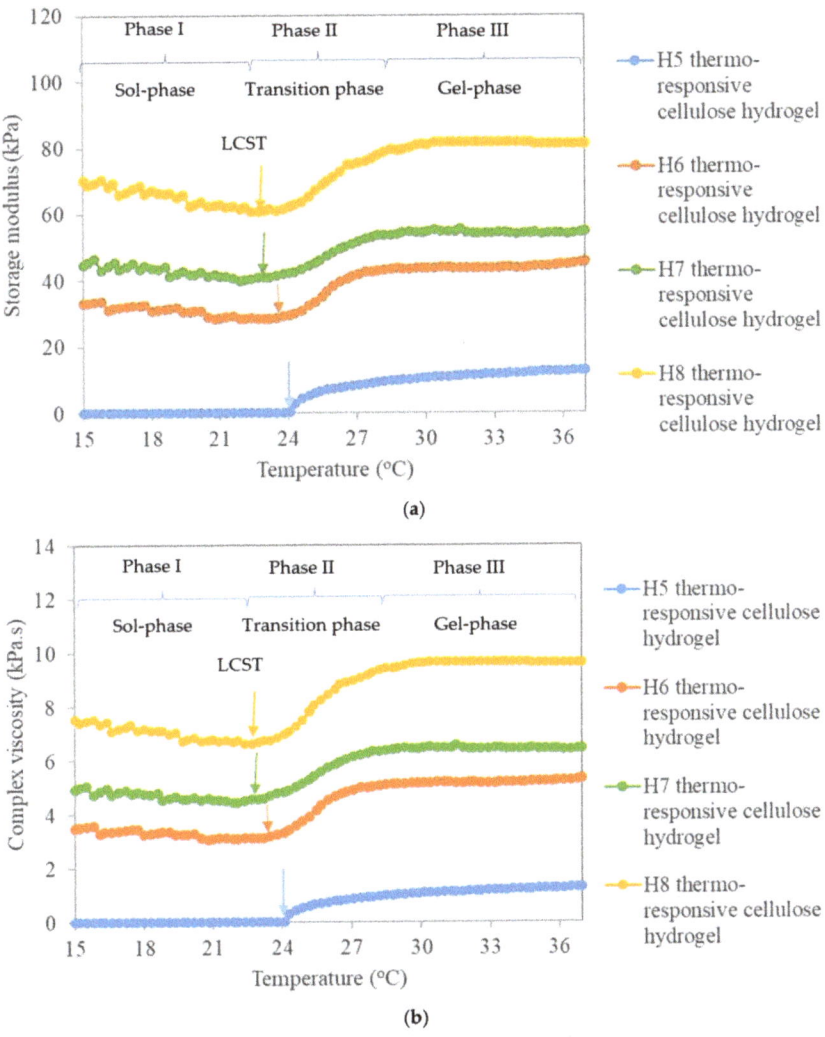

Figure 4. (a) Storage modulus (G′) and (b) complex viscosity (η^*) of thermo-responsive cellulose hydrogels as a function of temperature.

3.2. Performance of Thermo-Responsive Cellulose Hydrogel

3.2.1. Swelling and Degradation

Figure 5 shows the swelling ratio and the percentage degradation of thermo-responsive cellulose hydrogels as a function of time after immersing them in DI water at room temperature. As presented in Figure 5a, the swelling ratio of thermo-responsive cellulose hydrogel was increased with time and reaches a maximum at 15 min for pristine thermo-responsive hydrogel, H5 and at 30 min for H6-H8 thermo-responsive cellulose hydrogels. At the beginning of swelling process, hydration of the thermo-responsive cellulose hydrogels occurred. Water molecules were bonded to hydrophilic O–H groups in the thermo-responsive cellulose hydrogels, either by adsorption into the pores and/or into the thermo-responsive cellulose hydrogel's matrix [68]. Besides, the swelling ratio was increased with the increase of OPEFB-extracted cellulose percentage in the thermo-responsive hydrogels' formulation. Thermo-responsive cellulose hydrogels with higher weight percent of OPEFB-extracted cellulose exhibit a stronger affinity towards water absorption and are able to retain a higher fraction of water within their structure. This is mainly due to the presence of more O–H groups in hydrogel matrix as depicted by the higher O–H functional group intensity seen in the corresponding FTIR spectrum (Figure 2). A high swelling ratio is desirable for a thermo-responsive cellulose hydrogel to achieve a high drug loading capacity [69] and to maintain the moisture content on targeted skin area to provide cooling and soothing effects and reduce pain [70].

The swelling mechanism of thermo-responsive cellulose hydrogel could be explained by the Flory–Rehner theory [71], which posits that a cross-linked polymer swells in a solvent due to the swelling pressure, π. The swelling pressure is the summation of osmotic pressure ($\pi_{osmotic}$) and an elastic pressure ($\pi_{elastic}$), as described in Equation (10):

$$\pi = \pi_{osmotic} + \pi_{elastic} \tag{10}$$

As shown in Figure S2, when a dry thermo-responsive cellulose hydrogel was immersed in DI water, the thermo-responsive cellulose hydrogel started to swell by drawing in DI water due to the $\pi_{osmotic}$ difference between thermo-responsive cellulose hydrogel and DI water [72]. At first, $\pi_{osmotic}$ dominates the swelling pressure [72,73]. As swelling proceeds, incoming DI water molecules in the thermo-responsive cellulose hydrogel will exert pressure on the thermo-responsive cellulose hydrogel's chains. The pressure exerted on the thermo-responsive cellulose hydrogel's chains stretches the hydrogel's chains by pushing the cross-links apart [74]. Conversely, the thermo-responsive cellulose hydrogel's chains would resist the deformation, imposing a $\pi_{elastic}$ in opposite direction of $\pi_{osmotic}$ [74]. The swelling process was continued until $\pi_{elastic}$ and $\pi_{osmotic}$ were in equilibrium [72]. The swelling ratio reaches a maximum at this stage where the thermo-responsive cellulose hydrogel's chains will not further swell [75]. After reaching the maximum value, the swelling ratio starts to decrease and reaches saturation. A further increase of water content in thermo-responsive cellulose hydrogel would result in degradation and dissolution of the thermo-responsive cellulose hydrogel [68,76]. The swelling ratio values of the thermo-responsive cellulose hydrogels ranged between 1.20–4.22. This was higher than that of carbon nano-onions-reinforced natural protein nanocomposite hydrogels (swelling ratio value of 0.42–0.65) [10], validating the favorable use of thermo-responsive cellulose hydrogels as drug delivery media.

As shown in Figure 5b, percentage degradation of thermo-responsive cellulose hydrogels increased with time and eventually reached a plateau. The pristine thermo-responsive hydrogel H5 was degraded completely within 1 h. On the other hand, thermo-responsive cellulose hydrogels H6-H8 were only partially degraded after 4 h of degradation study. The degradation ability of the thermo-responsive cellulose hydrogels was decreased with the increase of OPEFB-extracted cellulose percentage in the thermo-responsive hydrogels' formulation. The higher cross-linking density of thermo-responsive hydrogel's attributed to a higher OPEFB-extracted cellulose percentage in its formulation produces strong a thermo-responsive cellulose hydrogel structure as proven by the rheological properties discussed in Section 3.1.4. The strong thermo-responsive cellulose hydrogel structure therefore prolonged the degradation process. Specifically, drug release from a drug delivery medium is controlled by the rate of degradation of the drug delivery medium. The duration of drug release can be regulated by tailoring the thermo-responsive cellulose hydrogel's degradation rate [77]. With this, H8 thermo-responsive cellulose hydrogel with high swelling ratio and low degradation rate is considered the most suitable drug delivery medium to retain high drug loading capacity with sustained drug release.

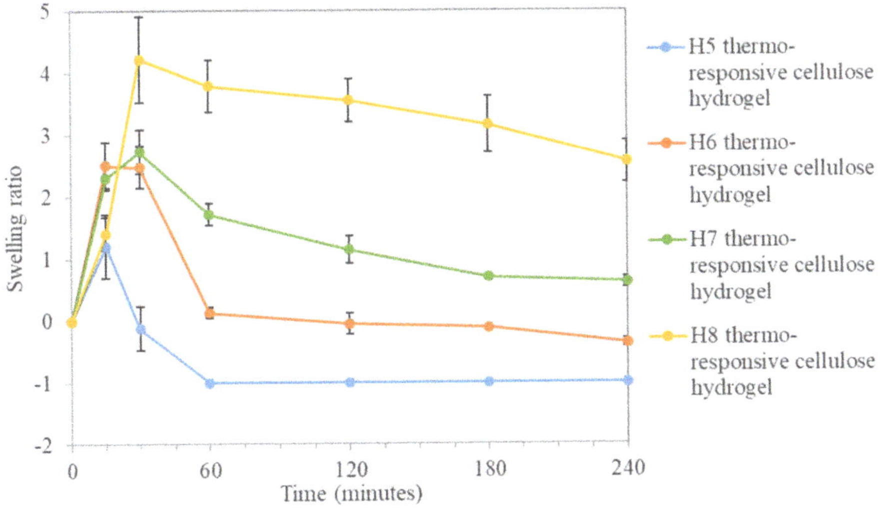

(a)

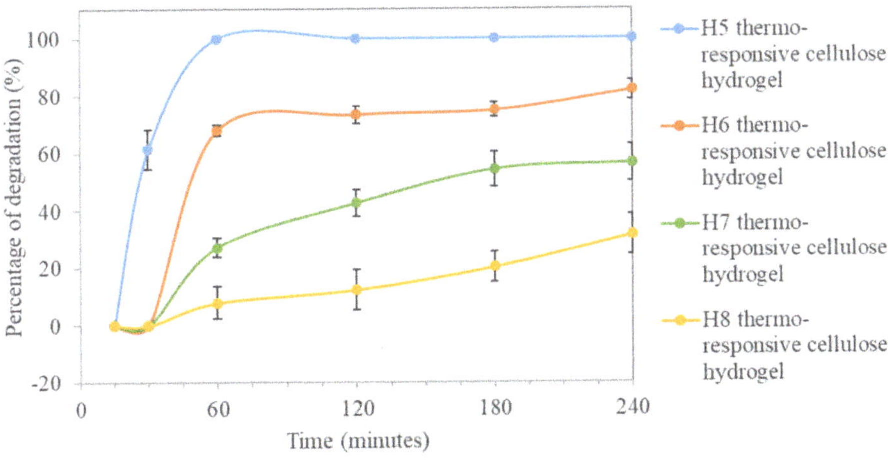

(b)

Figure 5. (a) Swelling ratio and (b) percentage degradation of thermo-responsive cellulose hydrogels.

3.2.2. In-Vitro Drug Delivery Study

The cumulative percentage of SSD released from the thermo-responsive cellulose hydrogels as a function of time is depicted in Figure 6. The half-life time of SSD release ($t_{50\%}$) is the time at which the mass fraction of SSD released reached 50%. $t_{50\%}$ for the pristine thermo-responsive hydrogel H5 was recorded at 8 h. It was increased to 17 h, 19 h, and 24 h for H6, H7, and H8 thermo-responsive cellulose hydrogels, respectively. Higher weight percent of OPEFB-extracted cellulose in the thermo-responsive cellulose hydrogels' formulation leads to strong interlocking of SSD within the thermo-responsive cellulose hydrogel network as a result of the greater interaction between PF127 and OPEFB-extracted cellulose, therefore, prolonging the sustained released of SSD. The $t_{50\%}$ values of thermo-responsive cellulose hydrogels range between 8 to 24 h were comparatively higher than commercial SSD cream and cubosomes aloe vera SSD hydrogel with $t_{50\%}$ value of 5 h and 8 h, respectively [78]. This is an impressive result proving that thermo-responsive cellulose hydrogel is a promising drug delivery medium which is able to accomplish sustained drug release reducing the fluctuation of drug levels during administration. Extended-release drug delivery media are an attractive therapeutic option for the treatment of complex chronic diseases, such as cancer [79].

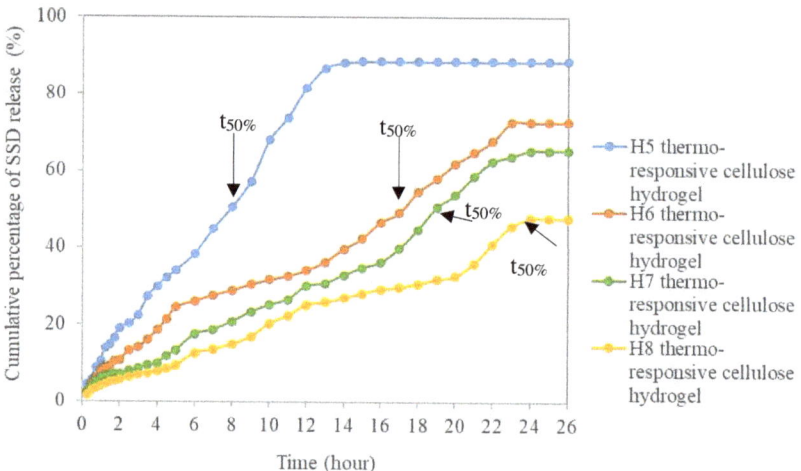

Figure 6. Cumulative percentage of SSD release from thermo-responsive cellulose hydrogels.

On top of that, the concentration of the released drug at the targeted tissue should attain the minimum inhibitory concentration (MIC) [68] where it is the minimum concentration of antimicrobial drug needed in plasma to inhibit the growth of micro-organisms [80]. Table S1 summarizes the concentration of SSD released from thermo-responsive cellulose hydrogels at different time intervals. As shown in Table S1, the concentration of SSD released from the thermo-responsive cellulose hydrogels was above 18 mg/L, the MIC for *Pseudomonas aeruginosa* [81] after 1, 2, 4.5, and 6 h of in-vitro release for H5, H6, H7, and H8 thermo-responsive cellulose hydrogels, respectively. Therefore, thermo-responsive cellulose hydrogel delivering SSD could be effective against *Pseudomonas aeruginosa*, one of the most common causes of burn wound infections [82].

3.2.3. Kinetic and Mechanism Study of Drug Release

Figure S3a–d show linear regressions of the kinetic models applied for our drug release study. The models' constants and correlation coefficient, R^2 are summarized in Table 3. A high R^2 value obtained from a specific kinetic model indicates the most appropriate kinetic model in explaining the release of the drug [82]. Among all kinetic models, the zero-order model has the highest R^2 value for all thermo-responsive cellulose hydrogel formulations in the drug release study. This indicates the zero-order model was the best fit in describing the SSD release mechanism from the thermo-responsive cellulose hydrogels. Based on the zero-order model, the release rate of thermo-responsive cellulose hydrogels is constant over time and independent of the drug concentration [83], in contrast to the other kinetic models, where the release rate is drug concentration dependent, and in which the higher the drug concentration, the faster the release rate [82]. The zero-order model constant, k_0, is used to reflect the release rate. As presented in Table 3, the k_0 value decreased with increasing OPEFB-extracted cellulose percent in the thermo-responsive hydrogel's formulation. H8 thermo-responsive cellulose hydrogel with the highest weight percent of OPEFB-extracted cellulose exhibited the lowest k_0, meaning it had a longer SSD release than other formulations. This agrees well with in-vitro drug delivery study outcomes described in Section 3.2.2.

Table 3. Kinetic and mechanism models' constants and correlation coefficient, R^2 of SSD release.

Hydrogel	Zero-Order		First-Order		Higuchi		Korsmeyer-Peppas		
	k_0	R^2	k_1	R^2	k_H	R^2	k_r	n	R^2
H5	5.583	0.980	0.140	0.954	26.662	0.968	10.839	0.727	0.994
H6	2.708	0.986	0.048	0.954	15.568	0.955	7.129	0.671	0.991
H7	2.585	0.985	0.041	0.946	14.330	0.923	5.023	0.710	0.967
H8	1.762	0.987	0.023	0.969	10.046	0.940	9.462	0.746	0.976

On the other hand, the linear regression of the Korsmeyer-Peppas model applied for drug release mechanism study is presented in Figure S3d. The diffusion exponents (n) of the Korsmeyer-Peppas model obtained from the plots are summarized in Table 3. The n values of the H5-H8 thermo-responsive cellulose hydrogels range between 0.671 and 0.746. An n value ranging between 0.5 and 1 indicates that the SSD release mechanism of thermo-responsive cellulose hydrogels followed a non-Fickian diffusion [84]. As explained by non-Fickian diffusion, the SSD release rate is controlled by both SSD drug diffusion and thermo-responsive cellulose hydrogel erosion [83]. The SSD drug release mechanism could be explained with the aid of Figure S4. Initially, SSD-loaded thermo-responsive cellulose hydrogel applied onto a wound surface is in close contact with wound exudates. Then, SSD molecules are transferred from the thermo-responsive cellulose hydrogel to the wound surface through diffusion due to this close contact. The driving force for SSD release by diffusion is the SSD concentration gradient between the thermo-responsive cellulose hydrogel and the wound surface [85]. Besides SSD diffusion, SSD release from thermo-responsive cellulose hydrogels is also contributed by physical dissolution of the thermo-responsive cellulose hydrogel as a result of its degradation [86]. As a thermo-responsive cellulose hydrogel's matrix is eroded, entrapped SSD molecules are set free and released to the wound surface. The degradation of thermo-responsive cellulose hydrogels contributing to SSD release rate had been validated in the degradation study presented in Section 3.2.1.

4. Conclusions

In conclusion, this study had successfully explored a green synthesis method for the development of thermo-responsive cellulose hydrogels from OPEFB, which is the most produced agricultural biomass in the palm oil industry. This study can not only solve the problem of OPEFB biomass waste, but also produce thermo-responsive cellulose hydrogels with superior performance in comparison with literature data and commercial drug delivery media. Among the different thermo-responsive cellulose hydrogel formulations, the H8 thermo-responsive cellulose hydrogel with 3 w/v% OPEFB-extracted cellulose is the best thermo-responsive cellulose hydrogel formulation in this study as it exhibited a high storage modulus and complex viscosity (81 kPa and 9.6 kPa.s, respectively), high swelling ratio (4.22 ± 0.70), and low degradation rate (31.3 ± 5.9%), in addition to high $t_{50\%}$ value of 24 h in SSD in-vitro drug release to accomplish sustained drug release which is comparatively higher than that of commercial SSD cream. The sustained drug release of the thermo-responsive cellulose hydrogel was also confirmed by best fitting of the drug release study results to a zero-order kinetic model. On the other hand, the SSD release mechanism of thermo-responsive cellulose hydrogels followed the non-Fickian diffusion, model, in which SSD release rate is controlled by both SSD drug diffusion and thermo-responsive cellulose hydrogel erosion. The impressive results obtained from this study confirm that thermo-responsive cellulose hydrogels are promising drug delivery media which are able to accomplish sustained drug release reducing the fluctuation of drug levels during administration while using an abundantly available agricultural biomass.

Supplementary Materials: The following are available online at https://www.mdpi.com/article/10.3390/polym13132153/s1, Figure S1: Schematic illustration of PF127 micelles formation and its interaction with OPEFB extracted cellulose for the synthesis of thermo-responsive cellulose hydrogel. Figure S2: Swelling mechanism of thermo-responsive cellulose hydrogel. Figure S3: Linear regression of kinetic and mechanism models (a) Zero-order model (b) First-order model (c) Higuchi model (d) Korsmeyer- peppas model. Figure S4: SSD release from thermo-responsive cellulose hydrogels by different mechanisms. Table S1: Concentration of SSD released from thermo-responsive cellulose hydrogels at different time interval.

Author Contributions: Conceptualization, Y.H.T. and M.M.A.-R.; methodology, Y.H.T. and M.M.A.-R.; formal analysis, M.M.A.-R.; writing—original draft preparation, M.M.A.-R.; writing—review and editing, Y.H.T.; supervision Y.H.T.; funding acquisition, Y.H.T. Both authors have read and agreed to the published version of the manuscript.

Funding: The authors wish to gratefully acknowledge the funding for this work by Dana Modal Insan (MI-2019-017) and Geran Universiti Penyelidikan (GUP-2017-098). The authors also wish to acknowledge Centre for Research and Instrumentation Management (CRIM), UKM for FESEM analysis.

Institutional Review Board Statement: Not applicable.

Informed Consent Statement: Not applicable.

Data Availability Statement: Not applicable.

Conflicts of Interest: The authors declare no conflict of interest.

References

1. Ribeiro, A.M.; Magalhães, M.; Veiga, F.; Figueiras, A. Cellulose-based hydrogels in topical drug delivery: A challenge in medical devices. In *Cellulose-Based Superabsorbent Hydrogels*; Springer: Berlin, Germany, 2018; pp. 1–29. [CrossRef]
2. Boateng, J.S.; Matthews, K.H.; Stevens, H.N.E.; Eccleston, G.M. Wound healing dressings and drug delivery systems: A review. *J. Pharm. Sci.* **2008**, *97*, 2892–2923. [CrossRef]
3. Kathe, K.; Kathpalia, H. Film forming systems for topical and transdermal drug delivery. *Asian J. Pharm. Sci.* **2017**, *12*, 487–497. [CrossRef] [PubMed]
4. Chang, R.; Raw, A.; Lionberger, R.; Yu, L. Generic development of topical dermatologic products: Formulation development, process development, and testing of topical dermatologic products. *AAPS J.* **2013**, *15*, 41–52. [CrossRef] [PubMed]
5. Parsa, M.; Trybała, A.; Malik, D.; Starov, V. Foam in pharmaceutical and medical applications. *Curr. Opin. Colloid Interface Sci.* **2019**, *44*, 153–167. [CrossRef]
6. Escobar-chávez, J.J.; Rodríguez-cruz, I.M.; Domínguez-delgado, C.L.; Díaz-torres, R.; Revilla-vázquez, A.L.; Aléncaster, N.C. Nanocarrier systems for transdermal drug delivery. In *Recent Advances in Novel Drug Carrier Systems*; IntechOpen: London, UK, 2012; pp. 201–239. [CrossRef]
7. Chai, Q.; Jiao, Y.; Yu, X. Hydrogels for biomedical applications: Their characteristics and the mechanisms behind them. *Gels* **2017**, *3*, 6. [CrossRef]
8. Ilochonwu, B.C.; Urtti, A.; Hennink, W.E.; Vermonden, T. Intravitreal hydrogels for sustained release of therapeutic proteins. *J. Control. Release* **2020**, *326*, 419–441. [CrossRef] [PubMed]
9. Zubik, K.; Singhsa, P.; Wang, Y.; Manuspiya, H.; Narain, R. Thermo-responsive poly(n-isopropylacrylamide)-cellulose nanocrystals hybrid hydrogels for wound dressing. *Polymers* **2017**, *9*, 119. [CrossRef]
10. Mamidi, N.; Villela Castrejón, J.; González-Ortiz, A. Rational design and engineering of carbon nano-onions reinforced natural protein nanocomposite hydrogels for biomedical applications. *J. Mech. Behav. Biomed. Mater.* **2020**, *104*. [CrossRef]
11. Park, S.H.; Shin, H.S.; Park, S.N. A novel pH-responsive hydrogel based on carboxymethyl cellulose/2-hydroxyethyl acrylate for transdermal delivery of naringenin. *Carbohydr. Polym.* **2018**, *200*, 341–352. [CrossRef]
12. Chang, C.; He, M.; Zhou, J.; Zhang, L. Swelling behaviors of pH- and salt-responsive cellulose-based hydrogels. *Macromolecules* **2011**, *44*, 1642–1648. [CrossRef]
13. Biyani, M.V.; Foster, E.J.; Weder, C. Light-healable supramolecular nanocomposites based on modified cellulose nanocrystals. *ACS Macro Lett.* **2013**, *2*, 236–240. [CrossRef]
14. Reddy, N.N.; Mohan, Y.M.; Varaprasad, K.; Ravindra, S.; Joy, P.A.; Raju, K.M. Magnetic and electric responsive hydrogel–magnetic nanocomposites for drug-delivery application. *J. Appl. Polym. Sci.* **2010**, *122*, 1364–1375. [CrossRef]
15. Kim, Y.; Matsunaga, Y.T. Thermo-responsive polymers and their application as smart biomaterials. *J. Mater. Chem. B* **2017**, *5*, 4307–4321. [CrossRef] [PubMed]

16. Sponchioni, M.; Capasso, U.; Moscatelli, D. Thermo-responsive polymers: Applications of smart materials in drug delivery and tissue engineering. *Mater. Sci. Eng. C* **2019**, *102*, 589–605. [CrossRef] [PubMed]
17. Escobar-Chávez, J.J.; López-Cervantes, M.; Naïk, A.; Kalia, Y.N.; Quintanar-Guerrero, D.; Ganem-Quintanar, A. Applications of thermo-reversible pluronic F-127 gels in pharmaceutical formulations. *J. Pharm. Pharm. Sci.* **2006**, *9*, 339–358.
18. Qiu, Y.; Hamilton, S.K.; Temenoff, J. Improving mechanical properties of injectable polymers and composites. In *Injectable Biomaterials Science and Applications. Woodhead Publishing Series in Biomaterials*; Elsevier: Amsdterdam, The Netherlands, 2011; pp. 61–91. [CrossRef]
19. Wang, W.; Wat, E.; Hui, P.C.L.; Chan, B.; Ng, F.S.F.; Kan, C.; Wang, X.; Hu, H.; Wong, E.C.W.; Lau, C.B.S.; et al. Dual-functional transdermal drug delivery system with controllable drug loading based on thermosensitive poloxamer hydrogel for atopic dermatitis treatment. *Sci. Rep.* **2016**, *6*, 24112. [CrossRef]
20. Derakhshandeh, K.; Fashi, M.; Seifoleslami, S. Thermosensitive pluronic® hydrogel: Prolonged injectable formulation for drug abuse. *Drug Des. Devel. Ther.* **2010**, *4*, 255–262. [CrossRef]
21. Kim, J.K.; Won, Y.W.; Lim, K.S.; Kim, Y.H. Low-molecular-weight methylcellulose-based thermo-reversible gel/pluronic micelle combination system for local and sustained docetaxel delivery. *Pharm. Res.* **2012**, *29*, 525–534. [CrossRef]
22. Chen, X.; Song, Z.; Li, S.; Tat Thang, N.; Gao, X.; Gong, X.; Guo, M. Facile one-pot synthesis of self-assembled nitrogen-doped carbon dots/cellulose nanofibril hydrogel with enhanced fluorescence and mechanical properties. *Green Chem.* **2020**, *22*, 3296–3308. [CrossRef]
23. Alammar, A.; Park, S.H.; Ibrahim, I.; Deepak, A.; Holtzl, T.; Dumée, L.F.; Lim, H.N.; Szekely, G. Architecting neonicotinoid-scavenging nanocomposite hydrogels for environmental remediation. *Appl. Mater. Today* **2020**, *21*, 100878. [CrossRef]
24. Yuan, J.; Yi, C.; Jiang, H.; Liu, F.; Cheng, G.J. Direct ink writing of hierarchically porous cellulose/alginate monolithic hydrogel as a highly effective adsorbent for environmental applications. *ACS Appl. Polym. Mater.* **2021**, *3*, 699–709. [CrossRef]
25. Haan, T.Y.; Mohd Syahmi Hafizi Ghani, M.A.W. Physical and Chemical Cleaning for Nanofiltration/Reverse Osmosis (NF/RO) Membranes in Treatment of Tertiary Palm Oil Mill Effluent (POME) for Water Reclamation. *J. Kejuruter.* **2018**, *1*, 51–58. [CrossRef]
26. Norwana, A.A.B.D.; Kunjappan, R.; Chin, M.; Schoneveld, G.; Potter, L.; Andriani, R. Center for International Forestry Research, Indonesia. The local impacts of oil palm expansion in Malaysia An assessment based on a case study in Sabah State. *CIFOR Working Paper* **2011**, *78*, 1–17.
27. Haan, T.Y.; Takriff, M.S. Zero waste technologies for sustainable development in palm oil mills. *J. Oil Palm. Environ. Health* **2021**, *12*, 55–68. [CrossRef]
28. Hamzah, N.; Tokimatsu, K.; Yoshikawa, K. Solid fuel from oil palm biomass residues and municipal solid waste by hydrothermal treatment for electrical power generation in Malaysia: A review. *Sustainability* **2019**, *11*, 1060. [CrossRef]
29. Haan, Y.; Norashiqin, S.; Chun, K. Sustainable approach to the synthesis of cellulose membrane from oil palm empty fruit bunch for dye wastewater treatment. *J. Water Process Eng.* **2020**, *34*, 1–9. [CrossRef]
30. Sudiyani, Y.; Styarini, D.; Triwahyuni, E. Utilization of biomass waste empty fruit bunch fiber of palm oil for bioethanol production using pilot–scale unit. *Energy Procedia* **2013**, *32*, 31–38. [CrossRef]
31. Nazir, M.S.; Wahjoedi, B.A.; Yussof, A.W.; Abdullah, M.A. Eco-friendly extraction and characterization of cellulose from oil palm empty fruit bunches. *BioResources* **2013**, *8*, 2161–2172. [CrossRef]
32. Schmolka, I.R. Artificial Skin, I. Preparation and Properties of Pluronic F-127 Gels for Treatment of Burns. *J. Biomed. Mater. Res.* **1972**, *6*, 571–582. [CrossRef]
33. Nie, S.; Hsiao, W.W.; Pan, W.; Yang, Z. Thermoreversible pluronic® f127-based hydrogel containing liposomes for the controlled delivery of paclitaxel: In vitro drug release, cell cytotoxicity, and uptake studies. *Int. J. Nanomedicine* **2011**, *6*, 151–166. [CrossRef]
34. Dewan, M.; Sarkar, G.; Bhowmik, M.; Das, B.; Chattoapadhyay, A.K.; Rana, D.; Chattopadhyay, D. Effect of gellan gum on the thermogelation property and drug release profile of poloxamer 407 based ophthalmic formulation. *Int. J. Biol. Macromol.* **2017**, *102*, 258–265. [CrossRef] [PubMed]
35. Carvalho, F.C.; Calixto, G.; Hatakeyama, I.N.; Luz, G.M.; Gremião, M.P.D.; Chorilli, M. Rheological, mechanical, and bioadhesive behavior of hydrogels to optimize skin delivery systems. *Drug Dev. Ind. Pharm.* **2013**, *39*, 1750–1757. [CrossRef] [PubMed]
36. Sun, M.; Sun, H.; Wang, Y.; Sánchez-Soto, M.; Schiraldi, D. The relation between the rheological properties of gels and the mechanical properties of their corresponding aerogels. *Gels* **2018**, *4*, 33. [CrossRef]
37. Sudipta, C.; Hui, P.C.; Kan, C.; Wan, W. Dual-responsive (pH/temperature) pluronic f-127 hydrogel drug delivery system for textile-based transdermal therapy. *Sci. Rep.* **2019**, *9*, 1–13. [CrossRef]
38. Stojkovska, J.; Djurdjevic, Z.; Jancic, I.; Bufan, B.; Milenkovic, M.; Jankovic, R.; Miskovic-Stankovic, V.; Obradovic, B. Comparative in vivo evaluation of novel formulations based on alginate and silver nanoparticles for wound treatments. *J. Biomater. Appl.* **2018**, *32*, 1197–1211. [CrossRef] [PubMed]
39. Morsi, N.M.; Abdelbary, G.A.; Ahmed, M.A. Silver sulfadiazine based cubosome hydrogels for topical treatment of burns: Development and in vitro/in vivo characterization. *Eur. J. Pharm. Biopharm.* **2014**, *86*, 178–189. [CrossRef] [PubMed]
40. Al-Rajabi, M.M.; Haan, T.Y. Influence of vertical diffusion cell set-up on in vitro silver sulfadiazine drug release from thermo-responsive cellulose hydrogel. *Mater. Sci. Forum* **2021**, *1030*, 19–26. [CrossRef]
41. Marques, M.; Ueda, C.T.; Shah, V.P.; Derdzinski, K.; Ewing, G.; Flynn, G.; Maibach, H.; Rytting, H.; Shaw, S.; Thakker, K.; et al. Topical and transdermal drug products. *Pharmacopeial Forum* **2009**, *35*, 750–764.

42. Deo, S.S.; Inam, F.; Karmarkar, N.P. Analytical method development for determination of performance of adapalene in adapalene 0.1% gel formulation using manual diffusion cell. *Chem. Sci. Trans.* **2012**, *2*, 251–257. [CrossRef]
43. Mustafa, F.M.; Hodali, H.A. Use of mesoporous silicate nanoparticles as drug carrier for mefenamic acid. *IOP Conf. Ser. Mater. Sci. Eng.* **2015**, *92*. [CrossRef]
44. Varelas, C.G.; Dixon, D.G.; Steiner, C.A. Zero-order release from biphasic polymer hydrogels. *J. Control. Release* **1995**, *34*, 185–192. [CrossRef]
45. Bruschi, M.L. Mathematical models of drug release. In *Strategies to Modify the Drug Release from Pharmaceutical Systems*; Elsevier: Amsterdam, The Netherlands, 2015; pp. 63–86. [CrossRef]
46. Siepmann, J.; Peppas, N.A. Higuchi equation: Derivation, applications, use and misuse. *Int. J. Pharm.* **2011**, *418*, 6–12. [CrossRef]
47. Korsmeyer, R.W.; Gurny, R.; Doelker, E.; Buri, P.; Peppas, N.A. Mechanisms of solute release from porous hydrophilic polymers. *Int. J. Pharm.* **1983**, *15*, 25–35. [CrossRef]
48. Gioffredi, E.; Boffito, M.; Calzone, S.; Giannitelli, S.M.; Rainer, A.; Trombetta, M.; Mozetic, P.; Chiono, V. Pluronic F127 hydrogel characterization and biofabrication in cellularized constructs for tissue engineering applications. *Procedia CIRP* **2016**, *49*, 125–132. [CrossRef]
49. Jung, Y.; Seok, P.W.; Park, H.; Lee, D.K.; Na, K. Thermo-sensitive injectable hydrogel based on the physical mixing of hyaluronic acid and pluronic f-127 for sustained NSAID delivery. *Carbohydr. Polym.* **2017**, *156*, 403–408. [CrossRef] [PubMed]
50. Pereira, G.G.; Dimer, F.A.; Guterres, S.S.; Kechinski, C.P.; Granada, J.E.; Cardozo, N.S.M. Formulation and characterization of poloxamer 407®: Thermoreversible gel containing polymeric microparticles and hyaluronic acid. *Quim. Nov.* **2013**, *36*, 1121–1125. [CrossRef]
51. Dragicevic, N.; Maibach, H.I. Polymeric micelles in dermal and transdermal delivery. In *Percutaneous Penetration Enhancers Chemical Methods in Penetration Enhancement: Nanocarriers*; Springer: Berlin, Germany, 2016; pp. 1–384. [CrossRef]
52. Katas, H.; Thian Sian, T.; Abdul Ghaf, M. Topical temperature-sensitive gel containing DsiRNA-chitosan nanoparticles for potential treatment of skin cancer. *Trends Med. Res.* **2017**, *12*, 1–13. [CrossRef]
53. Hopkins, C.C.; de Bruyn, J.R. Gelation and long-time relaxation of aqueous solutions of pluronic f127. *J. Rheol.* **2019**, *63*, 191–201. [CrossRef]
54. Tae, G.; Won, D. Composition for Forming Pluronic-Based Hydrogel With Improved Stability. US Patent 2015/0231246 A1, 19 February 2015.
55. Garala, K.; Joshi, P.; Patel, J.; Ramkishan, A.; Shah, M. Formulation and evaluation of periodontal in situ gel. *Int. J. Pharm. Investig.* **2013**, *3*, 29. [CrossRef]
56. Orasugh, J.T.; Sarkar, G.; Saha, N.R.; Das, B.; Bhattacharyya, A.; Das, S.; Mishra, R.; Roy, I.; Chattoapadhyay, A.; Ghosh, S.K.; et al. Effect of cellulose nanocrystals on the performance of drug loaded in situ gelling thermo-responsive ophthalmic formulations. *Int. J. Biol. Macromol.* **2019**, *124*, 235–245. [CrossRef]
57. Gong, C.Y.; Shi, S.; Dong, P.W.; Zheng, X.L.; Fu, S.Z.; Guo, G.; Yang, J.L.; Wei, Y.Q.; Qian, Z.Y. In vitro drug release behavior from a novel thermosensitive composite hydrogel based on pluronic f127 and poly(ethylene glycol)-poly(ε-caprolactone)-poly(ethylene glycol) copolymer. *BMC Biotechnol.* **2009**, *9*, 1–13. [CrossRef]
58. Omri, M.A.; Triki, A.; Guicha, M.; Hassen, M.; Ben, A.M.; Ahmed El Hamzaoui, H.; Bulou, A. Effect of wool and thermo-binder fibers on adhesion of alfa fibers in polyester composite. *J. Appl. Phys.* **2013**, *114*. [CrossRef]
59. Teow, Y.; Ming, K.; Mohammad, A. Synthesis of cellulose hydrogel for copper (II) ions adsorption. *J. Environ. Chem. Eng.* **2018**, *6*, 4588–4597. [CrossRef]
60. Prasad, S.G.; De, A.; De, U. Structural and optical investigations of radiation damage in transparent PET polymer films. *Int. J. Spectrosc.* **2011**, *2011*. [CrossRef]
61. Liu, T.; Jiao, C.; Peng, X.; Chen, Y.-N.; Chen, Y.; He, C.; Liu, R.; Wang, H. Super-strong and tough poly(vinyl alcohol)/poly(acrylic acid) hydrogels reinforced by hydrogen bonding. *J. Mater. Chem. B* **2018**, 8105–8114. [CrossRef]
62. Rasoulzadeh, M.; Namazi, H. Carboxymethyl cellulose/graphene oxide bio-nanocomposite hydrogel beads as anticancer drug carrier agent. *Carbohydr. Polym.* **2017**, *168*, 320–326. [CrossRef]
63. Mastropietro, D.J.; Nimroozi, R.; Omidian, H. Rheology in pharmaceutical formulations-A perspective. *J. Dev. Drugs* **2013**, *2*, 2–7. [CrossRef]
64. Grassi, G.; Crevatin, A.; Farra, R.; Guarnieri, G.; Pascotto, A.; Rehimers, B.; Lapasin, R.; Grassi, M. Rheological properties of aqueous pluronic-alginate systems containing liposomes. *J. Colloid Interface Sci.* **2006**, *301*, 282–290. [CrossRef] [PubMed]
65. Owczarz, P.; Zi, P.; Modrzejewska, Z.; Kuberski, S.; Dziubinski, M. Rheo-kinetic study of sol-gel phase transition of chitosan colloidal systems. *Polymers* **2018**, *10*, 47. [CrossRef] [PubMed]
66. Fakhari, A.; Corcoran, M.; Schwarz, A. Thermogelling properties of purified poloxamer 407. *Heliyon* **2017**, *3*, e00390. [CrossRef]
67. Bobbala, S.; Tamboli, V.; McDowell, A.; Mitra, A.K.; Hook, S. Novel injectable pentablock copolymer based thermoresponsive hydrogels for sustained release vaccines. *AAPS J.* **2016**, *18*, 261–269. [CrossRef]
68. Hsu, S.H.; Leu, Y.L.; Hu, J.W.; Fang, J.Y. Physicochemical characterization and drug release of thermosensitive hydrogels composed of a hyaluronic acid/pluronic f127 graft. *Chem. Pharm. Bull.* **2009**, *57*, 453–458. [CrossRef]
69. Wani, T.U.; Rashid, M.; Kumar, M.; Chaudhary, S.; Kumar, P.; Mishra, N. Targeting aspects of nanogels: An overview. *Int. J. Pharm. Sci. Nanotech.* **2014**, *7*, 2612–2630.

70. Gupta, A.; Kowalczuk, M.; Heaselgrave, W.; Britland, S.T.; Martin, C.; Radecka, I. The production and application of hydrogels for wound management: A review. *Eur. Polym. J.* **2018**, *111*, 134–151. [CrossRef]
71. Flory, P.J.; Rehner, J. Statistical mechanics of cross-linked polymer networks II. Swelling. *J. Chem. Phys.* **1943**, *11*, 521–526. [CrossRef]
72. Vervoort, S. Behaviour of hydrogels swollen in polymer solutions under mechanical action. Ph.D. Thesis, École Nationale Supérieure des Mines de Paris, Paris, France, 2006.
73. Sakai, T. Swelling and deswelling. In *Physics of Polymer Gels*; Wiley-VCHVerlag GmbH& Co. KGaA.: Weinheim, Germany, 2020; pp. 77–107.
74. Hillery, A.M.; Park, K. *Drug Delivery: Fundamentals and Applications*, 2nd ed.; CRC Press: Boca Raton, FL, USA, 2016. [CrossRef]
75. Siqueira, N.M.; Cirne, M.F.R.; Immich, M.F.; Poletto, F. Stimuli-responsive polymeric hydrogels and nanogels for drug delivery applications. In *Stimuli Responsive Polymeric Nanocarriers for Drug Delivery Applications*; Elsevier Ltd.: Amsdterdam, The Netherlands, 2018; Volume 1, pp. 343–374. [CrossRef]
76. Jodar, K.S.P.; Balcão, V.M.; Chaud, M.V.; Tubino, M.; Yoshida, V.M.H.; Oliveira, J.M.; Vila, M.M.D.C. Development and characterization of a hydrogel containing silver sulfadiazine for antimicrobial topical applications. *J. Pharm. Sci.* **2015**, *104*, 2241–2254. [CrossRef]
77. Huang, S.; Fu, X. Naturally derived materials-based cell and drug delivery systems in skin regeneration. *J. Control. Release* **2010**, *142*, 149–159. [CrossRef]
78. Thakkar, V.; Korat, V.; Baldaniya, L.; Gohel, M.; Gandhi, T.; Patel, N. Development and characterization of novel hydrogel containing antimicrobial drug for treatment of burns. *Int. J. Pharm. Investig.* **2016**, *6*, 158. [CrossRef] [PubMed]
79. Kalaydina, R.V.; Bajwa, K.; Qorri, B.; Decarlo, A.; Szewczuk, M.R. Recent advances in "smart" delivery systems for extended drug release in cancer therapy. *Int. J. Nanomedicine* **2018**, *13*, 4727–4745. [CrossRef] [PubMed]
80. Gupta, P.K. Principles and basic concepts of toxicokinetics. In *Fundamentals of Toxicology*; Academic Press: Waltham, MA, USA, 2016; pp. 87–107. [CrossRef]
81. Nascimento, E.G.D.; Sampaio, T.B.M.; Medeiros, A.C.; Azevedo, E.P.D. Evaluation of chitosan gel with 1% silver sulfadiazine as an alternative for burn wound treatment in rats. *Acta Cirúrgica Bras.* **2009**, *24*, 2009–2460.
82. Church, D.; Elsayed, S.; Reid, O.; Winston, B.; Lindsay, R. Burn wound infections. *Clin. Microbiol. Rev.* **2006**, *19*, 403–434. [CrossRef] [PubMed]
83. Dorota, W.; Krzak, J.; Macikowski, B.; Berkowski, R.; Osinski, B.; Musiał, W. Evaluation of the release kinetics of a pharmacologically active substance from model intra-articular implants replacing the cruciate ligaments of the knee. *Materials (Basel)* **2019**, *12*, 1202. [CrossRef]
84. Wong, R.S.H.; Dodou, K. Effect of drug loading method and drug physicochemical properties on the material and drug release properties of poly (ethylene oxide) hydrogels for transdermal delivery. *Polymers* **2017**, *9*, 286. [CrossRef] [PubMed]
85. Olejnik, A.; Glowka, A.; Nowak, I. Release studies of undecylenoyl phenylalanine from topical formulations. *Saudi Pharm. J.* **2018**, *26*, 709–718. [CrossRef]
86. Rabin, C.R.; Siegel, S.J. Delivery systems and dosing for antipsychotics. In *Handbook of Experimental Pharmacology*; Springer: Berlin/Heidelberg, Germany, 2012; Volume 212, pp. 267–291.

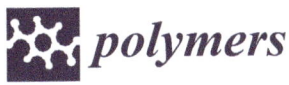

Review

Cotton Wastes Functionalized Biomaterials from Micro to Nano: A Cleaner Approach for a Sustainable Environmental Application

Samsul Rizal [1,*], Abdul Khalil H. P. S. [2,*], Adeleke A. Oyekanmi [2,*], Olaiya N. Gideon [2], Che K. Abdullah [2], Esam B. Yahya [2], Tata Alfatah [2], Fatimah A. Sabaruddin [2] and Azhar A. Rahman [3]

1. Department of Mechanical Engineering, Universitas Syiah Kuala, Banda Aceh 23111, Indonesia
2. School of Industrial Technology, Universiti Sains Malaysia (USM), Penang 11800, Malaysia; ngolaiya@futa.edu.ng (O.N.G.); ck_abdullah@usm.my (C.K.A.); essam912013@gmail.com (E.B.Y.); tataalfatah83@gmail.com (T.A.); atiyah88@gmail.com (F.A.S.)
3. School of Physics, Universiti Sains Malaysia (USM), Penang 11800, Malaysia; arazhar@usm.my
* Correspondence: samsul_r@yahoo.com (S.R.); akhalilhps@gmail.com (A.K.H.P.S.); abdulkan2000@yahoo.com (A.A.O.)

Abstract: The exponential increase in textile cotton wastes generation and the ineffective processing mechanism to mitigate its environmental impact by developing functional materials with unique properties for geotechnical applications, wastewater, packaging, and biomedical engineering have become emerging global concerns among researchers. A comprehensive study of a processed cotton fibres isolation technique and their applications are highlighted in this review. Surface modification of cotton wastes fibre increases the adsorption of dyes and heavy metals removal from wastewater. Cotton wastes fibres have demonstrated high adsorption capacity for the removal of recalcitrant pollutants in wastewater. Cotton wastes fibres have found remarkable application in slope amendments, reinforcement of expansive soils and building materials, and a proven source for isolation of cellulose nanocrystals (CNCs). Several research work on the use of cotton waste for functional application rather than disposal has been done. However, no review study has discussed the potentials of cotton wastes from source (Micro-Nano) to application. This review critically analyses novel isolation techniques of CNC from cotton wastes with an in-depth study of a parameter variation effect on their yield. Different pretreatment techniques and efficiency were discussed. From the analysis, chemical pretreatment is considered the most efficient extraction of CNCs from cotton wastes. The pretreatment strategies can suffer variation in process conditions, resulting in distortion in the extracted cellulose's crystallinity. Acid hydrolysis using sulfuric acid is the most used extraction process for cotton wastes-based CNC. A combined pretreatment process, such as sonication and hydrolysis, increases the crystallinity of cotton-based CNCs. The improvement of the reinforced matrix interface of textile fibres is required for improved packaging and biomedical applications for the sustainability of cotton-based CNCs.

Keywords: cotton wastes; textile; nanomaterials; cellulose nanocrystal; extraction methods; environmental application

1. Introduction

The demand for textile materials is increasing globally due to population growth and economic development [1]. Textile materials have been widely used in many fields of life [2]. Reference [3] stated that, among world fibre consumption, cotton is the most consumed fibre after polyester. According to global statistics, Reference [4] reported that more than 24 million tons of cotton cloths are produced and used. Textile industry contributes tons of cotton wastes, which affects the environment [5]. As a result of continuous production and utilization of cotton fibres, the wastes generated have proportionately increased over the years, causing significant environmental issues. Reference [6] categorized cotton wastes as pre-consumers and post-consumer wastes. Reference [7] regarded pre-consumer wastes as wastes from processed fibres, including finished yarns, technical textiles, garments off-cuts, nonwovens, and other rejected processing cotton materials. Reference [8] also added that pre-consumer cotton wastes are generated during yarn, fabric, and garment product manufacturing. Reference [9] classified post-consumer wastes as dirty wastes, which include both natural and synthetic materials. Reference [8] further classified post-consumer cotton wastes as discarded wastes at the end of the product's service life. Reference [10] stated that most of the cotton waste streams are post-consumer. The environmental impact, as a result of the ineffective processing mechanism of cotton wastes, is a major concern. Reference [11] reported that only 15% of cotton garments are recycled in the United States, 19% are incinerated, and 66% are landfilled. The processing and utilization of cotton wastes have the propensity to reduce environmental impact associated with its discharge. In other words, processed cotton wastes can be utilized as adsorbent media for the removal of recalcitrant pollutants from wastewater, as reinforcement materials for packaging and biomedical applications. Furthermore, the various forms of cotton wastes, especially post-consumer wastes, can be utilized as a promising source of cellulose nanocrystals (CNCs) due to its high cellulose contents including its inherent prospects for advanced applications toward environmental sustainability [12]. CNCs are the derivatives of cellulose used for various applications in nanoscience due to its high mechanical strength, high aspect ratio, large surface area, high modulus, and biodegradability [13,14]. CNCs are obtained in the form of rod-like crystalline domain consisting of diameters in the range of 1 to 100 nm and length around 10 to 100 nm [15]. The extent of the crystallinity of CNCs as a functional material depends on the precursor used for the extraction of cellulose and the intensity of the extraction methods. In nanotechnology, the utilization of CNCs as reinforcements for bio-based polymers are attracting remarkable prospects to enhance the properties of biomaterials. Due to its unique properties, many studies have discussed the extraction of CNCs from biomass and have reported the potential applications in varieties of broad-based areas, such as in environmental, packaging, construction, biomedicals, and automobiles [16]. The extraction of CNCs from cotton wastes have remarkable prospects and application in various areas of nanotechnology. In this review, based on current knowledge on the fundamental aspects and properties of cotton wastes, a comparative study of the pretreatment methods for the extraction of CNCs from cotton wastes are comprehensively discussed. The application of cotton wastes as a functional material in different fields of biotechnology for environmental sustainability and health safety have been discussed. This includes the advanced application and future prospects of nanocrystals from cotton wastes. This review will likely increase research interest in cotton wastes as a new source of nanocrystals.

Cotton wastes generated have either been incinerated or landfilled. Few research works on the use of cotton waste for functional application rather than disposal have been done. However, no review has critically discussed research findings and analyses of the state of cotton waste use. In this review, the relevance of our study is benchmarked on three critical aspects including cotton wastes generation and the associated environmental impacts, and functional properties of cotton fibres into functional materials to reduce the environmental impact as a result of disposal and functional biomaterials for sustainable environmental applications. A critical review on this subject is rare. The awareness of functional usage of cotton waste for CNC production has not been fully explored because many studies focus on isolation of CNC from a wood source. Therefore, no review has been written in this respect, which has limited the valorisation of cotton waste for a functional application. This review critically discussed the functional application of cotton waste at micro to nano sizes. This covers cotton fibre and CNC isolation, their application, limitations, and proposed research to combat these limitations. To the best of our knowledge, no review study has discussed the potentials of cotton wastes from source (Micro-Nano) to application. The utilization of the cotton wastes based on its intrinsic functional properties have found wide application in environmental sustainability through dye and heavy metals removal from wastewater as well as geotechnical, packaging, and biomedical applications. The improvement of the reinforced matrix interface of textile fibres is required for improved packaging and biomedical applications for the sustainability of cotton-based CNCs.

2. Environmental Impact of Cotton Wastes

Over the last decade, 40% of global fibre production is attributed to cotton [17]. China is the world's largest producer and consumer of cotton [18]. Reference [19] gave an instance that the Keqiao industrial park, which is one of the largest textile industrial parks in China, has the capacity for more than 3 million tons of varieties of cotton fibres, including 19 billion clothing fabrics. As the global demand for cotton increases, there is an exponential increase in wastes generated. Cotton wastes generated have either been incinerated or landfilled, creating many environmental problems. On a yearly basis, about 15.1 million tons and 1.7 million tons of cotton wastes, are generated in the United States and the United Kingdom, respectively. Reference [20] reported that, in Australia, 76% of cotton wastes were deposited in the landfill in 2016. In recent years, according to Zhang, Wu [19], cotton wastes in China have increased to more than 100 million tons annually, most of which ends in the landfill. Most cotton wastes from the plants add up to the waste streams and are mainly deposited in the landfill without effective management. Reference [17] stated that cotton diseases and pests are disposal issues attributed to cotton wastes discharge. The authors also added that the slow decomposition of cotton in the soil causes cultivation problems. The dyeing of textile products utilizes a lot of energy, water, and chemical bleaching agents. In the process, a high amount of wastewater is generated with harmful effects on the environment [21]. Furthermore, the emission of greenhouse gases during cotton production as a result of the utilization of gaseous fuel contributes to climate change. This includes the potential risks that could be detrimental to freshwater ecosystems. This is because rivers, streams, lakes, and other receiving waters are often the primary receivers of generated wastes during and after the production of textiles materials [22]. The need for sustainable solutions to cotton wastes accumulation is becoming more urgent as world fibre consumption increases. Reference [1] have reviewed by conducting a comparative study and concluded that wastes recycling is a more beneficial and sustainable approach to reduce environmental impact compared to incineration or landfilling. In addition, to reduce the environmental impact as a result of cotton wastes generation, there should be a proper mechanism to process these wastes into functional bio-materials for environmental sustainability. This includes a global waste summit to enact laws, regulations, and legislation with a penal code to promote strict compliance toward creating a circular economy that will minimize the impact of environmental pollution.

2.1. Functional Properties of Cotton Fibres

The strength of cotton fibres as a functional material is attributed to its rigidity of a cellulose chain, crystalline structure that allows intramolecular and intermolecular hydrogen bonding [23]. Wang, Farooq [24] stated that cotton fibres have high crystallinity, predominantly composed of cellulose. Reference [25] stated that molecules of cellulose in cotton typically are within 800 and 10,000 units. The strength of cotton fibre depends on the degree of orientation of the molecules in the fibre axis [26]. Reference [24] reported that the tensile strength of cotton fibres is influenced by the degree of orientation of molecular chains and the fibre axis. The authors stated further that the more the molecules are oriented, the higher is the fibre strength. Reference [27] have stated that the breaking strength and modulus of cotton fibres are less than the theoretical value of cellulose crystal due to the molecules shorter than the fibre. The authors revealed that the degree of orientation of the chain molecules and fibre axis determines the extent of its tensile strength for reinforcements. Reference [28] investigated the modulus of elasticity in yarn and revealed that the torsional rigidity of cotton fibre in the yarn is approximately twice as high as a single fibre. According to Reference [29], cotton fibres exhibit a tensile strength of 287–840 MPa, a tensile modulus of 9.4 to 22 GPa, 3–10% elongation, and 7–25% of moisture absorption.

The adsorption capacity of cotton fibre depends on the structure of the fibre, the fibre density, immersion, solid-liquid interface, adhesion, and wetting potential [30]. Prakash [31] revealed that cotton fibre has high moisture absorption potential, which makes it better suited for dye application. Reference [32] reported that pure cotton, cotton-hemp blends are inherently suitable as UV absorbers and concluded that naturally pigmented cotton combined with lignin composition in hemp fibres demonstrate better UV absorption capacity. In a related study, Gorade, Chaudhary [33] revealed that cotton fibre is hygroscopic and are non-abrasive with a low mass density, which makes it more suitable for liquid transport applications, such as for medical products. The functional properties of cotton fibre wastes have a remarkable prospect for various applications, such as environmental, packaging, and biomedicine. Reference [34] have reported that cotton wastes contain more than 95% of cellulose. The intrinsic properties of cotton wastes, including its abundance of cellulosic composition, have increased the prospect of its utilization as a precursor for the extraction of nanocrystals.

2.2. Applications of Cotton Wastes

During the processing and production of cotton fibres, cotton wastes are generated. Synthesis of these waste fibres have created potentials and a wide application due to their intrinsic properties. The functionalization of polymers for improved surface properties such as adhesion, surface properties including porosity roughness, surface adsorption, and wettability potentials through active functional groups determines its applicability [2,5]. The recycling of wastes into functional materials and application for water treatment, food packaging, biomedical applications, and textiles have been reviewed by Reference [35]. Some functional applications of polymers are illustrated in Figure 1.

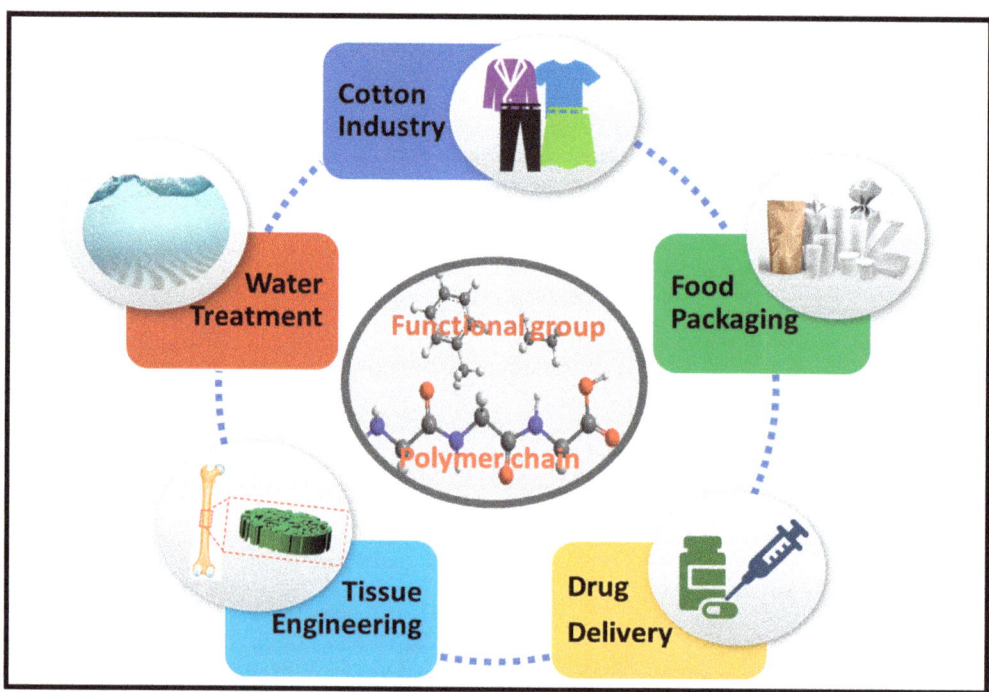

Figure 1. Polymer functionalization and application [35].

The application of cellulose-derived cotton wastes for a wide range of applications, such as environmental including dyes and heavy metal remediation from wastewater, geotechnical, packaging, and biomedical applications, with the potentials of forming composites, is presented in this section.

2.2.1. Environmental Applications

The cost-effectiveness and abundance of cotton wastes can be utilized as the ideal material for the fabrication of highly efficient catalysts and adsorbents for dye and heavy metal removal and for geotechnical applications. For the benefit of the environment, recycling these wastes into value-added products is very significant for environmental sustainability. Çay, Yanık [36] have recycled cotton wastes into biochar to improve the properties of the material. Carbonization of the wastes at a low temperature was applied to improve cotton fabric through the conventional printing process. The produced biochar demonstrated hydrophobic characteristics on the surface of the material, thereby, creating an improved structure with hydrophobic-hydrophilic properties. Cotton waste has been utilized for various applications, such as dyes and heavy metals pollutant removal, due to high sorption mechanisms and synergistic interactions as a suitable adsorption media. The application of cotton wastes for the removal of recalcitrant pollutants from wastewater, soil amendments, and reinforcements are very significant. Cotton fibres is applicable as reinforcement fillers in the matrix for building applications and as a functional material for reinforcement of composites. Cotton fibre properties and techniques of cellulose extraction from cotton waste have been applied for environmental sustainability.

- Water Treatment

Cotton waste is a very significant textile waste that has found application in water purification and wastewater treatment, especially for the removal of emerging and recalcitrant pollutants from wastewater. The removal of pollutants of emerging concerns such as pharmaceuticals, agrochemical personal care products (PPCPs), and dyes have gained interest in recent times. From an environmental perspective, PPCPs are the excreted drugs by humans, sediments, or pollutants from the treatment plant. Meng, Liu [37] have conducted a review study on the potential toxicity of PPCPs. Sathishkumar, Binupriya [38] stated that the disposal of these wastes could be lethargic and hazardous to the environment and the receiving water. However, cotton wastes have been synthesized as adsorbents for the removal of these wastes, which contain organic and inorganic pollutants. Reference [39] stated that the porous nature of cotton wastes determines the adsorption efficiency. Czech, Shirvanimoghaddam [40] studied the adsorption efficiency of carbon microtubes using cotton strips as a precursor for the removal of PPCPs from water. The investigated pollutants from the PPCP were naproxen (NAP), caffeine (CAF), and triclosan (TCS). The efficiency of the carbon microtube achieved 69% of the maximum removal capacity for NAP. In addition, 89.9% removal of TCS and 98% removal of CAF from PPCP was achieved. The removal of recalcitrant pollutants, such as dyes and heavy metals, using cotton wastes have been reported for dye and heavy metal removal.

- Dye Removal

The application of cotton wastes for the adsorption of synthetic dyes is economically viable because of the availability and ease of processing the wastes. Cotton wastes have been widely applied for the treatment of cationic dyes [41,42]. Qin, Guo [43] have reviewed methods of removal of dyes from aqueous solution. Some of the methods including Fenton oxidation, advanced oxidation, and ozonation have been applied widely in wastewater treatment. A treatment system using adsorption and a photocatalytic process on waste cotton composite film for the removal of dye is illustrated in Figure 2. Qiu, Shu [44] reported that, in a photocatalytic process, TiO_2 has good photocurrent activity and chemical stability. Although the light absorption of TiO_2 is limited to ultraviolet light, the energy of ultraviolet light accounts for about 4% of solar energy, and, therefore, restricts its practical application. A schematic illustration of adsorption and a photocatalytic process of pollutant removal in solution.

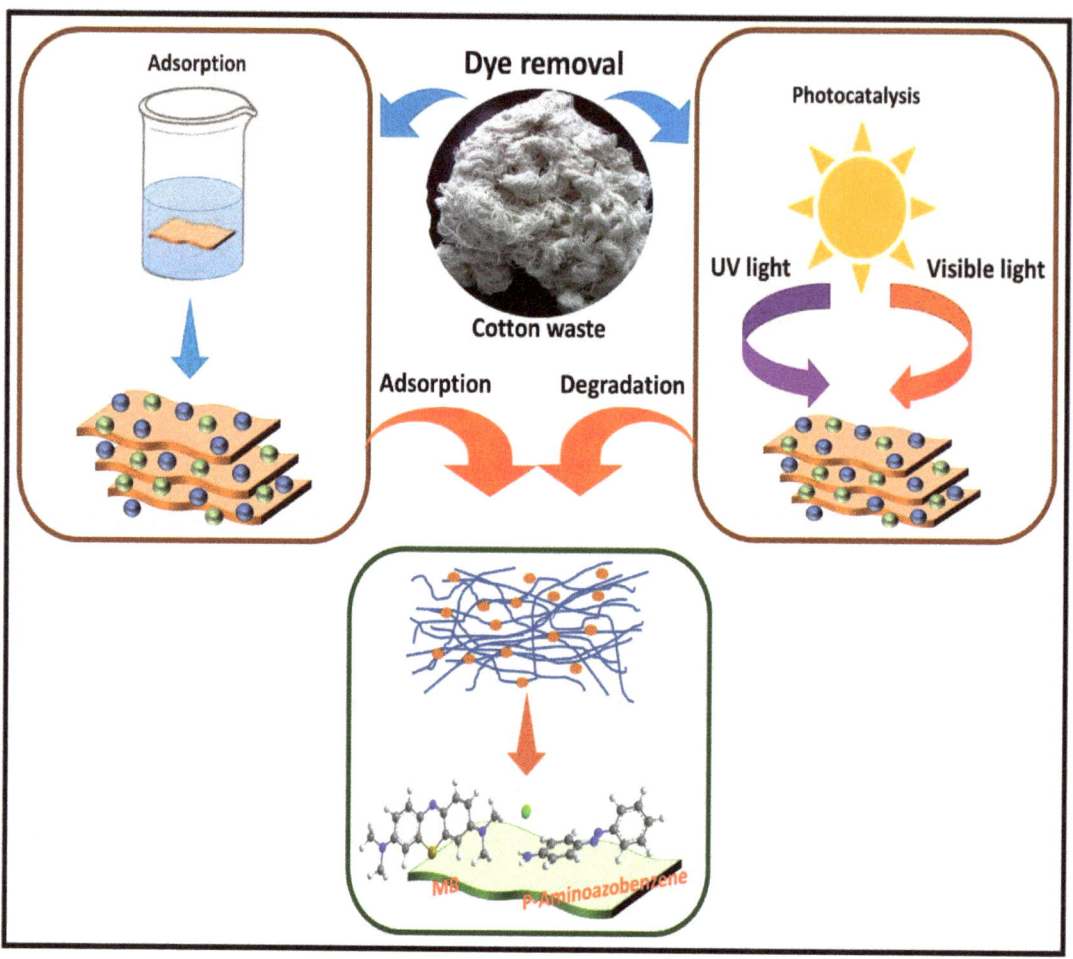

Figure 2. Adsorption and photocatalysis of dye removal using a waste cotton cloth fibre composite.

The efficiency of the processed cotton wastes for the removal of different forms of dyes with a focus on the adsorption method is presented in this review due to its low cost of treatment and high removal capacity from the solution. Table 1 shows the most recent application of cotton wastes for dye adsorption. Song, Li [45] investigated the adsorptive capacity of poly acrylethyl trimethyl ammonium chloride cotton (PA-cotton) for the removal of reactive scarlet 3BS dye. The result indicated that an increase in dosage and temperature favoured the adsorbent, which signified the applicability of the modified cotton wastes for dye removal. Maximum adsorption capacity (Qmax) from the Langmuir isotherm model for the PA-cotton was 540.54 mg/g, which proved to be higher than the widely applied activated carbon and unmodified cotton by 292.18 and 2702.70 times, respectively. The significance of increase in temperature and dosage improved on the adsorption capacity and regeneration potential of the cotton wastes for reuse. Sivarajasekar, Baskar [46] also indicated the prospect of cotton wastes for dye adsorption. The effect of chemisorption influenced maximum adsorption uptake of basic red 2 (BR2) and basic violet 3 (BR3), even though the decrease in adsorption efficiency may likely occur. The increase of initial dye concentrations may likely inhibit the uptake of dyes due to the boundary layer resistance of active pores on the surface of the waste cotton seed. The decrease in the adsorption capacity may also occur due to the unavailability of active sites and pores on the waste cotton seed particles for the attachment of solutes of the dye molecules. Haque, Remadevi [47] investigated the adsorptive capacity of cotton gin trash for the uptake of methylene blue. The result was compared with the uptake of methylene blue onto polypyrole-coated cotton fabrics [48]. Higher sorption capacity was achieved on the polypyrole-coated cotton fabrics as against cotton gin wastes. Comparatively, cotton gin wastes demonstrated lower uptake of methylene blue (MB). This is because the sorption of methylene blue on cotton gin wastes had less attraction toward the cellulosic surface due to the anionic nature of the dye. Tunc, Tanacı [49] used cotton stalk and cotton hull for the adsorption of Ramazol Black B from an aqueous solution.

The investigation revealed that higher sorption capacity was observed on the cotton hull adsorbent than cotton stalk even though the adsorption efficiency for both adsorbents was not very high. The pH of the adsorbate affected the aqueous chemistry and the binding surface of the adsorbents. It was found that cotton wastes, which are predominantly natural cellulosic fibres, are negatively charged as a result of the presence of hydroxyl groups in cellulose. The higher uptakes may likely be obtained at a very acidic pH due to the electrostatic interaction between the positively charged surface of the adsorbent and the negative charge of the Ramazol black B dye anions. Furthermore, Deng, Lu [50] reported that cotton stalk demonstrated less sorption capacity for methylene blue, but the effect of modified sulphuric acid-treated cotton stalk achieved very high adsorption efficiency, which is similar to phosphorus-treated cotton stalk. Cotton wastes bio-adsorbents are potentially suitable for the remediation of dyes from aqueous solutions. Cotton wastes adsorbents have demonstrated promising potentials for the removal of hazardous wastes. However, desorption of dyes on the spent adsorbents without compromising the quality of air safety through incineration requires a sound and sustainable technological approach. This requires intensive research.

Table 1. Adsorption efficiency of textile wastes for dyes.

Precursor	Adsorbate	Qmax (mg/g)	References
Cotton wastes	Methylene blue solution	76.0	[47]
Cotton waste	Aqueous solution of Alizarin red dye	73.8	[51]
Cotton wastes	Dissolution of $Bi(NO_3)_3 5H_2O$ in a solution of Rhodamine B and Methylene Blue	93.7 / 97.04	[43]
Cotton fibre	Aqueous solution of Methylene blue dye	95.6	[48]
Cotton stalk	Aqueous solution of Methylene blue	147.06	[50]
Sulphuric acid-treated cotton stalk	Aqueous solution of Methylene blue	555.56	[50]
Phosphorus acid-treated cotton stalk	Aqueous solution of Methylene blue	222.22	[50]
Cotton waste fabrics	Methylene blue and Rhodamine 6G (R6G) solution	185.63 / 118.21	[52]
Waste cottonseed	Basic red 2 (BR2) and solutions Basic violet 3 (BV3)	50.11 / 66.69	[46]
Cotton fibre	Concentration of crystal red and methylene blue	175.1 / 113.1	[53]
cotton hull waste	Reactive blue, red concentration		
CNC from cotton waste	Methylene blue solution	12.91	[54]

- Heavy metals removal

Heavy metal ions are persistent environmental pollutants. They are toxic and hazardous at a low concentration [55]. Heavy metals constitute environmental problems when they exceed the required threshold. The precipitation of metal ions on the surface of activated carbon is the conventional method of metal ion sequestration, while being quite expensive. To obtain cheaper adsorbents, the synthesis and application of cotton wastes for heavy metals removal from wastewater have been studied [56–59]. Table 2 illustrates the most recent studies of heavy metals adsorption on cotton wastes as an adsorbent media. Mihajlović, Vukčević [58] investigated the sorption capacity of waste cotton yarn for the removal of heavy metal ions from an aqueous solution. High removal of Pb ions was achieved on waste cotton yarn in a binary mixture, unlike in a single solution. This is because the influence of other ions species in the adsorbate favourably induced the removal of Cr. Although, slight removal of Pb ions was achieved. The sorption of Cd in the binary mixture was not affected by the presence of Cr, which implies that there was no competitive interaction of ions in the binary mixture. Despite the high uptake capacity in the binary mixture, there was low removal efficiency of the adsorbent for arsenic removal. This problem is caused by the repulsion of a negatively charged surface and negative ion surface. Another problem is that ions bind on different active sites. This may likely cause a different mechanism of adsorption.

Ma, Liu [60] studied the adsorption of Cd, Cu, Pb, Fe, and Zn onto waste cotton fabrics. The results indicated a high adsorption efficiency of the adsorbent for the investigated heavy metals. The result revealed that the adsorption of all the metal ions increases as the dosage grows. For the metal ions to be effectively removed from the industrial effluent, the treatment volume should be under control below 1932 mL. To achieve this benchmark in this work, 420-min bed contact time was achieved to process 1932 mL effluent. Yahya, Yohanna [56] reported an adsorption capacity of 27.65 mg/g and suggested that the monolayer surface influenced the removal of Pb (II) onto the cotton hull. Mendoza-Castillo et al. (2014) reported the sorption of Pb (II), Cd (II), Zn (II), and As (II) and attributed the high adsorption efficiency to the active functional groups and complex compounds when compared with other studies. The presence of Pb (II) on the surface of the denim waste influenced the removal of arsenic (II).

Table 2. Adsorption of heavy metals on textile wastes.

Precursor	Adsorbate	Qmax (mg/g)	Reference
Waste cotton Yarn	Aqueous solution of Pb, Cd, Cr, As ions	890.8 191.7 236.2 72.90	[58]
Waste cotton fibres	Inorganic pollutants As (III) Inorganic arsenic Arsanilic acid (ASA) $C_6H_8AsNO_3$ Roxarsone (Rox $C_6H_6AsNO_6$)	126.5 164 261.4 427.5	[61]
Char-FeCl$_3$ from cotton waste Char-FeCl$_2$ from cotton waste Char-FeCit from cotton waste	Cr(VI) solution, HNO$_3$, NaOH Concentration	73.79 68.87 43.84	[57]
Cotton hull	Pb(II) from gold mining liquid effluent	27.65	[56]
Cellulose from waste cotton fabrics	Heavy metal nitrate salts (CuNO$_3$)$_2$, Cd(NO$_3$)$_2$, and Pb(NO$_3$)$_2$ Cd, Cu, Pb, Fe, and Zn investigated	Cd: 99.8 Cu: 99.8 Pb: 99.7 Fe: 98.0 Zn: 99.9	[60]
Cotton fibre	Aqueous solution of Cu(II) ions Aqueous solution of Pb(II) ions Aqueous solution of Cr(III) ions	81.97 123.46 72.99	[59]
Denim fibre scraps	Aqueous solution of Pb(II) Aqueous solution of Cd(II) Aqueous solution of Zn(II) Aqueous solution of As(II)	9.83 2.71 2.69 1.23	[62]
Cotton fibre	Heavy metal solution of Cu(II), Zn(II), Pb(II), and Cd(II)	6.12 4.53 8.22 21.62	[63]
Cotton fibre	Aqueous solution of Cu(II) and Pb(II)	88.9 70.6	[53]

This synergistic interaction implies that competitive adsorption better described the sorption mechanism. Reference [59] investigated the use of cotton fibre for the sorption of Cu (II), Pb (II), and Cr (III) and attributed the sorption capacity of the adsorbent to the effect of increase in pH, contact time, and initial concentration of the adsorbate. The result from the sorption capacity of cotton wastes from the most recent studies indicated the effectiveness of the wastes as an alternative to commercial activated carbon for the adsorption of heavy metals from aqueous solution. The application of cotton wastes for the removal of metal ions has the prospect of ensuring safety for both the environment and the receiving water.

The removal of dyes and heavy metals depends on the chemical composition of the analyte and the nature of the wastes. Spent cotton adsorbents contain toxic substances, which have a severe environmental impact. Desorption of these hazardous wastes is required to ensure environmental sustainability. At the end of the adsorption of dyes on cotton wastes, not all reactive dyes are adsorbed. Consequently, the dyeing baths could generate enormous hazardous effluents.

Hu, Shang [64] have studied the use of catalytic ozonation for effluent reuse and reported that catalytic ozonation could regenerate cotton effluent and can be reused twice without compromising fabric quality. In order to regenerate spent cotton wastes for dyes and heavy metals removal, Yousef, Tatariants [65] have reported that leaching treatment is favourable when compared to incineration, which may further constitute a pollution effect. The authors carried out a leaching test for the removal of dyes from cotton wastes using nitric acid leaching and regenerated the used acid by activated carbon. This technique could recover used cotton fabrics by desorption including the spent acid. The desorption of dyes was attributed to the significance of sound waves during leaching, which influenced breaking down of bonds held by a polar interaction or Van-der Waals forces. This implies that the reduction of carbon footprints is achievable by recycling million tons of cotton wastes that could end up in the landfills and incinerators. Surface modification of highly efficient cotton wastes precursors have been effectively applied for wastewater treatment. Challenges such as surface modification of cotton fibres derived CNC for improved adsorption capacity using various pre-treatment methods have not been researched.

- Geotechnical applications

Processing of cotton waste fibres for the reinforcement is of huge environmental and economic benefits. For building applications, the processing of cotton fibres to form fibre-reinforced composites has been reported by Reference [66]. Reference [67] combined cotton wastes and limestone powder wastes for the production of low-cost, lightweight composite for building applications. The new material exhibited high energy absorption capacity and mechanical properties compared to the conventional concrete bricks. In another study, Reference [68] utilized cotton fibres for the reinforcement of composite sheets for a building application while polyester resin served as the matrix. The ultimate tensile strength and flexural strength increases as the reinforced cotton fibre filler was increased in the composite. This suggests that the incorporation of cotton fibre in a matrix improved the mechanical properties of the composite. The properties of the prepared bio-composites are very significant for the insulation of building applications. A schematic representation of sustainable composite properties for effective application of the cotton fibre composite is illustrated in Figure 3.

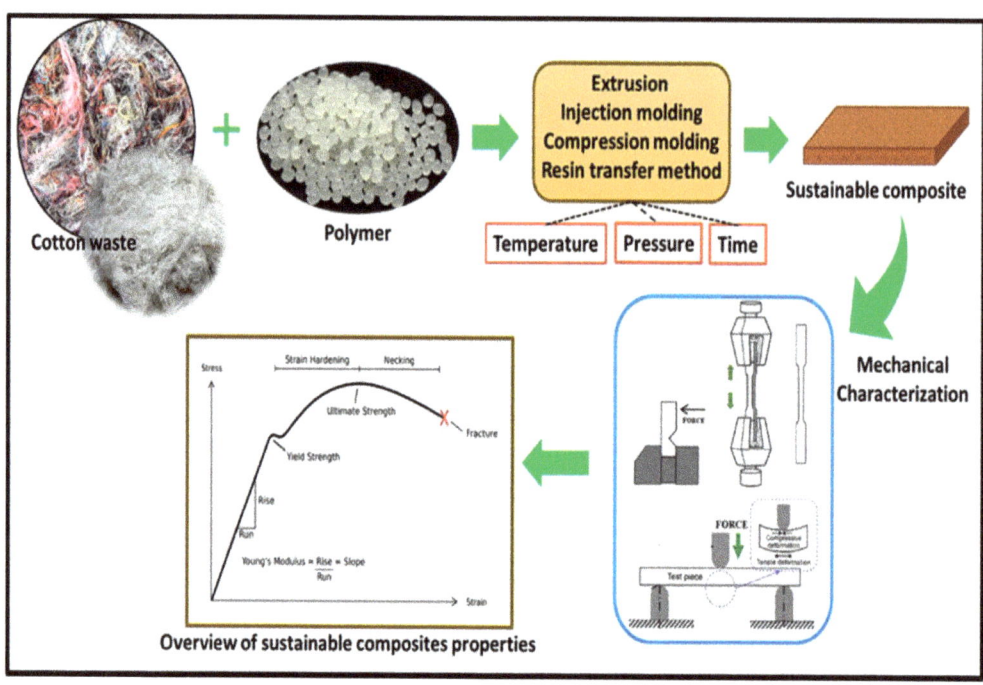

Figure 3. Sustainable composite properties for effective application [69].

Holt, Chow [70] developed bio-composites from cotton carpel blended with Kenaf, flax, cotton stalks, and southern yellow pine and revealed that composite blends with cotton carpel demonstrated improved mechanical properties when compared to an unblended composite. Authors attributed the behaviour of the cotton carpel to its low modulus of rapture. The strength properties of the bio-composite depends on the physical and mechanical properties of the cotton fibers as filler, the fiber orientation, and the manner the components were combined in the composite structure. Reference [71] reinforced the geopolymer composite using cotton fabrics via a lay-up technique. An increase of cotton fibre further improved the flexural strength of the composite. The authors attributed the improvement of the flexural strength to the increase in the number of woven fibres in the composite. It can be suggested that the lower layer of the cotton fabrics in the composite resisted shear failure and sustains the applied load to the composite. Reference [72] reported the reinforcement of fly ash-based geopolymer using cotton fibre. The porosity of the geopolymer increased as the percentage weight of cotton fibre increased in the composite. The improved flexural strength of the composite could be attributed to the good adhesion of the cotton fibre in the matrix. This could be as a result of the good dispersion of the cotton fibres in the matrix. Improvement of geosynthetic materials using recycled cotton wastes have found application as filler for reinforced structures. Reference [73] investigated the effect of mechanical properties of cotton fibre as filler in polyvinyl chloride (PVC) composites under quasi-loading. The increase in stiffness of the composite is attributed to the increase in the filler, suggesting improved mechanical properties. Reference [74] reported that biopolymers have found useful applications in soil amendment. The strength of cotton fibre-reinforced soil was found to increase compared with unreinforced soil [75]. Sadrolodabaeea, Claramunt [76] fabricated composite materials from short fibre textile materials from cotton wastes for building insulations. The compressive strength, stiffness, and durability properties were found to enhance the composite. This can be attributed to the post cracking performance of the cotton waste composite due to the compressive strength and flexural stiffness, which indicated an improvement in energy absorption and suitability as building components.

In another study, Muthuraj, Lacoste [77] reported the incorporation of cotton fibres for the improvement of the properties of the polylactic acid (PLA) composite. Improved mechanical properties of the composite can be attributed to the addition of fibre, which enhanced the crystallization ability of the PLA in the composite. Improved mechanical properties can be achieved using cotton fibre as a filler in composite applications.

2.2.2. Packaging Applications

Technology and engineering sectors are key drivers of innovative waste management strategies to ensure environmental sustainability. Waste engineering and an effective technological approach may well create waste management solutions through recycling of industrial wastes, especially cotton wastes, which are the most generated textile wastes. The concept of bioeconomy has created a sustainable approach to processing cotton wastes into bio-products. Leal Filho, Saari [78] have reported that effective circular design limits the utilization of additives and chemicals for cotton wastes recycling for packaging materials. Reichert, Bugnicourt [79] have reviewed the potentials and significance of processing cellulose from biobased materials for bio-packaging applications. The concept of a bioeconomy can be derived from cotton wastes processing for sustainable packaging application, as illustrated in Figure 4. Circular design toward bio-packaging using cotton wastes as a functional material has a promising prospect.

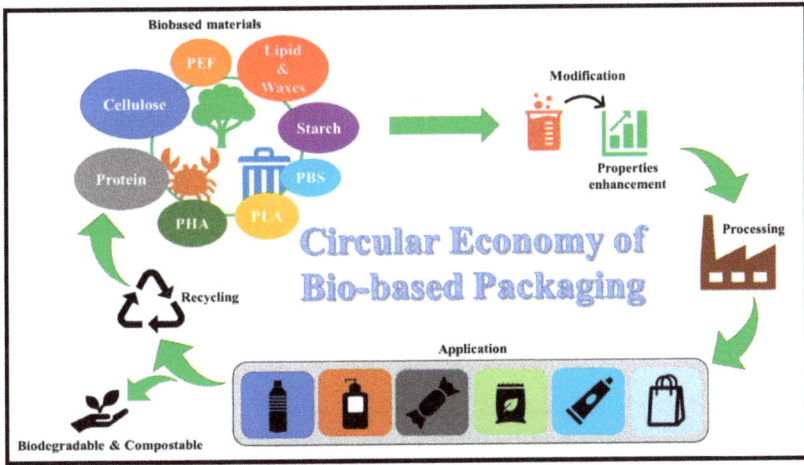

Figure 4. Schematic of bio-derived materials for packaging applications [79]. (PEF: polyethylene furanoate, PLA: polylactic acid, PBS: polybutylene succinate, PHA: polyhydrodyalkanoate).

At a structural level, cellulose derivatives from bio-cellulose composites are an attractive prospect for sheet coating and binders [12]. Fibres of cellulosic composition have a high degree of polymerization, including high modulus and tensile strength for the selection and fabrication of bio-composites. According to Reddy [80], cotton-derived polymer composites can be blended with fillers and additives. After then, they can be reprocessed into different forms. Ramamoorthy, Skrifvars [81] reported that handspun yarns could be made from cotton wastes blended with fibres of feathers for packaging applications. Montava-Jordà, Torres-Giner [82] stated that cotton fibres with a diameter in the range of 10–30 µm and a high aspect ratio improved the mechanical properties in a polymer composite for non-food packaging applications. High aspect ratio cotton fibers in composite reinforcement provides high tensile strength, which implies that cotton fibre reinforcement in a composite have good mechanical properties suitable for packaging applications.

The processing of cotton fibres provides an interesting phase in bio-composite applications due to their elasticity and high tensile strength. The application of bio-composites has focused on synthetic thermoplastics or thermosets and other non-biodegradable polymers as a matrix. Cotton fibres have been applied with other fibres to form bio-composites of good thermal and mechanical properties. Reference [83] have incorporated cotton waste-derived CNF as a reinforcement of PLA/chitin bio-composite and suggested that the improved mechanical, thermal, and wettability property of the bio-composite was attributed to the addition of the filler in the bio-composite.

Meekum and Kingchang [84] demonstrated that the PLA bio-composite with cotton fibre and empty fruit bunch (EFB) from oil palm showed high biodegradability, even though there was a reduction in the flowability of the material when the cotton fraction was increased. This challenge can be eliminated by injecting reactive plasticizers into the bio-composite. de Oliveira, de Macedo [85] have reported that composites in which cotton fibres were used as fillers demonstrated effective packaging potentials when compared to PLA or thermoplastic starch. The authors concluded from the result that the incorporation of 20% cotton fibre demonstrated better mechanical properties when compared to PLA. Furthermore, Montava-Jordà, Torres-Giner [82] have revealed that the injection of recycled cotton fibre improved the thermal and mechanical strength of a bio-poly (ethylene terephthalate) (bio-PET) biopolymer composite for food packaging applications. Hou, Sun [86] have conducted a comparative study of cotton fibre blended with polypropylene fibres by compression moulding under optimized conditions. The cotton stalks were treated using alkali treatment by steam flash explosion and a combination of both methods. The authors concluded that the polypropylene-reinforced composite with the cotton stalk by a combined alkali-steam explosion exhibited excellent mechanical properties including stability in water due to the large surface area attributed to the highest cellulose content obtained from the combined process. Cotton wastes have demonstrated inherent properties that favour its utilization as a packaging material.

2.2.3. Biomedical Applications

Biopolymers are often processed for bio-medical packaging, anti-microbial materials, biosensing, and tissue engineered applications [87]. Cotton is a typical example of natural fibre, which contains more than 95% cellulose and have been widely used for bio-medical packaging, anti-microbial materials, biosensing, and tissue engineered applications. Abdul Khalil, Adnan [88] reviewed the biomedical application for cellulose nanofiber and revealed numerous potentials for cellulose nanofiber materials. This includes drug delivery, biosensing, antibacterial, medical implant, tissue engineering, and so on (Figure 5).

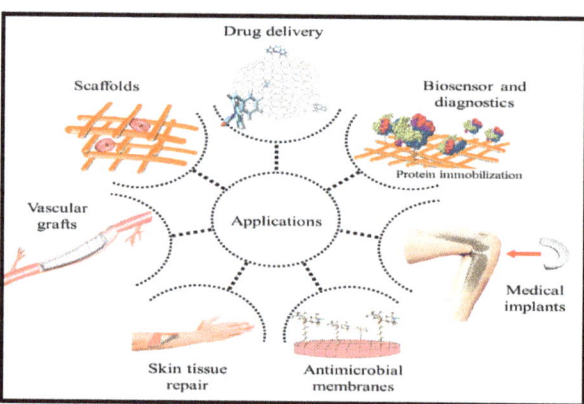

Figure 5. Biomedical application of cellulose-based materials [88].

- Antimicrobial Materials

The unique physicochemical characteristics of cotton fibres are attributed to the available highly reactive functional groups, such as the hydroxyl groups on the surface of the cellulose-based material, which have the tendency to react with other functional groups such as carboxylic acids amines, and aldehydes, resulting in diverse properties. Hence, the prospect of producing antimicrobial biofilm can be achieved for biomedical packaging applications [89]. The production of antimicrobial films from antimicrobial raw materials is shown in Figure 6.

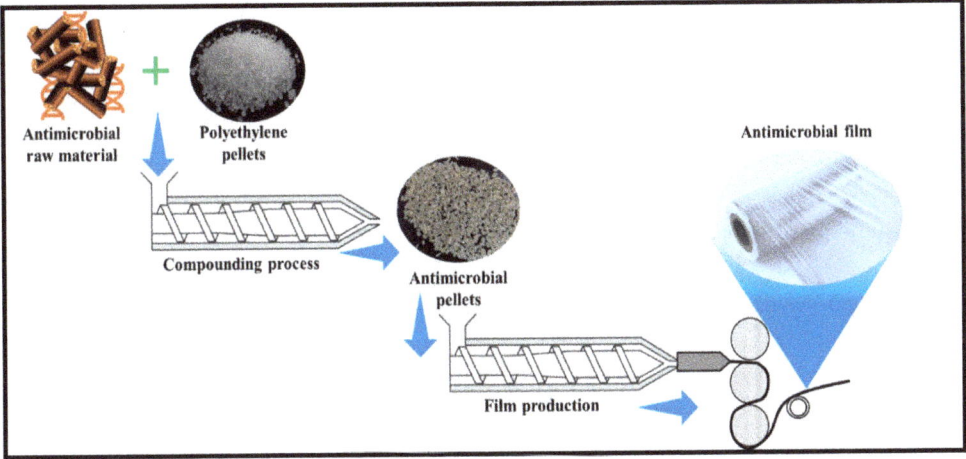

Figure 6. Production of antimicrobial film from antimicrobial raw material.

Tavakolian, Jafari [90] have reported that, through surface modification, cellulose material creates a wide range of compounds that can be grafted on its structure. Such compounds include proteins, antibiotics, and nanoparticles. The prevention of cross-infection and cross-contamination have been effective by incorporating antimicrobial agents into fabric materials. Vignesh, Suriyaraj [91] have produced a membrane from cotton micro-dust wastes via thermochemical treatment and a spin coating method. The authors reported that the membrane demonstrated good tensile and rheological properties. The membrane was effective for the absorption of antibiotics and antimicrobial activities against *E. coli*. Coradi, Zanetti [92] have synthesized cotton fabrics as a precursor for the production of antimicrobial textiles. The investigation of antimicrobial activity was carried out using strains of gram-positive bacteria staphylococcus aureus and staphylococcus epidermis, strains of gram-negative bacteria escherichia coli and pseudomonas aeruginosa, and yeast strain of candida Albicans. The enzymatic immobilization and characterization were conducted using pre-bleached cotton fabrics. The immobilization assays indicated that there was enzymatic immobilization on the modified cotton fabrics. It was suggested that oxidation of cotton fabrics via periodate reaction could be an effective alternative to functionalization and immobilization of the investigated enzymes. Zhang, Yang [93] reported that finished cotton fabrics demonstrated excellent bacterial and anti-fouling properties without affecting their water and air permeability when compared to raw cotton. The intrinsic behaviour was attributed to their breaking strength, tearing strength, bursting strength, and hydrophilic behaviour. Vinod, Sanjay [94] stated that, when natural fibres are used as reinforcements in a composite, the hydrophilic behaviour of the fibres and the hydrophobicity of the matrix are the major challenge. Dassanayake, Wansapura [95] fabricated a composite membrane by incorporating cellulose-cadmium (Cd)-tellurium (TE) quantum dots on a cellulose matrix obtained from a waste cotton linter and investigated the antibacterial activity of the composite using staphylococcus aureus. The in vitro analysis revealed that the composite membrane inhibited biofilm formation [96]. The synthesis of chitosan-silver nanoparticles on cotton fibre was reported to achieve 100% antibacterial activity against Gram-positive and Gram-negative bacteria [97]. Wang, Yin [98] evaluated the antimicrobial efficiency of synthesized and grafted polymeric N-halamine precursors (PSPH) onto cotton fabrics by investigating the hemostatic behaviour and the adhesion characteristics of platelets and red blood cells using the modified mesoporous cotton-based materials.

The authors revealed that the bioactive behaviour of the mesoporous surface has the potentials for attaching red blood cells to the surface of the materials. The modified N-halamine polymer was found to have a higher surface area and pore volume, resulting in the increase of a blood clot index. The synthesis of the membrane from cotton wastes is potential candidates for biomaterial applications, such as coating. The incorporated, extracted cellulose from the cotton into membranes have prospects for biomedical packaging, wound dressing applications, and drug delivery systems.

- Biosensing and Diagnostic

Cotton fibres have promising potentials for the integration of biosensors [99]. Idumah [100] have reported that the integration of sensors within cotton fibres has improved physicochemical properties such as flame resistance, UV protection, and antimicrobial behaviour. In view of fabricating a biosensor from cotton, Subbiah, Mani [101] have developed a biosensor from nanostructured ZnO-modified cotton fabrics by modifying the surface of the carbon-cellulose fabric via a sol-gel technique toward enhanced ultraviolet protection. It was reported that the modified functionalized cotton fabric exhibited maximum UV protection. Authors attributed it to the scattering effect within the layers of ZnO, which influenced the enhanced absorption of UV light. Khattab, Fouda [102] have produced a biosensor from cotton gauze by creating a thin layer of alginate via a dip-coating method.

The obtained cotton sensor was co-encapsulated in a biopolymer shell by loading the sensor on a pH probe. The sensor system monitored the enzymatic reaction of urease through a pH-responsive tricyanofuran hydrazone chromophore. Under the atmospheric conditions, the encapsulated sensor recorded a detection limit in the range of 0.1 ppm and 250 ppm. The findings revealed that cotton gauze strips that coated the sensor system can be applied for the implantation of a smart bandage. A cotton-derived biosensor was applied for wound dressing. Reference [103] fabricated a cotton-based protease sensor and was applied for chronic wound dressings. A schematic representation of the fabricated sensor is presented in Figure 7.

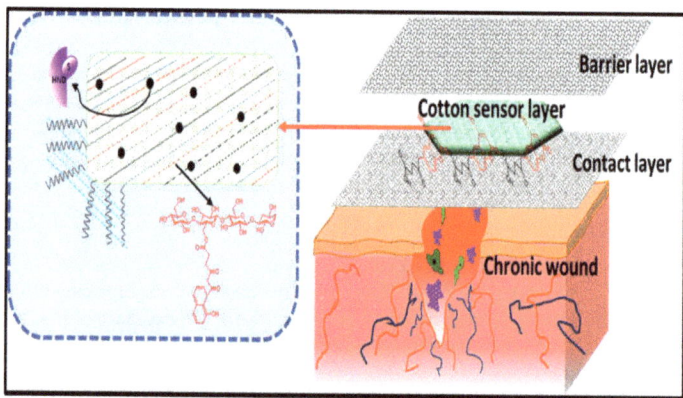

Figure 7. Cotton-based biosensor for wound dressing [103].

Abdelrahman, Fouda [104] have developed a colourimetric swab using a molecular switching hydrazone probe in calcium alginate by encapsulating the probe immobilized onto cotton. The micro-encapsulated cotton probe exhibited a colour shift from light yellow to purple. The result of the morphology study of the colourimetric sensor on a pad of dry cotton fibre indicated the colour-fastness and air permeability. Hence, the sensor can be applied for real-time identification of the status of sweat fluid and for other real-time drug test assessments. Song, Xu [105] produced a flexible sensor from polyurethane silver nanowires on cotton yarn via a dip-coating technique. The pressure sensor could effectively be tuned by changing the amount of polyurethane and silver nanowires. The pressure sensor was reported by authors to be conductive, durable, highly sensitive, and stable. It indicated that a conductive pressure sensor with properties such as a low detection limit, working stability, and high sensitivity have the potential to distinguish signals and frequencies and is applicable for real time assessments.

- Tissue Engineering

Cotton fibres have been explored for wide applications in tissue engineering through the regeneration of tissues from cell growth supported by biomaterials. Cotton materials have been explored for the fabrication of scaffolds for tissue-engineered applications. King, Chen [106] reported that woven fabrics have effective mechanical strength and structural stability due to properties such as porosity and thickness. A porous structure of a tissue-engineered scaffold derived for fibre is shown in Figure 8.

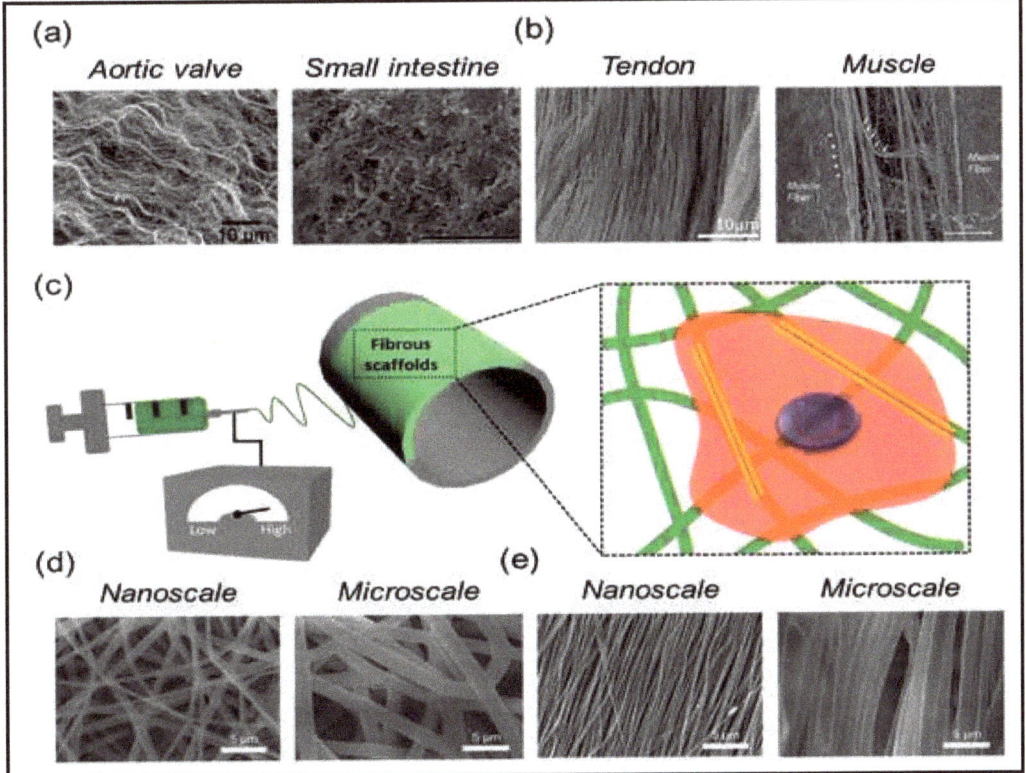

Figure 8. Pore structures of the tissue-engineered scaffold from fibre (**a**) aortic valve & small intestine (**b**) tendon & muscle (**c**) fibrous scaffold (**d**) nanoscale & microscale (**e**) nanoscale & microscale [107].

Singh, Dutt [108] stated that cotton cellulose exhibits functional material properties for scaffold designs. The authors fabricated the scaffold using cotton microfiber, and they investigated the mechanical and physicochemical properties, including the osteogenic properties of the derived cotton scaffold, by crosslinking with citric acid and modification thereafter with gelatin. The porosity and swelling tests revealed the hydrophilicity of the scaffold and also its nontoxicity, as observed in the micrograph. The authors attributed the water absorption potential of the scaffold to the hydrophilic functional groups [109]. Xu, Wu [110] stated that the hydrophilic properties of scaffold influence the interaction between the materials and the cells. The increase in the hydrophilic property of the citric acid crosslinked gelatin scaffold influenced the rate of biodegradability and the adhesion property of the cell in vivo. Nelson, Tallia [111] have developed electrospun, cotton, wool-like silica/gelatin with covalent coupling applicable as a regenerative medicine scaffold. The investigation included the effect of electrospinning process variables of sol viscosity (and ageing time) and amount of coupling agent on the 3D morphology of the fibres, including their structure and dissolution. Authors attributed the 3D structure of the cotton wool fibre to a slow solvent evaporation rate as a result of high humidity and the elongation of the fibres to form optimal sol viscosity. The observable less congruence in the dissolution of silica and gelatin could be a result of the higher surface area of the spun fibres attributed to the evolution of the hybrid structure. The breaking down of the silica and the crosslinking and release of the gelatin molecules significantly influenced the application of 3D fibres as scaffolds. Liu, Liao [112] synthesized a nanofiber core-spun yarn with nylon filament as core and poly (L-lactide-co-caprolactone) (PLCL) nanofibers as the shell via core-spun electrospinning technology. This was designed to obtain a nanofiber vascular scaffold. The nanofibers of core-spun yarn nanofiber vascular scaffolds demonstrated a directional orientation when compared to conventional nanofiber vascular scaffolds. This indicated that core-spun yarn nanofiber vascular scaffolds have the potential to overcome the weakness of the cohesive force between nanofibers while maintaining effective mechanical properties in the axial and radial directions. The applications of nanomaterials create promising applications in biological and bio-medicine for health safety. The use of nanoparticles via surface modification of cotton wastes with multifunctional properties has promising potential in nanotechnology. The functional properties of cotton fibre can further be explored on a nanoscale for the extraction of CNC.

3. Extraction of CNC from Cotton Wastes and Applications

The conversion of cotton wastes into valuable nanomaterials is attracting research attention. From an environmental point-of-view, the recycling of cotton wastes, which is the most generated textile waste, is very significant for the protection of the ecosystem. Cotton fibre is an abundant renewable resource. Therefore, cotton waste remains a suitable candidate for the extraction of cellulose and a precursor for the production of CNC. Cotton fibres consist of three major components, namely, cellulose, hemicellulose, and lignin [113]. The crystalline region contains an abundance of cellulose content. The effect of bleaching could further increase the cellulose composition in cotton fibre [114]. The hemicellulose and lignin are the other components of cotton fibre and are amorphous in nature [3]. A schematic illustrating extraction of CNC from a cellulosic precursor is illustrated in Figure 9.

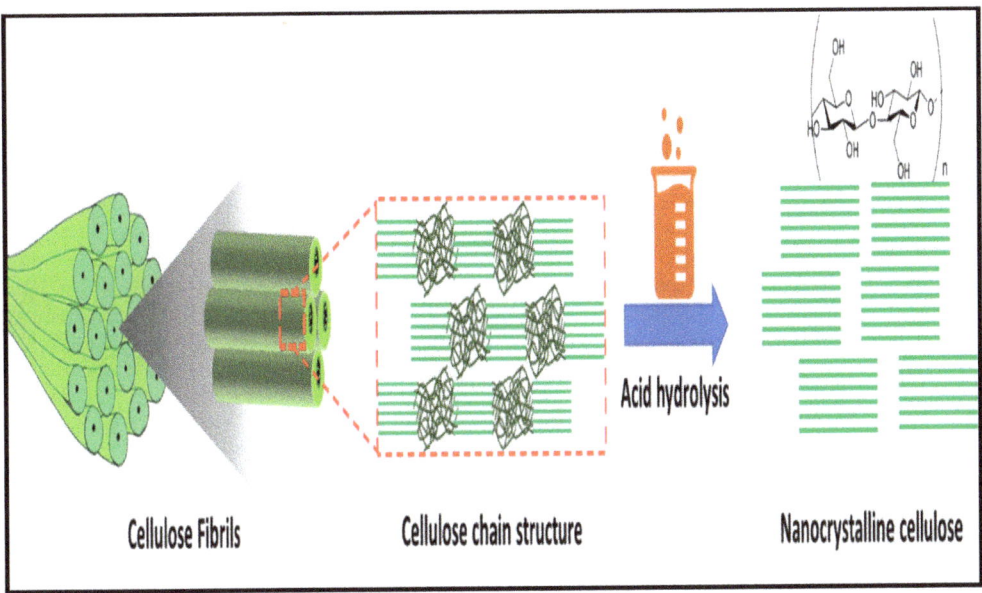

Figure 9. Schematic of nanocrystalline cellulose extraction from cellulose chain using acid hydrolysis for the removal of the amorphous region [115].

However, the abundance of cellulose composition in cotton waste has increased the prospect of its utilization for the production of CNC. Due to its inherent properties such as high strength, large surface area, high modulus, unique optical properties, and biodegradability potentials, several extraction methods have been explored for the production of CNC from cotton wastes. The degree and extent of the crystallinity of the isolated CNC from cotton wastes depend on the pretreatment methods and the processing conditions utilized. CNC is often good biodegradable reinforcement materials and has become a focus of attention for producing biodegradable composites due to their excellent mechanical properties. However, an effective processing mechanism of cotton wastes into functional materials for the extraction of CNC is rarely reported. Isolation of CNC from cotton wastes have been rarely explored. Much focus is on wood-based plants for commercialization purposes. The prospect of cotton-derived CNC has been explored as reinforcement material to improve the functional properties of composite films for packaging applications [116]. Similarly, the advanced applications of cotton-based CNC have been reported for improved biomedical applications [117].

3.1. Chronological Study of Cellulose Extraction from Cotton Wastes

Cotton wastes have been explored as a suitable precursor for the extraction of cellulose. In the early years of textile production, the extraction of cellulose from cotton wastes was achieved by the removal of nanocellulose (NC) compounds. A significant evolutionary trend has been achieved over the years for the extraction of cellulose from cotton wastes. In earlier studies, several pretreatment methods have been used for the extraction, such as chemical treatment using sulphuric compounds, acids, alkali, thermal treatment using pressing, heating, steam puffing, and decompressing. In some cases, pretreatment methods were combined. The sources and the method of extraction are the two major factors attributed to the production of cellulose from cotton wastes. However, there is still much more progress that is required. In recent years, methods of extraction have focused on different pretreatment methods, such as alkali and acid hydrolysis, including ultrasonication. Cotton production has increased over the years, resulting in the increase in cotton wastes that have been applied in different areas, such as environmental pollution control, soil amendment, reinforcements, and biomedicine. Studies have advanced on the increase in the crystallinity and yield of CNC produced from extracted cellulose from cotton wastes.

Recently, hydrolysis pretreatment has been combined with sonication to improve the cellulose content in waste cotton fibre. CNCs from cotton wastes have been used in the biopolymer matrix for reinforcements. The bio-composites have been used for packaging, building materials, and biomedical purposes. The chronological order of cellulose extraction from cotton wastes is presented in Table 3. Earlier studies of cellulose separation from cotton linter were reported by Talbot [118]. The separation was achieved by cooking cotton linter at a uniform temperature distribution in order to avoid the effect of poor liquid penetration. This approach was exhibited to remove the nanocellulose component in the cotton. Speakman [119] separated cellulose from cotton wool using sulphuric acid hydrolysis. The authors revealed that the filaments formed after pretreatment with sodium sulphide were attributed to the effect of formaldehyde. Masselli and Burford [120] extracted cellulose from cotton waste by bleaching using hypochlorite and peroxide. After then, sodium bisulphate was applied. The authors revealed that an increase in cellulose was achieved from the waste due to scouring. Kramar, Obradović [121] have attributed the increase in cellulose extraction for raw cotton silver waste to the influence of chemical pretreatment, bleaching, and scouring. Masselli [122] attributed the effect of scouring to an increase in the extracted cellulose for the removal of natural wax, pefins, alcohol, and impurities. The caustic kiering method was achieved using a mixture of sodium carbonate, caustic soda, and sodium silicate, resulting in the formation of the white fibre. The removal of natural wax from cotton was done by the addition of pine oil soap. Gallagher and Elliott [123] obtained cellulose from waste cotton cloths using alkaline scouring and peroxide bleaching of grey cotton cloth. The alkaline scouring was achieved via steaming of the cotton cloth impregnated with an alkaline medium. This resulted in the removal of the nanocellulose component of the cotton cloths. In 1980, enzymatic hydrolysis was used to obtain cellulose from cotton waste by compression-milling [124].

The energy requirement and enzymatic hydrolysis were affected by the milling time. The authors reported that the relationship between milling time, enzymatic hydrolysis, and the surface area of biomass determines the extent of crystallinity of cellulose materials. Maciel, Kolodziejski [125] separated cellulose from cotton linters by ball milling. The authors revealed that prolonged milling increased the cellulose chain due to the presence of carbon monomer units. Two years later, Bertran and Dale [126] extracted cellulose from cotton wastes by enzymatic hydrolysis. The authors attributed the low crystallinity from the cotton wastes to the susceptibility of the material to enzymatic attack. However, in a related study by Mokeev, Iljin [127], the effect of enzyme activities influenced the degradation and, by extension, the extraction of cellulose. Ludwig and Fengel [128] extracted cellulose from cotton linters by nitration using a mixture of nitric acid and sulphuric acid. The authors concluded that the significance of nitration was vital to the interaction between the cell wall, swelling capacity of the medium, and esterification on the surface of the cotton linters. Brooks and Moore [129] have extracted cellulose from cotton fabric waste by alkaline hydrogen peroxide bleaching. The authors emphasized that the hydroxyl anion was primarily responsible for the bleaching moiety in alkaline hydrogen peroxide.

Acid Hydrolysis has been reported for the separation of cellulose from cotton fabric waste. Elazzouzi-Hafraoui, Nishiyama [130] and Kuo, Lin [131] reported the extraction of cellulose from cotton wastes by enzyme saccharification. Wu, Zhang [132] extracted cellulose from cotton flax yarn waste using the alkali cooking method. The influence of an increase in temperature resulted in the water absorbency and bimolecular collision, which resulted in the increase in the extracted cellulose. The authors suggested that a cellulose macro-radical can be increased due to the effect of the collision. Bidgoli, Zamani [133] extracted cellulose from cotton and viscose waste (VW) by the dissolution of the wastes in water, followed by carboxymethylation. It was revealed that super adsorbent prepared from the extracted cellulose demonstrated effective water-binding capacity. Wang, Yao [4] have extracted cellulose by subjecting waste cotton cloth to alkali and bleaching treatments. Rough surface and disruption of the outer layer of the fibres and exposure of the fibril strands as a result of cracks in the inner structure of fibre were observed. Authors attributed it to the effect of bleaching and alkali treatment, causing the removal of major components, which include hemicellulose, waxes, and impurities. At present, the ease of cellulose extraction from cotton wastes has been improved by advanced technologies.

Ultrasonication is one method that has been used for the isolation of micro and macro fibrils and for the extraction of cellulose. A few studies have combined the process of hydrolysis and sonication for cellulose extraction. Pandi, Sonawane [134] stated that cellulose extracted from cotton wastes had been used in various applications, such as packaging and reinforcement for biomedical applications. This includes the production of CNCs of high crystallinity and thermal stability.

Table 3. Summary of the evolution of cellulose from cotton wastes.

Source	Method of Cellulose Extraction	Reference
Cotton linter	Cooking cotton linter, depolymerization	[118,135]
Cotton wool waste	Hydrolysis using concentrated sulphuric acid	[119]
Cotton Waste	Sodium bisulphite concentration	[120]
Cotton waste	Caustic kiering method, Geiger counter spectrometer	[122,136]
Cotton cloth waste	Kiering bleaching method	[123]
Cotton waste	Graft copolymerization	[137,138]
Cotton waste	Enzymatic hydrolysis by compressing milling	[124]
Cotton linters	Ball milling	[125]
Cotton waste	Enzymatic hydrolysis	[126]
Cotton linters	Nitration, organic solvent	[128,139]
Cotton waste	Strains of trichoderma and Aspergillus bleaching	[127]
Cotton fibre waste	Alkaline hydrogen peroxide bleaching	[129]
Cotton fabric waste	Hydrolysis using hydrochloric acid	[130]
Cotton waste	Ultrasonication, hydrolysis	[127,133,134]

3.2. Extraction of Nanocrystals from Cotton Wastes

The synthesis of cotton wastes on a nanoscale toward pollution management and environmental protection is drawing attention for diverse applications. The characterization of nanocrystals from cotton wastes is attracting significant interest. This can be attributed to its abundance of cellulose, environmentally-friendliness, high crystallinity, biodegradability, low toxicity, and high surface area, including its inherent optical properties and renewability potentials. The extraction of CNCs from cotton wastes can be achieved via chemical, mechanical, or enzymatic processes.

The isolation of CNCs from semi-crystalline cellulose is usually achieved through a hydrolytic process whereby the amorphous phase undergoes digestion while, in the process of mechanical disintegration, the crystallites are released. Bahloul, Kassab [140] reported that the characteristics of CNC are affected by the source of cellulose, hydrolysis condition, and reaction temperature. References [141,142] reported that CNCs could be extracted from sources such as cotton stalks, cotton linters, and cotton slivers. Khalil, Davoudpour [142] have reported that extraction of CNC is mostly achieved using mineral acids such as hydrochloric, phosphoric, and sulfuric acids. The extraction of the cellulose chain by the removal of the amorphous region from cellulose-composed materials is shown in Figure 10.

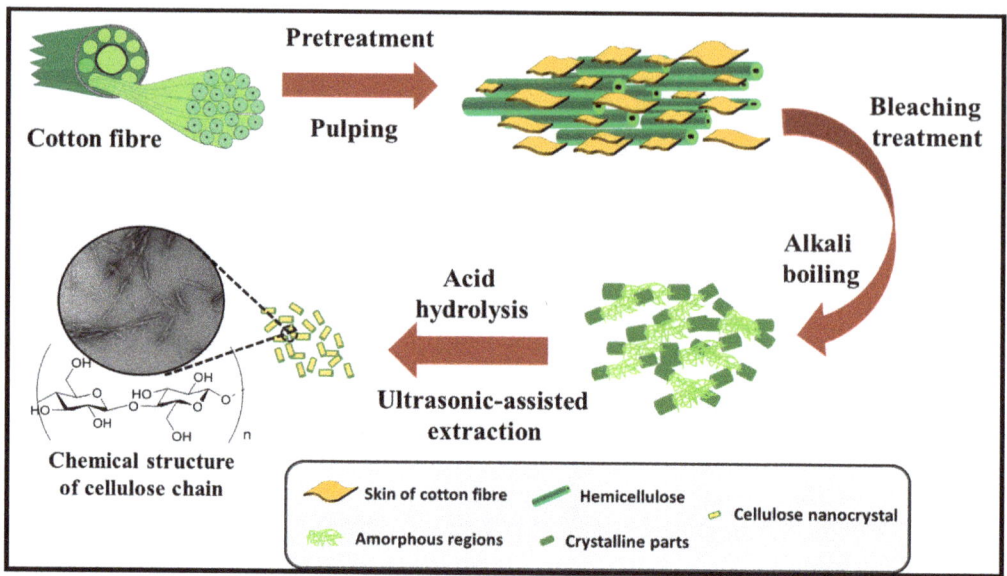

Figure 10. Schematic of nanocrystalline cellulose extraction from the cellulose chain using acid hydrolysis for the removal of the amorphous region [4].

Maciel, de Carvalho Benini [143] have reported the extraction of CNC from cotton waste fibres by hydrolysis. It was revealed that the chemical treatment was responsible for the decrease in the amounts of hemicellulose and total lignin (soluble + insoluble), including the extractives and ash. As a result, the relative amount of cellulose was found to increase for bleached fibres compared to the untreated ones. Kalia, Thakur [144] stated that the effect of chemical treatments did not only decrease the amorphous components but also were responsible for the colour changes from brown to white. The findings revealed that cotton wastes provided a good source of cellulose and for the extraction of nanocellulose. Pandi, Sonawane [134] extracted CNC from cotton wastes by bleaching, hydrolysis, and sonication pretreatment methods. High stability and crystallinity of CNC were found higher than the raw material. The combined effect of hydrolysis and sonication can possibly isolate more CNC from cotton. Thambiraj and Shankaran [34] have isolated CNC from cotton wastes by hydrolysis using H_2SO_4. The produced CNC demonstrated high crystalline, biocompatible, and sustainable behaviour. The type of acid used for extracting CNCs can affect the characteristics of the CNC. When sulfuric acid was used, sulphate groups from the acid esterifies the free hydroxyl groups on the surface of the CNC. Conversely, CNCs hydrolyzed by HCl have relatively low concentrations of strong and weak acid groups bound on the surface that allow the crystals to aggregate and flocculate due to van der Waals attraction in aqueous solutions.

Wang et al. (2019) extracted CNC from cotton wastes by alkali treatment, which is followed by acid hydrolysis. The extracted CNC demonstrated smooth and flat surface morphology, including a high aspect ratio. CNC exhibiting a high aspect ratio have the reinforcing potentials in composites due to good mechanical properties. Reference [145] stated that the efficiency of CNC for reinforcement applications is influenced by the crystallinity and the aspect ratio. This implies that cotton-derived CNC can serve as an effective reinforcement material for various applications. According to Kassab, Kassem [146], a high aspect ratio CNC provides a reinforcing effect due to the good physical and mechanical properties. Hemmati, Jafari [147] reported the extraction of CNC from cotton linters by alkali treatment using NaOH. High crystallinity of CNC was achieved. Although, the thermal stability was lower than the raw material. This problem was caused by an increase in the hydrolysis concentration, leading to fragmentation of cellulose particles. Culsum, Melinda [148] extracted CNC from denim wastes using alkali treatment and ammonium persulfate (APS) oxidation at 60 °C within 5 to 15 h. Zhong, Dhandapani [141] have synthesized CNC from indigo-dyed denim fabrics by bleaching 2 m long cotton fabrics using contra selector mill, and then hydrolysis using 64 wt% of H_2SO_4 at 45 °C for 1 h. Higher crystallinity of CNC was achieved. This was attributed to the susceptibility of the amorphous region as a result of the effect of hydrolysis, which influenced the removal of amorphous cellulose. Reference [149] extracted CNC from viscose yarn via a facile one-step process (Figure 11). Higher crystallinity of CNC was achieved. This was attributed to the susceptibility of amorphous region as a result of the effect of hydrolysis, which influenced the removal of amorphous cellulose.

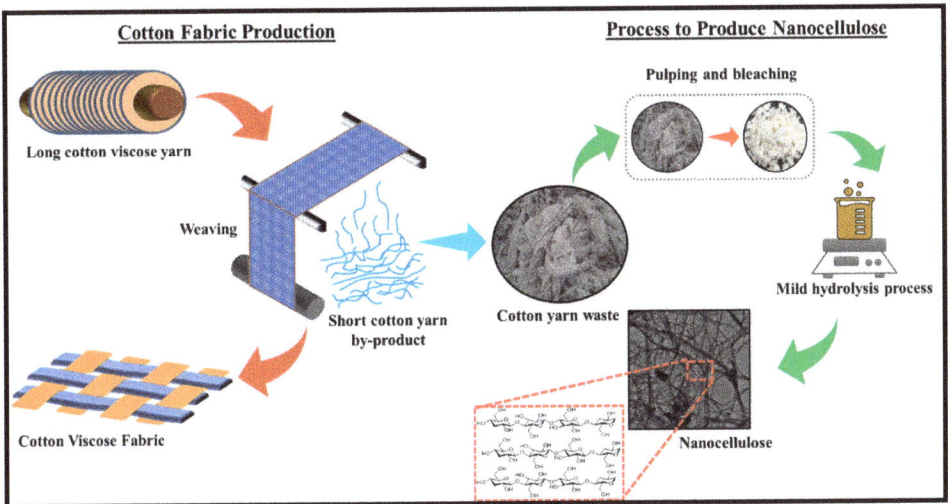

Figure 11. Viscose yarn waste production and one-step extraction of NC from VW [149].

The extraction of CNC from cotton wastes, pretreatment methods, and process conditions, including their strength and limitation from the previous studies, is presented in Table 4.

Table 4. Cotton wastes and method of extraction of CNC.

Sources	Pretreatment Methods	Process Condition	Average Size of CNC	Advantages	Disadvantages	Reference
Waste cotton cloth	Alkaline treatment and acid hydrolysis	Alkali treatment 10% (wt%) of NaOH for 2 h Hydrolysis: H_2SO_4 and HCl at 55 °C for 7 h using ultrasonic waves	Length 28 to 470 nm Diameter: 35 nm	High crystallinity of CNC.	Low thermal stability compared to raw material	[4]
Cotton wastes	Acid hydrolysis	Hydrolysis: 60% H_2SO_4 at 50 °C for 8 h	Diameter: 6.5 nm Length: 180 ± 60 nm	Unique fluorescence properties for bioimaging and biosensing applications	Energy time consumption to determine fluorescence property	[34]
Waste cotton cloth	Alkali treatment and acid hydrolysis	Alkali treatment NaOH at 70 °C for 2 h Hydrolysis: H_2SO_4, HCl, at 65 °C for 5 h	Length: 38 nm to 424 nm Diameter: 2 to 17 nm	Smooth and dense surface with high crystallinity	Excess CNC will inhibit the formation of the transparent film in composite	[150]
Cotton waste	Bleaching, acid hydrolysis, and sonication	Bleaching at 60 °C for 4 h using NaOCl Hydrolysis: H_2SO_4 (30–50%) for 4 h Sonication: Ultrasound probe sonication for 45 min	Length: 20–100 nm Diameter: 10–50 nm	High thermal stability than the raw material	The combined effect of hydrolysis and ultrasound treatment is expensive	[134]

Table 4. Cont.

Sources	Pretreatment Methods	Process Condition	Average Size of CNC	Advantages	Disadvantages	Reference
Cotton wastes	Alkali treatment, bleaching, and acid hydrolysis	Alkali treatment: NaOH solution for 1 h at 70 °C. Bleaching: Added H_2O_2 NaOH for 1 h at 50 °C Hydrolysis: H_2SO_4 at 50 °C. Pulp: Solution of 1:20 (g/mL)	Length: 105–5880 nm Diameter 23.8 ± 5.6	Effective removal of amorphous compounds before hydrolysis	Increase in reaction time not feasible at 15 min due to an increase in processing cost	[143]
Indigo Denim fabrics	Bleaching and acid hydrolysis	Bleaching cotton fabrics using cotton selection mill Hydrolysis: H_2SO_4 at 45 °C for 1 h	Length: 197 Diameter: 7	High crystallinity and thermal stability	Sulphuric acid hydrolysis could not degrade indigo dyes	[141]

The extraction of CNC from cotton wastes is achieved at a low cost. The wastes are harnessed into value added products with promising potentials for a wide range of application. The functional properties of cotton wastes as a precursor for the extraction of CNCs depend on the extent of its crystallinity. The crystallinity of a material is a function of the hydrolysis reaction time, which is determined by the crystalline index [151]. Ilyas, Sapuan [152] stated that the formation mechanism of the nanostructure is based on the extent of susceptibility of the amorphous phase to acid hydrolysis in comparison with the crystalline phases. The extent of the removal of the amorphous phase relative to the amorphous region is affected by the hydrolytic time in view of extraction of cellulose from cotton wastes. Zhong, Dhandapani [141] and other studies have reported extraction of CNCs from cellulose derived from cotton wastes (Table 4). The functionality and efficiency of the produced CNCs were determined from the yield and crystallinity index. The result of a comparative study of the crystallinity and yield of CNC extracted from cotton wastes is presented in Table 5.

Table 5. Comparative study of material properties of cellulose nanocrystals obtained from cotton wastes.

Source	Size Width (nm)	Length (nm)	Crystallinity Index (%)	Yield (%)	Reference
Cotton waste	221	20–100	81.23	45	[134]
Denim waste	80–120	76.14 ± 8.56	86	24.14	[148]
Cotton fibre from denim fabrics	11.9 ± 6.7	127.7 ± 43.8	86.4	39.6	[141]
Cotton gin motes and cotton gin wastes	78–247	100–300	78	29.3–48.6	[153]
Cotton waste	-	105–5880	75–81	80–89	[143]
Cotton linter	133	229 ± 97	82	59	[147]
Cotton waste	10 ± 1	180 ± 60	-	45	[34]
Waste cotton cloth	-	28–470	55.76 ± 7.82	46.7 ± 1.8	[150]
Cotton waste	40–90	70–200	82.80	25.21	[154]
Cotton waste	-	76–159	79.	30–35	[155]

The findings of the crystallinity indexes in Table 5 is discussed on the basis of the extraction process in Table 4, which is relative to the source of the waste and the pretreatment methods. Maciel, de Carvalho Benini [143] reported a high crystallinity index and yield from the CNC processed from cotton waste. The high crystallinity index was attributed to the effective removal of amorphous compounds before hydrolysis. Reference [148] achieved higher crystallinity similar to the value obtained for CNC derived denim wastes from cotton fibres [141]. Low hydrolytic time of extraction of CNC from indigo denim fabrics produced CNC of high crystallinity and thermal stability. The significance of high crystallinity at low hydrolytic time indicated that the highly crystalline CNC produced could be attributed to the effect of bleaching since sulfuric acid hydrolysis could not degrade indigo dyes. The inability of hydrolysis by sulphuric acid to degrade indigo dyes was attributed to the poor yield of CNC obtained from cotton wastes [141]. Morais, de Freitas Rosa [156] have extracted nanocellulose from cotton linters. The authors revealed that the CNC have prospects for the synthesis of hydrophilic nanocomposites. The linter have potentials for the formation of CNC. Hemmati, Jafari [147] achieved high crystallinity and yield of CNC from cotton linters. The CNC obtained have high water holding capacity compared to the raw cellulose. The water holding capacity is attributed to the larger pore width size of the extracted CNC. The amount of amorphous region is expressly removed at the expense of crystalline cellulose as a result of the removal of hydrogen bonding in the network structure. Therefore, the extracted CNC have higher water sorption capacity as a result of the increased width of the CNC. The high crystallinity can be attributed to the preferential hydrolysis of the cellulose amorphous regions, which degraded during hydrolysis of sulphuric acid into soluble products [157].

Furthermore, Wang, Yao [4] reported that the crystallinity of CNC derived from cotton cloth waste was not too high. Low thermal stability of CNC was achieved when compared to the raw material. This is because the crystalline domains were found to be intact due to the arrangement of a highly ordered molecular chain that resisted acid penetration. Higher stability of suspension in the CNC produced is attributed to the high crystallinity of the CNC. Although there was evidence of poor thermal stability in some of the produced CNC, this can be attributed to the pretreatment methods and the nature of the surface of the wastes. In addition, the hydrolysis reaction affected the amorphous and crystalline region. In a related study, Reference [158] used a mechanical, high-pressure homogenizer for the extraction of CNC from the cotton fibre after pretreatment using sulphuric acid hydrolysis. The concentrations of sulfuric acid, hydrolysis time, and homogenizer speed were the optimum conditions that resulted in CNC of high crystallinity and thermal stability. High crystallinity and good thermal stability of the CNC could be attributed to the mercerization process. Authors suggested that the mercerization process increased the size of the crystallites.

Thambiraj and Shankaran [34] synthesized cellulose microcrystals (CMCs) from industrial waste cotton by hydrolysis to produce CNC. The prepared CNCs from waste cotton was performed via a two-step procedure. First, CMCs were extracted from the cotton wastes by the removal of pectin including hemicellulose by alkaline hydrolysis and then lignin in an acidic condition. The effect of alkali treatment in the first stage was to remove and hydrolyze the hemicellulose contents including ash, minerals, and other forms of impurities. The impact of alkaline hydrolysis influences the morphology of the produced material. The result revealed that cotton fibre treated at 70 °C have a smooth and well-aligned surface. The removal of the amorphous region was achieved in the second process by acid hydrolysis in order to obtain a crystalline region of CNCs. The result from the findings indicated that the isolated CNCs demonstrated high crystallinity, excellent optical properties, and thermal stability. It is suggested that the improved properties of the isolated CNC is attributed to a high aspect ratio. This is because a high aspect ratio plays a prominent role in increasing the path length of permanent molecules. As a result, the barrier properties are improved. This is shown in the improved mechanical properties of the extracted CNC.

Previous studies have indicated the significance of the hydrolysis reaction in the extraction of CNC. The effect of chemical hydrolysis in the removal of an amorphous region can be observed in the difference in the morphology of treated and untreated cotton waste fibres, as illustrated in Figure 12.

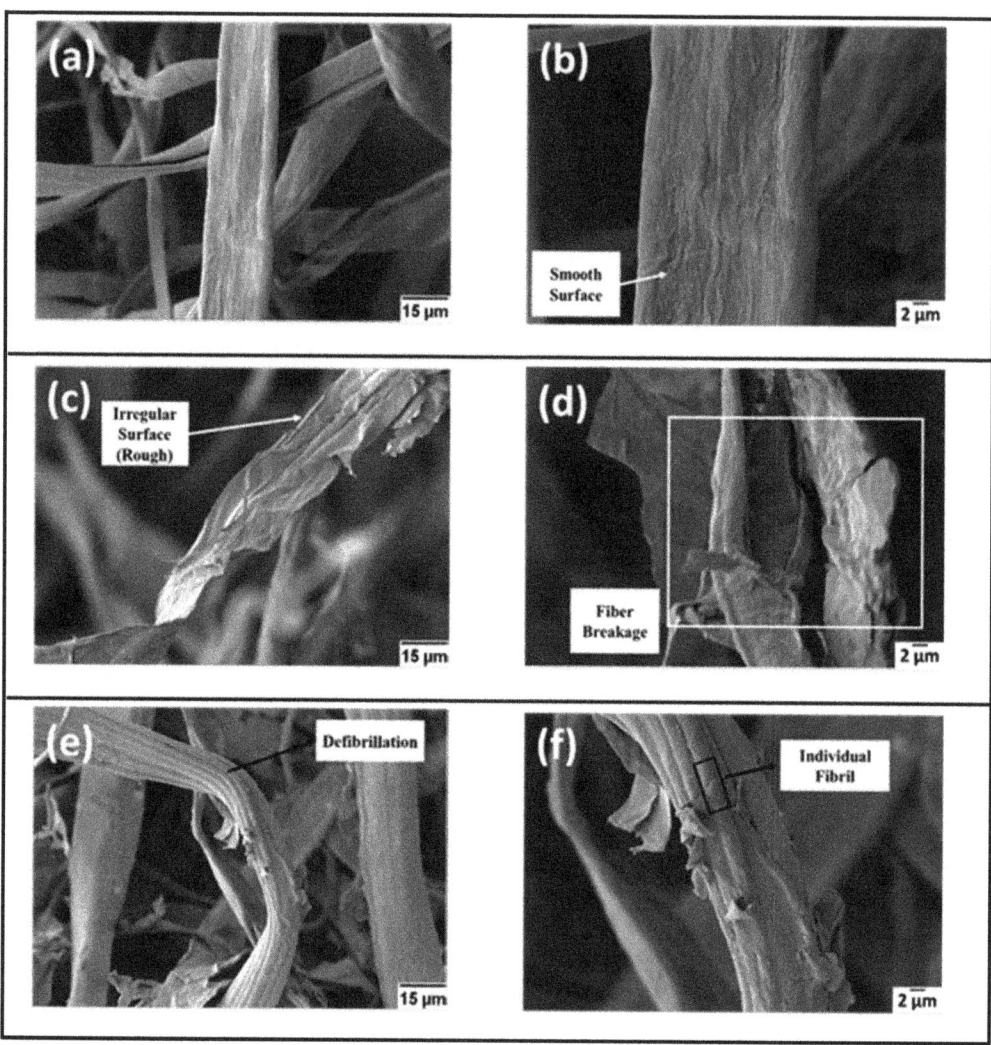

Figure 12. SEM Micrograph of (**a,b**) untreated, (**c,d**) alkali-treated, and (**e,f**) bleached waste cotton fibres [143].

In Figure 12a,b, the morphology of the untreated fibres indicated a smooth surface with predominance of fibre bundles. The effect of alkali hydrolysis affected the morphology of the fibre in Figure 12c,d. Very rough and irregular surfaces that include a breakage of fibre can be observed. This is attributed to the removal of the superficial layer, which was formed by wax [159]. After the bleaching of fibre, more expressive removal of amorphous contents was achieved. The removal of the amorphous region including hemicellulose and lignin resulted in the production of fibrils via fibrillation (Figure 12e,f). A well-structured superfine fibril of cotton fibers consisting of high cellulose content is an ideal precursor for the extraction of CNC. An observable trend of residue weight decreases as a result of the progression of chemical treatment noticed, suggesting that the effect of alkali and bleaching treatment on the cotton fibres effectively removed the amorphous content and other impurities. This process is very significant due to the effect of the components on the crystallinity and thermal stability of the extracted CNC. A summary of the extraction processes of CNCs derived from cotton wastes is presented in Figure 13.

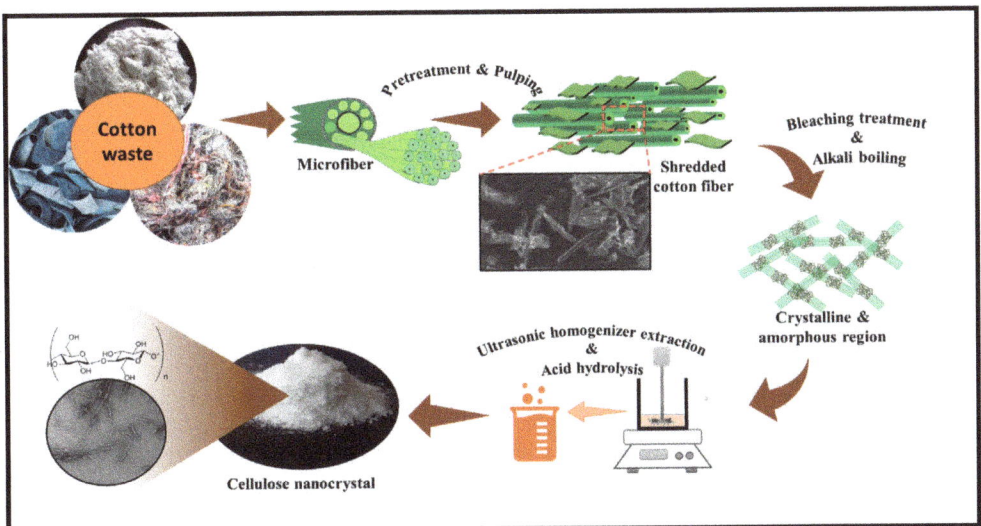

Figure 13. A summary of extraction methods of CNC derived from cotton wastes.

Chemical pretreatment is the most used extraction method for CNC from cotton wastes precursor. However, the techniques applied to extract CNC depends on the chemical composition of the precursor and the pretreatment process conditions of extraction. The extraction procedures involve four major steps, which are washing to remove compounds from cotton wastes, pulping of cotton wastes, then bleaching, and, finally, hydrolysis. The extraction of CNC from cotton wastes using chemical pretreatment mostly applied sulfuric acid hydrolysis to achieve CNC of high crystallinity. Bleaching and sulfuric acid hydrolysis is a highly efficient process of producing CNC of high purity and a high aspect ratio, which can suitably be utilized as reinforcement material in a polymer composite. A combined acid hydrolysis using H_2SO_4 and HCl produced a very dense, flat, and smooth surface of CNC with good thermal stability. The extracted CNC when used as a reinforcement agent exhibited good surface area, high crystallinity, tensile strength, and modulus of elasticity. However, the CNCs were not well distributed in the polymer matrix at a high concentration. This could be improved using ultrasonic assisted hydrolysis. This process could produce CNC of high crystallinity, a high aspect ratio, and higher thermal stability. However, the dispersibilty for CNCs is caused by hydrolysis of HCl. This problem could be improved by the electrostatic repulsion of sulphate ions during hydrolysis with a mixed acid. The combined effect of sonication and hydrolysis using ultrasonic assisted H_2SO_4 hydrolysis could enhance the structural and thermal properties of CNC extracted from cotton wastes. This included increased crystallinity due to a combined effect. Enzymatic hydrolysis and sonication could produce high crystallinity CNC from cotton wastes. Low power and short sonication could disrupt the weak linkages of the hydrolyzed cotton for CNC extraction. Lastly, sulfuric acid hydrolysis assisted high pressure of a mechanical homogenizer that could increase crystallinity and a high aspect ratio of CNC from cotton wastes, which has shown to produce composites of high tensile strength.

3.3. Advanced Applications of Cotton Derived CNC

With recent advances in extraction of CNC from cotton wastes using different isolation techniques, there is remarkable interest in advancing its potentials in various applications, which will be a significant benefit in nanotechnology. The utilization of the intrinsic properties of derived CNC from cotton wastes have prospects in advancing applications in areas such as biomedical and packaging. The design of nanocomposite films using cotton-based CNC as reinforcement to enhance films' properties have attracted significant attention in recent years to improve properties or as a replacement for existing applications.

3.3.1. Biomedical Engineering

The potential of CNC has been explored for biomedical application, such as anti-cancer activity, drug delivery, and tissue engineering in previous studies [160]. Traditionally, carbon nanotubes have been utilized for drug and gene delivery due to their effective penetrability and their encapsulation efficiency. Reference [161] reported the potentials of cotton-derived CNC as replacement of carbon nanotubes for drug delivery. The authors investigated advanced application of cotton-based CNC functionalized with disulfide bond-linked poly (2-(dimethylamino) ethyl methacrylate. The condensation ability of the gene including reduction sensitivity, cytotoxicity, gene transfection, and in vivo anti-tumor activities were investigated. It was revealed that the cotton-based CNC demonstrated very effective transfection efficiencies and low cyto-toxicities. The toxicity of CNCs through pathways such as inhalation into the lungs and cellular uptake is a critical property that determines the utility of the nanoparticles in the biomedical sector [162]. Reference [163] fabricated a gelatin composite film using cotton as reinforcement for a wound dressing application. The developed bio-composite film exhibited homogenous dispersion of the CNC within the gelatin matrix including a strong interfacial interaction between the matrix and the reinforcement. The evidence of excellent biocompatibility indicated improved mechanical properties of the composite film. Authors found no cytotoxicity from the in vitro analysis, which suggested the significance of the reinforcement in advanced application of the composite film for wound dressing. Reference [117] compared the interaction of CNC derived from cotton with CNC extracted from tunicate using a multicellular in vitro model of the epithelial airway barrier after aerosol exposure. Authors reported that an increase in the doses of short cotton CNCs were readily cleared from the surface by internalization within 24 h after exposure when compared to tunicate CNC. However, both extracted CNC exhibited no translocation and no clearance pathway. The fibres alone were able to pass through the insert membrane. This gives a valuable insight when considering the long-term effect and consequences of CNC exposure to humans. However, CNC is found to be non-cytotoxic. Further in-depth studies are required, such as immunogenicity and blood compatibility, which are currently lacking in the literature. A non-cytotoxic CNC could interact with biomolecules and evoke other responses in the biological systems. Therefore, advanced studies are required to further explore the potentials of cotton-based CNC as a replacement for carbon nanotubes for medical applications.

3.3.2. Packaging Applications

The reinforcement potentials of CNCs have often been explored for the improvement of mechanical properties of composites for packaging applications. Reference [164] compared the potentialities of CNC isolated from cotton linters under two different processing conditions as coating for Poly (ethylene terephthalate) films. The pretreatment processes were sulphuric acid hydrolysis and ammonium persulfate (APS) oxidation. It was revealed that CNC isolated using APS oxidation exhibited a higher charge density, higher crystallinity, and higher transparency of the coating compared to sulphuric acid that extracted CNC. Higher oxygen barrier was exhibited by the sulphuric acid-treated CNC. This indicated that the potential surface modification or grafting is required for improved properties of the APS-treated CNC. Although both CNCs demonstrated a lower oxygen permeability coefficient compared to commonly used synthetic resin for packaging. There is a need for coating the layer of the APS-treated CNC to enhance the final properties of the packaging application and as an alternative to other food packaging materials. Reference [165] investigated the mechanical, optical, and anti-barrier properties of polyethylene terephthalate (PET), oriented polypropylene (OPP), oriented polyamide (OPA), and cellophane films coated using CNC extracted from cotton linters. The authors reported that CNC-coated PET and OPA exhibited the best performance when compared to other coating films. The less effective performance of OPP was attributed to the weak adhesion interaction between CNC coating and the OPP surface, which resulted in the removal of CNC from the substrate during dynamic measurements. A well-dispersed CNC coating on the substrate is likely to improve the surface property of the OPP for the packaging application. Jiang et al., 2021, extracted CNC from cotton fibre and the extracted CNC was applied as reinforcement for composite films. The addition of the CNC as filler improved the tensile strength of the fabricated film. Oun and Rhim, 2015 reported that cotton linters have been used as a precursor for the extraction of CNC and investigated its effect on films' property for the production of the composite. The improved tensile strength of the film including water vapour barrier properties indicated the effectiveness of the CNC as filler for the reinforcement of the composite. Huang et al., 2020, reported the extraction of CNC from cotton wastes and was applied for the reinforcement of soy protein film. The incorporated CNC improved the tensile properties, Young modulus, and water vapour barrier property of the composite film. This implies that cotton wastes can be successfully applied as an effective alternative in blended composites and can complement other fibres for the improvement of mechanical and thermal properties for packaging applications.

4. Challenges of Cotton Derived CNC and Prospects

Cotton wastes is the most generated wastes during textile production. The ineffective disposal mechanism of the wastes causes environmental problems. The valorization of cotton wastes into functional materials is essential, but its sustainability and the potentials of extracted cellulose from the waste streams have not been fully explored even though they have been utilized for some geotechnical, environmental, and biomedical applications. A cotton fiber–biopolymer blend for functional and advanced applications such as biomedical, tissue engineering, etc. still requires further research. Issues of biocompatibility and toxicity of the cotton fibre are still pending. The extraction of CNC from cotton wastes have been widely reported in the literature. However, there still exists a drawback that could limit their applications. The pretreatment strategies can suffer variation in terms of process conditions, which could result in a distortion in the crystallinity of the extracted cellulose. Chemical pretreatment is considered the most efficient process. In some instances, an increase of acid concentration affects the crystalline region of the CNC. The use of ultrasonic techniques for the extraction of CNC from cotton wastes is attracting attention in the area of sonochemistry.

In an ultrasonic technique, the exchange of ultrasonic energy to the cellulose chain requires energy for the breaking down of cellulose into nanofibers. Sonication depends on cavitation, whereby the ultrasound energy may likely cause shear-in resulting in breakage of intra-fibrillar bonds due to scission. On a commercial scale, ultrasonication is effective for the extraction of nanofibers from the original, natural fibre by altering the duration and power of the ultrasound. The diameter of the nanocrystal can be increased in the process. Many studies have combined pretreatment methods, such as hydrolysis, alkali treatment, and sonication, to reduce production cost. An increase in the acid hydrolysis affects the crystalline region in some cotton wastes. Therefore, more studies should focus on the best approach to minimize the effect of hydrolysis for an optimal benefit. Mass production of CNCs from cotton wastes have rarely been explored on industrial level even though there are promising potentials for various industrial applications, such as biocompatibility, biodegradability, and high thermal conductivity. The issues and challenges that affect the production of CNC from cotton wastes documented in this review is illustrated in Figure 14.

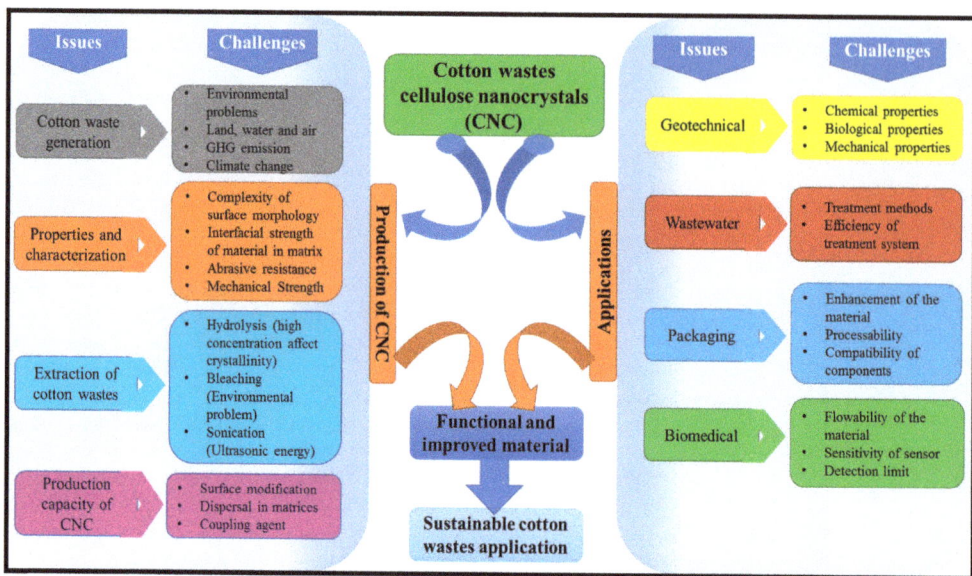

Figure 14. Issue challenges of extraction methods for the production of CNC.

Furthermore, cotton wastes as a precursor for activated carbon have demonstrated high adsorption capacity for the removal of recalcitrant pollutants for dyes and heavy metals including organic pollutants in solution. However, the sorption capacity of the wastes adsorbent depends on the operational conditions of the treatment system and the surface structure of the wastes adsorbent. Surface modification of cotton wastes will improve the extent of attachment of pollutants on the active sites of the adsorbent. The challenge of improvement of reinforced matrix interface using cotton fibre as reinforcement in composites is still a subject of intensive research. According to Claramunt, Ventura [166], the reinforcement of the composite with natural fibres are usually dispersed randomly on the matrix in textile structures, such as woven or nonwoven fabrics. This is because short fibres are used for reinforcements. Even though the flexural strength and ductility of the composite are improved by the randomly dispersed fibres, the improvement is limited due to the short length of the fibres. In some other cases, nonwoven cotton fabrics are rarely applied in cement composites due to the problem of infiltration of the cement matrix as a result of the closed structures. These challenges need to be eliminated through comprehensive research. Furthermore, the mechanical properties depend on the fibre type, orientation, and degree of entanglement. In view of producing cotton fibre-reinforced cement composite for structural application, studies on cotton fibres in a cement matrix with a high degree of fibre entanglement are still very limited. However, more research needs to be done for improved properties of cotton fibres for reinforcement of composites. Cellulose derived from cotton wastes have demonstrated remarkable potential for packaging and biomedical applications.

The possibility of combining biopolymers and cellulose from cotton wastes to obtain a biodegradable plastic can be very useful for packaging applications. The effect of cellulose as filler can significantly impact the mechanical properties of the biopolymer. Masmoudi, Bessadok [167] reported that an increase in cellulose in the biopolymer increases the elongation rate. There was minimal elongation at break as a result of strong adhesion at a saturation between the filler and the matrix. This can be attributed to the increasing forces of the interaction of density and interfacial tension. To a large extent, through innovative ideas, cotton wastes have been synthesized to nanocellulose materials and have found wide application in various areas of nanotechnology, ranging from non-food packaging, bio-nanocomposites, environmental, and biomedical applications. The availability of the technologies for producing CNCs from cotton wastes with enhanced properties having stable, fluid aqueous dispersion, abundant surface functional groups, and a strong interaction with a polymer matrix are subject to future development, considering its socio and economic benefits.

5. Propositions for Future Research on Nanocellulose Crystals from Cotton Wastes

The development of CNCs from cotton wastes have demonstrated vast potentials for various application in our earlier discussions. For the advancement of nanotechnology, enhancement by chemical modification, adding compatibilizers to improve CNC is significant since hydrophilic behaviour of CNC inhibits mixing and dispersal in hydrophobic matrices. PLA obtained from agricultural sources combined with nanocellulose produced cellulose extracted from cotton wastes as reinforcement have the potential to create a functional CNC. The new biomaterials have the prospect of creating new areas of applications in construction and automotive engineering including a sustainable water treatment system, biomedicine, and food packaging. However, the challenge of the dispersal of the reinforcement in the polymer matrix without causing degradation of the biopolymer may limit the application. This limitation can be improved through the nanofibers-matrix compatibility by an interaction or by adopting suitable processing techniques. The technique to be adopted to manufacture a biopolymer/cotton-derived nanocellulose composite should be on the basis of the desired application. This includes the physico-chemical characteristics and the thermal stability of the nanocellulose. For example, the use of nanocellulose as an additive for reinforcement of the biopolymer must have effective binding potential, high molecular strength, and functional surface chemistry that allows reactivity. Cellulose nanocomposites are fibrous in nature with effective mechanical properties including their biocompatibility. They are potentially suitable as components for improved membranes for water treatment systems, conductive thin films, wound dressing, and paper barriers. The incorporation of CNC within the polymer matrix alters the membrane properties. More research attention should focus on the enhancement of the tensile strength from the added CNC and the hydrophilic behaviour, permeability, and resistance to biofouling of the produced bio-nanomaterial.

6. Conclusions

Cotton wastes have been effectively applied due to the high adsorption capacity of the precursor for the removal of dyes and heavy metals from solution. The functional properties of cotton wastes have been utilized for the production of lightweight composite-building materials. Cotton wastes have found application as filler for reinforced structures for geotechnical applications for slope amendments as well as reinforcement of soils and composites with effective binding properties. The cellulosic properties of cotton wastes with other pozolanic materials could be an effective alternative binder to improved flexural strength and physical properties of materials for packaging. Cotton fibres have also been utilized as effective reinforcement materials for packaging, antimicrobial materials, and for the fabrication of scaffolds for tissue-engineered applications. The extraction of CNCs from cotton wastes offer remarkable prospects to produce functional materials for reinforcement applications of composites. The processing of cotton wastes for the extraction of CNCs holds promising potentials for advanced application due to the high crystallinity and thermal stability. Several novel extraction methods including a combination of processes have been reported. Acid hydrolysis using sulphuric acid remains the most used pretreatment method. Cotton gin motes and cotton gin wastes could extract CNCs of high thermal stability and high crystallinity using sulfuric acid hydrolysis. CNCs of high crystallinity could be extracted from cotton wastes by alkali treatment and acid hydrolysis. A combined process of bleaching, acid hydrolysis, and sonication could extract CNC of high thermal conductivity from cotton wastes compared to the raw material. The crystallinity of cotton wastes derived CNC that could increase by 81.23% and the yield by 45%. Reduction in hydrolysis time and enhanced yield of CNC could be achieved by combined extraction methods. Physical and chemical modification of cellulosic sources derived cotton wastes significant for improved functional materials for improved packaging and biomedical applications through improved bonding of the fibre and the matrix of bionanomaterial products. Cotton wastes-extracted CNC have been incorporated as a reinforcement agent in composite films with improved compatibility, homogenous dispersion, and effective fiber-matrix interaction for packaging. CNC derived from cotton wastes have proven to exhibit a non-cytotoxic effect and could be applied as a replacement for carbon nanotubes for biomedical applications. The high crystallinity of CNCs derived from cotton wastes and their properties including high tensile strength and high-water holding capacity create prospects for the production of CNC aerogels/hydrogels for drug delivery, building insulations, and oil and water separation applications. Previous researchers have explored the ideas of producing CNCs. However, effort toward advancing the commercialization potential remains a challenge. Future works to improve on the dispersibility of CNC in the polymer matrix as reinforcements, including the potentials of CNCs from cotton wastes to produce novel bionanomaterials, need to be explored.

Author Contributions: Conceptualization, S.R., A.K.H.P.S., and A.A.O. Data curation, O.N.G., C.K.A., E.B.Y., and T.A. Methodology, A.K.H.P.S. and O.N.G. Project administration, S.R., A.K.H.P.S., and A.A.O. Resources, A.A.O., F.A.S., and A.A.R. Software, C.K.A., E.B.Y., F.A.S. and A.A.R. Writing–original draft, A.A.O. Writing–review & editing, A.A.O., O.N.G., C.K.A., E.B.Y., and T.A. All authors have read and agreed to the published version of the manuscript.

Funding: This research was funded by the Ministry of Education grant number RUI 1001/PTEKIND 8014119 and the APC was funded by the Ministry of Culture and Education of the Republic of Indonesia by World Class Professor (WCP), Program 2020 Contract number: 101.26/E4.3/KU/2020.

Institutional Review Board Statement: Not applicable.

Informed Consent Statement: Not applicable.

Data Availability Statement: Not applicable.

Acknowledgments: This work was financially supported by the Ministry of Culture and Education of the Republic of Indonesia by World Class Professor (WCP), Program 2020 Contract number: 101.26/E4.3/KU/2020. The researchers would like to thank the collaboration between Universitas Syiah Kuala, Banda Aceh, Indonesia, and Universiti Sains Malaysia, Penang, Malaysia, that has made this work possible.

Conflicts of Interest: The authors declare no conflict of interest.

References

1. Sandin, G.; Peters, G.M. Environmental impact of textile reuse and recycling–A review. *J. Clean. Prod.* **2018**, *184*, 353–365. [CrossRef]
2. Xiao, B.; Zhang, Y.; Wang, Y.; Jiang, G.; Liang, M.; Chen, X.; Long, G. A fractal model for kozeny–carman constant and dimensionless permeability of fibrous porous media with roughened surfaces. *Fractals* **2019**, *27*, 1950116. [CrossRef]
3. Ütebay, B.; Çelik, P.; Çay, A. Effects of cotton textile waste properties on recycled fibre quality. *J. Clean. Prod.* **2019**, *222*, 29–35. [CrossRef]
4. Wang, Z.; Yao, Z.; Zhou, J.; Zhang, Y. Reuse of waste cotton cloth for the extraction of cellulose nanocrystals. *Carbohydr. Polym.* **2017**, *157*, 945–952. [CrossRef]
5. Xiao, B.; Huang, Q.; Chen, H.; Chen, X.; Long, G. A fractal model for capillary flow through a single tortuous capillary with roughened surfaces in fibrous porous media. *Fractals* **2021**, *29*, 2150017. [CrossRef]
6. Dobilaite, V.; Mileriene, G.; Juciene, M.; Saceviciene, V. Investigation of current state of pre-consumer textile waste generated at Lithuanian enterprises. *Int. J. Cloth. Sci. Technol.* **2017**, *29*, 491–502. [CrossRef]
7. Rani, S.; Jamal, Z. Recycling of textiles waste for environmental protection. *Int. J. Home Sci.* **2018**, *4*, 164–168.
8. Liu, W.; Liu, S.; Liu, T.; Liu, T.; Zhang, J.; Liu, H. Eco-friendly post-consumer cotton waste recycling for regenerated cellulose fibers. *Carbohydr. Polym.* **2019**, *206*, 141–148. [CrossRef] [PubMed]
9. Ahmad, S.S.; Mulyadi, I.M.M.; Ibrahim, N.; Othman, A.R. The application of recycled textile and innovative spatial design strategies for a recycling centre exhibition space. *Procedia Soc. Behav. Sci.* **2016**, *234*, 525–535. [CrossRef]

10. Aronsson, J.; Persson, A. Tearing of post-consumer cotton T-shirts and jeans of varying degree of wear. *J. Eng. Fibers Fabr.* **2020**, *15*, 1558925020901322. [CrossRef]
11. Johnson, S.; Echeverria, D.; Venditti, R.; Jameel, H.; Yao, Y. Supply Chain of Waste Cotton Recycling and Reuse: A Review. *Aatcc J. Res.* **2020**, *7*, 19–31. [CrossRef]
12. Rizal, S.; Alfatah, T.; Abdul Khalil, H.P.S.; Mistar, E.; Abdullah, C.; Olaiya, F.G.; Sabaruddin, F.; Muksin, U. Properties and Characterization of Lignin Nanoparticles Functionalized in Macroalgae Biopolymer Films. *Nanomaterials* **2021**, *11*, 637. [CrossRef]
13. Jawaid, M.; Abdul Khalil, H.P.S. Cellulosic/synthetic fibre reinforced polymer hybrid composites: A review. *Carbohydr. Polym.* **2011**, *86*, 1–18. [CrossRef]
14. Majeed, K.; Jawaid, M.; Hassan, A.; Bakar, A.A.; Abdul Khalil, H.P.S.; Salema, A.; Inuwa, I. Potential materials for food packaging from nanoclay/natural fibres filled hybrid composites. *Mater. Des.* **2013**, *46*, 391–410. [CrossRef]
15. Fortunati, E.; Luzi, F.; Puglia, D.; Petrucci, R.; Kenny, J.M.; Torre, L. Processing of PLA nanocomposites with cellulose nanocrystals extracted from Posidonia oceanica waste: Innovative reuse of coastal plant. *Ind. Crop. Prod.* **2015**, *67*, 439–447. [CrossRef]
16. Rizal, S.; Lai, T.K.; Muksin, U.; Olaiya, N.; Abdullah, C.; Yahya, E.B.; Chong, E.; Abdul Khalil, H.P.S. Properties of Macroalgae Biopolymer Films Reinforcement with Polysaccharide Microfibre. *Polymers* **2020**, *12*, 2554. [CrossRef]
17. Sharma-Shivappa, R.; Chen, Y. Conversion of cotton wastes to bioenergy and value-added products. *Trans. Asabe* **2008**, *51*, 2239–2246. [CrossRef]
18. Dai, J.; Dong, H. Intensive cotton farming technologies in China: Achievements, challenges and countermeasures. *Field Crop. Res.* **2014**, *155*, 99–110. [CrossRef]
19. Zhang, L.; Wu, T.; Liu, S.; Jiang, S.; Wu, H.; Yang, J. Consumers' clothing disposal behaviors in Nanjing, China. *J. Clean. Prod.* **2020**, *276*, 123184. [CrossRef]
20. Ma, Y.; Rosson, L.; Wang, X.; Byrne, N. Upcycling of waste textiles into regenerated cellulose fibres: Impact of pretreatments. *J. Text. Inst.* **2020**, *111*, 630–638. [CrossRef]
21. Esteve-Turrillas, F.A.; de la Guardia, M. Environmental impact of Recover cotton in textile industry. *Resour. Conserv. Recycl.* **2017**, *116*, 107–115. [CrossRef]
22. Stone, C.; Windsor, F.M.; Munday, M.; Durance, I. Natural or synthetic–how global trends in textile usage threaten freshwater environments. *Sci. Total Environ.* **2020**, *718*, 134689. [CrossRef] [PubMed]
23. Gordon, S.; Hsieh, Y.-l. *Cotton: Science and Technology*; Woodhead Publishing Ltd.: Cambridge, UK, 2006.
24. Wang, H.; Farooq, A.; Memon, H. Influence of cotton fiber properties on the microstructural characteristics of mercerized fibers by regression analysis. *Wood Fiber Sci.* **2020**, *52*, 13–27. [CrossRef]
25. Grishanov, S. Structure and properties of textile materials. In *Handbook of Textile and Industrial Dyein*; Woodhead Publishing Series in Textiles; Woodhead Publishing Limited Elsevier: Cambridge, UK, 2011; Volume 1, pp. 28–63.
26. Ma, Y.; Rissanen, M.; You, X.; Moriam, K.; Hummel, M.; Sixta, H. New method for determining the degree of fibrillation of regenerated cellulose fibres. *Cellulose* **2020**, *28*, 31–44. [CrossRef]
27. French, A.D.; Kim, H.J. Cotton fiber structure. In *Cotton Fiber: Physics, Chemistry and Biology*; Springer: Berlin/Heidelberg, Germany; Cotton Fiber Bioscience Research UnitUSDA-ARS, Southern Regional Research Center: New Orleans, LA, USA, 2018; pp. 13–39.
28. Barkhotkin, Y.K. *Modulus of Elasticity of the Cotton Fibres in a Yarn*; Textile Industry Technology; Ivanovo State Polytechnic University: Ivanovo, Russia, 2003; Volume 3, pp. 50–53.
29. Varghese, A.M.; Mittal, V. Surface modification of natural fibers. *Biodegrad. Biocompatible Polym. Compos.* **2018**, *5*, 115–155.
30. Cruz, J.; Leitão, A.; Silveira, D.; Pichandi, S.; Pinto, M.; Fangueiro, R. Study of moisture absorption characteristics of cotton terry towel fabrics. *Procedia Eng.* **2017**, *200*, 389–398. [CrossRef]
31. Prakash, C. Bamboo fibre. In *Handbook of Natural Fibres*; Woodhead Publishing Series in Textiles; Elsevier Ltd.: Cambridge, UK, 2020; Volume 1, pp. 219–229.
32. Kocić, A.; Bizjak, M.; Popović, D.; Poparić, G.B.; Stanković, S.B. UV protection afforded by textile fabrics made of natural and regenerated cellulose fibres. *J. Clean. Prod.* **2019**, *228*, 1229–1237. [CrossRef]
33. Gorade, V.; Chaudhary, B.; Parmaj, O.; Kale, R. Preparation and Characterization of Chitosan/viscose Rayon Filament Biocomposite. *J. Nat. Fibers* **2020**, 1–12. [CrossRef]
34. Thambiraj, S.; Shankaran, D.R. Preparation and physicochemical characterization of cellulose nanocrystals from industrial waste cotton. *Appl. Surf. Sci.* **2017**, *412*, 405–416. [CrossRef]
35. Makvandi, P.; Iftekhar, S.; Pizzetti, F.; Zarepour, A.; Zare, E.N.; Ashrafizadeh, M.; Agarwal, T.; Padil, V.V.; Mohammadinejad, R.; Sillanpaa, M. Functionalization of polymers and nanomaterials for water treatment, food packaging, textile and biomedical applications: A review. *Environ. Chem. Lett.* **2020**, *19*, 1–29. [CrossRef]
36. Çay, A.; Yanık, J.; Akduman, Ç.; Duman, G.; Ertaş, H. Application of textile waste derived biochars onto cotton fabric for improved performance and functional properties. *J. Clean. Prod.* **2020**, *251*, 119664. [CrossRef]
37. Meng, Y.; Liu, W.; Liu, X.; Zhang, J.; Peng, M.; Zhang, T. A review on analytical methods for pharmaceutical and personal care products and their transformation products. *J. Environ. Sci.* **2021**, *101*, 260–281. [CrossRef] [PubMed]
38. Sathishkumar, M.; Binupriya, A.; Kavitha, D.; Selvakumar, R.; Sheema, K.; Choi, J.; Yun, S. Organic micro-pollutant removal in liquid-phase using carbonized silk cotton hull. *J. Environ. Sci.* **2008**, *20*, 1046–1054. [CrossRef]

39. Lu, S.; Liu, Q.; Han, R.; Guo, M.; Shi, J.; Song, C.; Ji, N.; Lu, X.; Ma, D. Potential applications of porous organic polymers as adsorbent for the adsorption of volatile organic compounds. *J. Environ. Sci.* **2021**, *105*, 184–203. [CrossRef]
40. Czech, B.; Shirvanimoghaddam, K.; Trojanowska, E. Sorption of pharmaceuticals and personal care products (PPCPs) onto a sustainable cotton based adsorbent. *Sustain. Chem. Pharm.* **2020**, *18*, 100324. [CrossRef]
41. Nayl, A.; Abd-Elhamid, A.; Abu-Saied, M.; El-Shanshory, A.A.; Soliman, H.M.; Akl, M.A.; Aly, H. A novel method for highly effective removal and determination of binary cationic dyes in aqueous media using a cotton–graphene oxide composite. *RSC Adv.* **2020**, *10*, 7791–7802. [CrossRef]
42. Rakruam, P.; Thuptimdang, P.; Siripattanakul-Ratpukdi, S.; Phungsai, P. Molecular dissolved organic matter removal by cotton-based adsorbents and characterization using high-resolution mass spectrometry. *Sci. Total Environ.* **2021**, *754*, 142074. [CrossRef]
43. Qin, Q.; Guo, R.; Lin, S.; Jiang, S.; Lan, J.; Lai, X.; Cui, C.; Xiao, H.; Zhang, Y. Waste cotton fiber/Bi 2 WO 6 composite film for dye removal. *Cellulose* **2019**, *26*, 3909–3922. [CrossRef]
44. Qiu, Z.; Shu, J.; Tang, D. Near-infrared-to-ultraviolet light-mediated photoelectrochemical aptasensing platform for cancer biomarker based on core–shell NaYF4: Yb, Tm@ TiO_2 upconversion microrods. *Anal. Chem.* **2018**, *90*, 1021–1028. [CrossRef]
45. Song, C.; Li, H.; Yu, Y. Synthesis, microstructure transformations, and long-distance inductive effect of poly (acrylethyltrimethylammonium chloride) cotton with super-high adsorption ability for purifying dyeing wastewater. *Cellulose* **2019**, *26*, 3987–4004. [CrossRef]
46. Sivarajasekar, N.; Baskar, R.; Ragu, T.; Sarika, K.; Preethi, N.; Radhika, T. Biosorption studies on waste cotton seed for cationic dyes sequestration: Equilibrium and thermodynamics. *Appl. Water Sci.* **2017**, *7*, 1987–1995. [CrossRef]
47. Haque, A.N.M.A.; Remadevi, R.; Wang, X.; Naebe, M. Sorption properties of fabricated film from cotton gin trash. *Mater. Today Proc.* **2020**, *31*, S221–S226. [CrossRef]
48. Fan, L.; Wei, C.; Xu, Q.; Xu, J. Polypyrrole-coated cotton fabrics used for removal of methylene blue from aqueous solution. *J. Text. Inst.* **2017**, *108*, 1847–1852. [CrossRef]
49. Tunc, Ö.; Tanacı, H.; Aksu, Z. Potential use of cotton plant wastes for the removal of Remazol Black B reactive dye. *J. Hazard. Mater.* **2009**, *163*, 187–198. [CrossRef] [PubMed]
50. Deng, H.; Lu, J.; Li, G.; Zhang, G.; Wang, X. Adsorption of methylene blue on adsorbent materials produced from cotton stalk. *Chem. Eng. J.* **2011**, *172*, 326–334. [CrossRef]
51. Wanassi, B.; Hariz, I.B.; Ghimbeu, C.M.; Vaulot, C.; Hassen, M.B.; Jeguirim, M. Carbonaceous adsorbents derived from textile cotton waste for the removal of Alizarin S dye from aqueous effluent: Kinetic and equilibrium studies. *Environ. Sci. Pollut. Res.* **2017**, *24*, 10041–10055. [CrossRef]
52. Tian, D.; Zhang, X.; Lu, C.; Yuan, G.; Zhang, W.; Zhou, Z. Solvent-free synthesis of carboxylate-functionalized cellulose from waste cotton fabrics for the removal of cationic dyes from aqueous solutions. *Cellulose* **2014**, *21*, 473–484. [CrossRef]
53. Xiong, J.; Jiao, C.; Li, C.; Zhang, D.; Lin, H.; Chen, Y. A versatile amphiprotic cotton fiber for the removal of dyes and metal ions. *Cellulose* **2014**, *21*, 3073–3087. [CrossRef]
54. Thangamani, K.; Sathishkumar, M.; Sameena, Y.; Vennilamani, N.; Kadirvelu, K.; Pattabhi, S.; Yun, S. Utilization of modified silk cotton hull waste as an adsorbent for the removal of textile dye (reactive blue MR) from aqueous solution. *Bioresour. Technol.* **2007**, *98*, 1265–1269. [CrossRef]
55. Wang, Q.; Zhang, Y.; Wangjin, X.; Wang, Y.; Meng, G.; Chen, Y. The adsorption behavior of metals in aqueous solution by microplastics effected by UV radiation. *J. Environ. Sci.* **2020**, *87*, 272–280. [CrossRef]
56. Yahya, M.; Yohanna, I.; Auta, M.; Obayomi, K. Remediation of Pb (II) ions from Kagara gold mining effluent using cotton hull adsorbent. *Sci. Afr.* **2020**, *8*, e00399.
57. Xu, Z.; Gu, S.; Sun, Z.; Zhang, D.; Zhou, Y.; Gao, Y.; Qi, R.; Chen, W. Synthesis of char-based adsorbents from cotton textile waste assisted by iron salts at low pyrolysis temperature for Cr (VI) removal. *Environ. Sci. Pollut. Res.* **2020**, *27*, 11012–11025. [CrossRef]
58. Mihajlović, S.; Vukčević, M.; Pejić, B.; Grujić, A.P.; Ristić, M. Application of waste cotton yarn as adsorbent of heavy metal ions from single and mixed solutions. *Environ. Sci. Pollut. Res.* **2020**, *27*, 35769–35781. [CrossRef] [PubMed]
59. Niu, Y.; Hu, W.; Guo, M.; Wang, Y.; Jia, J.; Hu, Z. Preparation of cotton-based fibrous adsorbents for the removal of heavy metal ions. *Carbohydr. Polym.* **2019**, *225*, 115218. [CrossRef] [PubMed]
60. Ma, J.; Liu, Y.; Ali, O.; Wei, Y.; Zhang, S.; Zhang, Y.; Cai, T.; Liu, C.; Luo, S. Fast adsorption of heavy metal ions by waste cotton fabrics based double network hydrogel and influencing factors insight. *J. Hazard. Mater.* **2018**, *344*, 1034–1042. [CrossRef] [PubMed]
61. Pang, D.; Wang, C.-C.; Wang, P.; Liu, W.; Fu, H.; Zhao, C. Superior removal of inorganic and organic arsenic pollutants from water with MIL-88A (Fe) decorated on cotton fibers. *Chemosphere* **2020**, *254*, 126829. [CrossRef] [PubMed]
62. Mendoza-Castillo, D.; Rojas-Mayorga, C.; García-Martínez, I.; Pérez-Cruz, M.; Hernández-Montoya, V.; Bonilla-Petriciolet, A.; Montes-Morán, M. Removal of heavy metals and arsenic from aqueous solution using textile wastes from denim industry. *Int. J. Environ. Sci. Technol.* **2015**, *12*, 1657–1668. [CrossRef]
63. Paulino, Á.G.; da Cunha, A.J.; da Silva Alfaya, R.V.; da Silva Alfaya, A.A. Chemically modified natural cotton fiber: A low-cost biosorbent for the removal of the Cu (II), Zn (II), Cd (II), and Pb (II) from natural water. *Desalination Water Treat.* **2014**, *52*, 4223–4233. [CrossRef]
64. Hu, E.; Shang, S.; Tao, X.-m.; Jiang, S.; Chiu, K.-l. Regeneration and reuse of highly polluting textile dyeing effluents through catalytic ozonation with carbon aerogel catalysts. *J. Clean. Prod.* **2016**, *137*, 1055–1065. [CrossRef]

65. Yousef, S.; Tatariants, M.; Tichonovas, M.; Kliucininkas, L.; Lukošiūtė, S.-I.; Yan, L. Sustainable green technology for recovery of cotton fibers and polyester from textile waste. *J. Clean. Prod.* **2020**, *254*, 120078. [CrossRef]
66. Echeverria, C.A.; Handoko, W.; Pahlevani, F.; Sahajwalla, V. Cascading use of textile waste for the advancement of fibre reinforced composites for building applications. *J. Clean. Prod.* **2019**, *208*, 1524–1536. [CrossRef]
67. Algin, H.M.; Turgut, P. Cotton and limestone powder wastes as brick material. *Constr. Build. Mater.* **2008**, *22*, 1074–1080. [CrossRef]
68. Uddin, M.M.; Karim, R.; Kaysar, M.; Dayan, M.A.R.; Islam, K.A. Low-Cost Jute-Cotton and Glass Fibre Reinforced Textile Composite Sheet. *Int. J. Mat. Math. Sci* **2020**, *2*, 1–7. [CrossRef]
69. Sanivada, U.K.; Mármol, G.; Brito, F.P.; Fangueiro, R. PLA Composites Reinforced with Flax and Jute Fibers—A Review of Recent Trends, Processing Parameters and Mechanical Properties. *Polymers* **2020**, *12*, 2373. [CrossRef] [PubMed]
70. Holt, G.; Chow, P.; Wanjura, J.; Pelletier, M.; Wedegaertner, T. Evaluation of thermal treatments to improve physical and mechanical properties of bio-composites made from cotton byproducts and other agricultural fibers. *Ind. Crop. Prod.* **2014**, *52*, 627–632. [CrossRef]
71. Alomayri, T.; Shaikh, F.U.A.; Low, I.M. Synthesis and mechanical properties of cotton fabric reinforced geopolymer composites. *Compos. Part B Eng.* **2014**, *60*, 36–42. [CrossRef]
72. Alomayri, T.; Shaikh, F.U.A.; Low, I.M. Characterisation of cotton fibre-reinforced geopolymer composites. *Compos. Part B Eng.* **2013**, *50*, 1–6. [CrossRef]
73. Mahdi, E.; Dean, A. The effect of filler content on the tensile behavior of polypropylene/cotton fiber and poly (vinyl chloride)/cotton fiber composites. *Materials* **2020**, *13*, 753. [CrossRef] [PubMed]
74. Chen, H.; Wang, F.; Chen, H.; Fang, H.; Feng, W.; Wei, Y.; Wang, F.; Su, H.; Mi, Y.; Zhou, M.; et al. Specific biotests to assess eco-toxicity of biodegradable polymer materials in soil. *J. Environ. Sci.* **2021**, *105*, 150–162. [CrossRef]
75. Liu, J.-l.; Hou, T.-s.; Luo, Y.-s.; Cui, Y.-x. Experimental Study on Unconsolidated Undrained Shear Strength Characteristics of Synthetic Cotton Fiber Reinforced Soil. *Geotech. Geol. Eng.* **2020**, *38*, 1773–1783. [CrossRef]
76. Sadrolodabaeea, P.; Claramunt, J.; Ardanuy, M.; de la Fuente, A. Mechanical and durability characterization of a new Textile Waste Micro-Fiber Reinforced Cement Composite for building applications. *Case Stud. Constr. Mater.* **2021**, *14*, e00492. [CrossRef]
77. Muthuraj, R.; Lacoste, C.; Lacroix, P.; Bergeret, A. Sustainable thermal insulation biocomposites from rice husk, wheat husk, wood fibers and textile waste fibers: Elaboration and performances evaluation. *Ind. Crop. Prod.* **2019**, *135*, 238–245. [CrossRef]
78. Leal Filho, W.; Saari, U.; Fedoruk, M.; Iital, A.; Moora, H.; Klöga, M.; Voronova, V. An overview of the problems posed by plastic products and the role of extended producer responsibility in Europe. *J. Clean. Prod.* **2019**, *214*, 550–558. [CrossRef]
79. Reichert, C.L.; Bugnicourt, E.; Coltelli, M.-B.; Cinelli, P.; Lazzeri, A.; Canesi, I.; Braca, F.; Martínez, B.M.; Alonso, R.; Agostinis, L. Bio-based packaging: Materials, modifications, industrial applications and sustainability. *Polymers* **2020**, *12*, 1558. [CrossRef]
80. Reddy, N. Non-food industrial applications of poultry feathers. *Waste Manag.* **2015**, *45*, 91–107. [CrossRef]
81. Ramamoorthy, S.K.; Skrifvars, M.; Persson, A. A review of natural fibers used in biocomposites: Plant, animal and regenerated cellulose fibers. *Polym. Rev.* **2015**, *55*, 107–162. [CrossRef]
82. Montava-Jordà, S.; Torres-Giner, S.; Ferrandiz-Bou, S.; Quiles-Carrillo, L.; Montanes, N. Development of sustainable and cost-competitive injection-molded pieces of partially bio-based polyethylene terephthalate through the valorization of cotton textile waste. *Int. J. Mol. Sci.* **2019**, *20*, 1378. [CrossRef] [PubMed]
83. Rizal, S.; Olaiya, F.G.; Saharudin, N.I.; Abdullah, C.K.; NG, O.; Mohamad Haafiz, M.K.; Yahya, E.B.; Sabaruddin, F.A.; Abdul Khalil, H.P.S. Isolation of Textile Waste Cellulose Nanofibrillated Fibre Reinforced in Polylactic Acid-Chitin Biodegradable Composite for Green Packaging Application. *Polymers* **2021**, *13*, 325. [CrossRef]
84. Meekum, U.; Kingchang, P. Compounding Oil Palm Empty Fruit Bunch/Cotton Fiber Hybrid Reinforced Poly (lactic acid) Biocomposites Aiming For High-temperature Packaging Applications. *BioResources* **2017**, *12*, 4670–4689. [CrossRef]
85. de Oliveira, S.A.; de Macedo, J.R.N.; dos Santos Rosa, D. Eco-efficiency of poly (lactic acid)-Starch-Cotton composite with high natural cotton fiber content: Environmental and functional value. *J. Clean. Prod.* **2019**, *217*, 32–41. [CrossRef]
86. Hou, X.; Sun, F.; Yan, D.; Xu, H.; Dong, Z.; Li, Q.; Yang, Y. Preparation of lightweight polypropylene composites reinforced by cotton stalk fibers from combined steam flash-explosion and alkaline treatment. *J. Clean. Prod.* **2014**, *83*, 454–462. [CrossRef]
87. Jummaat, F.; Yahya, E.B.; Abdul Khalil, H.P.S.; Adnan, A.; Alqadhi, A.M.; Abdullah, C.; AK, A.S.; Olaiya, N.; Abdat, M. The Role of Biopolymer-Based Materials in Obstetrics and Gynecology Applications: A Review. *Polymers* **2021**, *13*, 633. [CrossRef]
88. Abdul Khalil, H.P.S.; Adnan, A.; Yahya, E.B.; Olaiya, N.; Safrida, S.; Hossain, M.; Balakrishnan, V.; Gopakumar, D.A.; Abdullah, C.; Oyekanmi, A. A Review on plant cellulose nanofibre-based aerogels for biomedical applications. *Polymers* **2020**, *12*, 1759. [CrossRef]
89. Yahya, E.B.; Jummaat, F.; Amirul, A.; Adnan, A.; Olaiya, N.; Abdullah, C.; Rizal, S.; Mohamad Haafiz, M.; Abdul Khalil, H.P.S. review on revolutionary natural biopolymer-based aerogels for antibacterial delivery. *Antibiotics* **2020**, *9*, 648. [CrossRef] [PubMed]
90. Tavakolian, M.; Jafari, S.M.; van de Ven, T.G. A Review on Surface-Functionalized Cellulosic Nanostructures as Biocompatible Antibacterial Materials. *Nano Micro Lett.* **2020**, *12*, 1–23. [CrossRef]
91. Vignesh, N.; Suriyaraj, S.; Selvakumar, R.; Chandraraj, K. Facile Fabrication and Characterization of Zn Loaded Cellulose Membrane from Cotton Microdust Waste and its Antibacterial Properties—A Waste to Value Approach. *J. Polym. Environ.* **2021**, 1–12. [CrossRef]

92. Coradi, M.; Zanetti, M.; Valério, A.; de Oliveira, D.; da Silva, A.; Ulson, S.M.d.A.G.; de Souza, A.A.U. Production of antimicrobial textiles by cotton fabric functionalization and pectinolytic enzyme immobilization. *Mater. Chem. Phys.* **2018**, *208*, 28–34. [CrossRef]
93. Zhang, S.; Yang, X.; Tang, B.; Yuan, L.; Wang, K.; Liu, X.; Zhu, X.; Li, J.; Ge, Z.; Chen, S. New insights into synergistic antimicrobial and antifouling cotton fabrics via dually finished with quaternary ammonium salt and zwitterionic sulfobetaine. *Chem. Eng. J.* **2018**, *336*, 123–132. [CrossRef]
94. Vinod, A.; Sanjay, M.; Suchart, S.; Jyotishkumar, P. Renewable and sustainable biobased materials: An assessment on biofibers, biofilms, biopolymers and biocomposites. *J. Clean. Prod.* **2020**, *258*, 120978. [CrossRef]
95. Dassanayake, R.S.; Wansapura, P.T.; Tran, P.; Hamood, A.; Abidi, N. Cotton Cellulose-CdTe Quantum Dots Composite Films with Inhibition of Biofilm-Forming S. aureus. *Fibers* **2019**, *7*, 57. [CrossRef]
96. Jabra-Rizk, M.; Meiller, T.; James, C.; Shirtliff, M. Effect of farnesol on Staphylococcus aureus biofilm formation and antimicrobial susceptibility. *Antimicrob. Agents Chemother.* **2006**, *50*, 1463–1469. [CrossRef]
97. Gadkari, R.; Ali, S.W.; Joshi, M.; Rajendran, S.; Das, A.; Alagirusamy, R. Leveraging antibacterial efficacy of silver loaded chitosan nanoparticles on layer-by-layer self-assembled coated cotton fabric. *Int. J. Biol. Macromol.* **2020**, *162*, 548–560. [CrossRef] [PubMed]
98. Wang, Y.; Yin, M.; Li, Z.; Liu, Y.; Ren, X.; Huang, T.-S. Preparation of antimicrobial and hemostatic cotton with modified mesoporous particles for biomedical applications. *Colloids Surf. B Biointerfaces* **2018**, *165*, 199–206. [CrossRef] [PubMed]
99. Bastos, A.R.; da Silva, L.P.; Gomes, V.P.; Lopes, P.E.; Rodrigues, L.C.; Reis, R.L.; Correlo, V.M.; Souto, A.P. Electroactive polyamide/cotton fabrics for biomedical applications. *Org. Electron.* **2020**, *77*, 105401. [CrossRef]
100. Idumah, C.I. Influence of nanotechnology in polymeric textiles, applications, and fight against COVID-19. *J. Text. Inst.* **2020**, 1–21. [CrossRef]
101. Subbiah, D.K.; Mani, G.K.; Babu, K.J.; Das, A.; Rayappan, J.B.B. Nanostructured ZnO on cotton fabrics—A novel flexible gas sensor & UV filter. *J. Clean. Prod.* **2018**, *194*, 372–382. [CrossRef]
102. Khattab, T.A.; Fouda, M.M.; Abdelrahman, M.S.; Othman, S.I.; Bin-Jumah, M.; Alqaraawi, M.A.; Al Fassam, H.; Allam, A.A. Co-encapsulation of enzyme and tricyanofuran hydrazone into alginate microcapsules incorporated onto cotton fabric as a biosensor for colorimetric recognition of urea. *React. Funct. Polym.* **2019**, *142*, 199–206. [CrossRef]
103. Edwards, J.V.; Fontenot, K.R.; Liebner, F.; Condon, B.D. Peptide-Cellulose Conjugates on Cotton-Based Materials Have Protease Sensor/Sequestrant Activity. *Sensors* **2018**, *18*, 2334. [CrossRef] [PubMed]
104. Abdelrahman, M.S.; Fouda, M.M.; Ajarem, J.S.; Maodaa, S.N.; Allam, A.A.; Khattab, T.A. Development of colorimetric cotton swab using molecular switching hydrazone probe in calcium alginate. *J. Mol. Struct.* **2020**, *1216*, 128301. [CrossRef]
105. Song, Y.; Xu, W.; Rong, M.; Zhang, M. A sunlight self-healable fibrous flexible pressure sensor based on electrically conductive composite wool yarns. *Express Polym. Lett.* **2020**, *14*, 1089–1104. [CrossRef]
106. King, M.W.; Chen, J.; Deshpande, M.; He, T.; Ramakrishna, H.; Xie, Y.; Zhang, F.; Zhao, F. Structural Design, Fabrication and Evaluation of Resorbable Fiber-Based Tissue Engineering Scaffolds. *Biotechnol. Bioeng.* **2019**, 61–188. [CrossRef]
107. Jun, I.; Han, H.-S.; Edwards, J.R.; Jeon, H. Electrospun Fibrous Scaffolds for Tissue Engineering: Viewpoints on Architecture and Fabrication. *Int. J. Mol. Sci.* **2018**, *19*, 745. [CrossRef] [PubMed]
108. Singh, S.; Dutt, D.; Mishra, N.C. Cotton pulp for bone tissue engineering. *J. Biomater. Sci. Polym. Ed.* **2020**, *31*, 2094–2113. [CrossRef] [PubMed]
109. Khalil, H.; Jummaat, F.; Yahya, E.B.; Olaiya, N.; Adnan, A.; Abdat, M.; NAM, N.; Halim, A.S.; Kumar, U.; Bairwan, R. A review on micro-to nanocellulose biopolymer scaffold forming for tissue engineering applications. *Polymers* **2020**, *12*, 2043. [CrossRef] [PubMed]
110. Xu, Y.; Wu, P.; Feng, P.; Guo, W.; Yang, W.; Shuai, C. Interfacial reinforcement in a poly-l-lactic acid/mesoporous bioactive glass scaffold via polydopamine. *Colloids Surf. B Biointerfaces* **2018**, *170*, 45–53. [CrossRef]
111. Nelson, M.; Tallia, F.; Page, S.J.; Hanna, J.V.; Fujita, Y.; Obata, A.; Kasuga, T.; Jones, J.R. Electrospun cotton–wool-like silica/gelatin hybrids with covalent coupling. *J. Sol-Gel Sci. Technol.* **2021**, *97*, 11–26. [CrossRef]
112. Liu, F.; Liao, X.; Liu, C.; Li, M.; Chen, Y.; Shao, W.; Weng, K.; Li, F.; Ou, K.; He, J. Poly (l-lactide-co-caprolactone)/tussah silk fibroin nanofiber vascular scaffolds with small diameter fabricated by core-spun electrospinning technology. *J. Mater. Sci.* **2020**, *55*, 7106–7119. [CrossRef]
113. Reddy, N.; Yang, Y. Properties and potential applications of natural cellulose fibers from the bark of cotton stalks. *Bioresour. Technol.* **2009**, *100*, 3563–3569. [CrossRef]
114. Martins, M.A.; Teixeira, E.M.; Corrêa, A.C.; Ferreira, M.; Mattoso, L.H.C. Extraction and characterization of cellulose whiskers from commercial cotton fibers. *J. Mater. Sci.* **2011**, *46*, 7858–7864. [CrossRef]
115. Phanthong, P.; Reubroycharoen, P.; Hao, X.; Xu, G.; Abudula, A.; Guan, G. Nanocellulose: Extraction and application. *Carbon Resour. Convers.* **2018**, *1*, 32–43. [CrossRef]
116. Jiang, S.; Xia, Z.; Farooq, A.; Zhang, M.; Li, M.; Liu, L. Efficient recovery of the dyed cotton–polyester fabric: Cellulose nanocrystal extraction and its application in composite films. *Cellulose* **2021**, 1–14. [CrossRef]
117. Endes, C.; Mueller, S.; Kinnear, C.; Vanhecke, D.; Foster, E.J.; Petri-Fink, A.; Weder, C.; Clift, M.J.D.; Rothen-Rutishauser, B. Fate of Cellulose Nanocrystal Aerosols Deposited on the Lung Cell Surface In Vitro. *Biomacromolecules* **2015**, *16*, 1267–1275. [CrossRef] [PubMed]
118. Talbot, K.W. Process for the Extraction of Cellulose Fibers from Lignified Fibrous Material. U.S. Patent 26,404,339, 28 April 1942.
119. Speakman, J. Present And Future Methods Of Separating Textile Fibres. *J. Text. Inst. Proc.* **1950**, *41*, P202–P207. [CrossRef]

120. Masselli, J.W.; Burford, M.G. Pollution Reduction in Cotton Finishing Wastes Through Process Chemical Changes. *Sew. Ind. Wastes* **1954**, *26*, 1109–1116.
121. Kramar, A.D.; Obradović, B.M.; Vesel, A.; Kuraica, M.M.; Kostić, M.M. Surface cleaning of raw cotton fibers with atmospheric pressure air plasma. *Cellulose* **2018**, *25*, 4199–4209. [CrossRef]
122. Masselli, J.W. *A Simplification of Textile Waste Survey and Treatment*; Wesleyan University (Middletown, Conn.) Hall Laboratory of Chemistry, New England Interstate Water Pollution Control Commission: Boston, MA, USA, 1959; p. 68.
123. Gallagher, G.T.; Elliott, E.J. Process of Alkaline Scouring and Peroxide Bleaching of Gray Cotton Cloth. U.S. Patent 3,148,019, 8 September 1964.
124. Tassinari, T.; Macy, C.; Spano, L.; Ryu, D.D. Energy requirements and process design considerations in compression-milling pretreatment of cellulosic wastes for enzymatic hydrolysis. *Biotechnol. Bioeng.* **1980**, *22*, 1689–1705. [CrossRef]
125. Maciel, G.E.; Kolodziejski, W.L.; Bertran, M.S.; Dale, B.E. Carbon-13 NMR and order in cellulose. *Macromolecules* **1982**, *15*, 686–687. [CrossRef]
126. Bertran, M.S.; Dale, B.E. Enzymatic hydrolysis and recrystallization behavior of initially amorphous cellulose. *Biotechnol. Bioeng.* **1985**, *27*, 177–181. [CrossRef]
127. Mokeev, A.; Iljin, V.; Gradova, N. Biotechnological degradation of the radioactive cellulose containing waste. *J. Mol. Catal. B Enzym.* **1998**, *5*, 441–445. [CrossRef]
128. Ludwig, M.; Fengel, D. Influence of the degree of substitution and pretreatments on the surface structure of cellulose nitrate fibres. *Wood Sci. Technol.* **1992**, *26*, 393–401. [CrossRef]
129. Brooks, R.E.; Moore, S.B. Alkaline hydrogen peroxide bleaching of cellulose. *Cellulose* **2000**, *7*, 263–286. [CrossRef]
130. Elazzouzi-Hafraoui, S.; Nishiyama, Y.; Putaux, J.-L.; Heux, L.; Dubreuil, F.; Rochas, C. The shape and size distribution of crystalline nanoparticles prepared by acid hydrolysis of native cellulose. *Biomacromolecules* **2008**, *9*, 57–65. [CrossRef]
131. Kuo, C.H.; Lin, P.J.; Lee, C.K. Enzymatic saccharification of dissolution pretreated waste cellulosic fabrics for bacterial cellulose production by Gluconacetobacter xylinus. *J. Chem. Technol. Biotechnol.* **2010**, *85*, 1346–1352. [CrossRef]
132. Wu, F.; Zhang, Y.; Liu, L.; Yao, J. Synthesis and characterization of a novel cellulose-g-poly (acrylic acid-co-acrylamide) superabsorbent composite based on flax yarn waste. *Carbohydr. Polym.* **2012**, *87*, 2519–2525. [CrossRef]
133. Bidgoli, H.; Zamani, A.; Jeihanipour, A.; Taherzadeh, M.J. Preparation of carboxymethyl cellulose superabsorbents from waste textiles. *Fibers Polym.* **2014**, *15*, 431–436. [CrossRef]
134. Pandi, N.; Sonawane, S.H.; Kishore, K.A. Synthesis of cellulose nanocrystals (CNCs) from cotton using ultrasound-assisted acid hydrolysis. *Ultrason. Sonochemistry* **2020**, *70*, 105353. [CrossRef]
135. Bartell, F.; Cowling, H. Depolymermiation of cellulose in viscose production. *Ind. Eng. Chem.* **1942**, *34*, 607–612. [CrossRef]
136. Wakelin, J.H.; Virgin, H.S.; Crystal, E. Development and comparison of two X-ray methods for determining the crystallinity of cotton cellulose. *J. Appl. Phys.* **1959**, *30*, 1654–1662. [CrossRef]
137. Calamari, T.A., Jr.; Schreiber, S.P.; Cooper, A.S., Jr.; Reeves, W.A. Liquid Ammonia Modification of Cellulose in Cotton and Polyester/Cotton Textiles. *Text. Chem. Colorist* **1971**, *3*, 10.
138. Hebeish, A.; Kantouch, A.; El-Rafie, M. Graft copolymerization of vinyl monomers with modified cotton. II. Grafting of acrylonitrile and methyl methacrylate on acetylated cotton. *J. Appl. Polym. Sci.* **1971**, *15*, 11–24.
139. Buschle-Diller, G.; Zeronian, S. Enhancing the reactivity and strength of cotton fibers. *J. Appl. Polym. Sci.* **1992**, *45*, 967–979. [CrossRef]
140. Bahloul, A.; Kassab, Z.; El Bouchti, M.; Hannache, H.; Oumam, M.; El Achaby, M. Micro-and nano-structures of cellulose from eggplant plant (*Solanum melongena* L.) agricultural residue. *Carbohydr. Polym.* **2020**, *253*, 117311. [CrossRef] [PubMed]
141. Zhong, T.; Dhandapani, R.; Liang, D.; Wang, J.; Wolcott, M.P.; Van Fossen, D.; Liu, H. Nanocellulose from recycled indigo-dyed denim fabric and its application in composite films. *Carbohydr. Polym.* **2020**, *240*, 116283. [CrossRef] [PubMed]
142. Abdul Khalil, H.P.S.; Davoudpour, Y.; Islam, M.N.; Mustapha, A.; Sudesh, K.; Dungani, R.; Jawaid, M. Production and modification of nanofibrillated cellulose using various mechanical processes: A review. *Carbohydr. Polym.* **2014**, *99*, 649–665. [CrossRef] [PubMed]
143. Maciel, M.M.Á.D.; de Carvalho Benini, K.C.C.; Voorwald, H.J.C.; Cioffi, M.O.H. Obtainment and characterization of nanocellulose from an unwoven industrial textile cotton waste: Effect of acid hydrolysis conditions. *Int. J. Biol. Macromol.* **2019**, *126*, 496–506. [CrossRef]
144. Kalia, S.; Thakur, K.; Celli, A.; Kiechel, M.A.; Schauer, C.L. Surface modification of plant fibers using environment friendly methods for their application in polymer composites, textile industry and antimicrobial activities: A review. *J. Environ. Chem. Eng.* **2013**, *1*, 97–112. [CrossRef]
145. Flauzino Neto, W.P.; Mariano, M.; da Silva, I.S.V.; Silvério, H.A.; Putaux, J.-L.; Otaguro, H.; Pasquini, D.; Dufresne, A. Mechanical properties of natural rubber nanocomposites reinforced with high aspect ratio cellulose nanocrystals isolated from soy hulls. *Carbohydr. Polym.* **2016**, *153*, 143–152. [CrossRef] [PubMed]
146. Kassab, Z.; Kassem, I.; Hannache, H.; Bouhfid, R.; El Achaby, M. Tomato plant residue as new renewable source for cellulose production: Extraction of cellulose nanocrystals with different surface functionalities. *Cellulose* **2020**, *27*, 4287–4303. [CrossRef]
147. Hemmati, F.; Jafari, S.M.; Taheri, R.A. Optimization of homogenization-sonication technique for the production of cellulose nanocrystals from cotton linter. *Int. J. Biol. Macromol.* **2019**, *137*, 374–381. [CrossRef]

148. Culsum, N.T.U.; Melinda, C.; Leman, I.; Wibowo, A.; Budhi, Y.W. Isolation and characterization of cellulose nanocrystals (CNCs) from industrial denim waste using ammonium persulfate. *Mater. Today Commun.* **2020**, 101817. [CrossRef]
149. Prado, K.S.; Gonzales, D.; Spinacé, M.A. Recycling of viscose yarn waste through one-step extraction of nanocellulose. *Int. J. Biol. Macromol.* **2019**, *136*, 729–737. [CrossRef] [PubMed]
150. Wang, Z.; Yao, Z.; Zhou, J.; He, M.; Jiang, Q.; Li, A.; Li, S.; Liu, M.; Luo, S.; Zhang, D. Improvement of polylactic acid film properties through the addition of cellulose nanocrystals isolated from waste cotton cloth. *Int. J. Biol. Macromol.* **2019**, *129*, 878–886. [CrossRef] [PubMed]
151. Ibarra-Díaz, N.; Castañón-Rodríguez, J.; Gómez-Rodríguez, J.; Aguilar-Uscanga, M. Optimization of peroxide-alkaline pretreatment and enzymatic hydrolysis of barley straw (Hordeum vulgare L.) to produce fermentable sugars using a Box–Behnken design. *Biomass Convers. Biorefinery* **2020**, 1–10. [CrossRef]
152. Ilyas, R.; Sapuan, S.; Atikah, M.; Asyraf, M.; Rafiqah, S.A.; Aisyah, H.; Nurazzi, N.M.; Norrrahim, M. Effect of hydrolysis time on the morphological, physical, chemical, and thermal behavior of sugar palm nanocrystalline cellulose (Arenga pinnata (Wurmb.) Merr). *Text. Res. J.* **2021**, *91*, 152–167. [CrossRef]
153. Jordan, J.H.; Easson, M.W.; Dien, B.; Thompson, S.; Condon, B.D. Extraction and characterization of nanocellulose crystals from cotton gin motes and cotton gin waste. *Cellulose* **2019**, *26*, 5959–5979. [CrossRef]
154. Meyabadi, T.F.; Dadashian, F.; Sadeghi, G.M.M.; Asl, H.E.Z. Spherical cellulose nanoparticles preparation from waste cotton using a green method. *Powder Technol.* **2014**, *261*, 232–240. [CrossRef]
155. Yue, Y.; Zhou, C.; French, A.D.; Xia, G.; Han, G.; Wang, Q.; Wu, Q. Comparative properties of cellulose nano-crystals from native and mercerized cotton fibers. *Cellulose* **2012**, *19*, 1173–1187. [CrossRef]
156. Morais, J.P.S.; de Freitas Rosa, M.; Nascimento, L.D.; do Nascimento, D.M.; Cassales, A.R. Extraction and characterization of nanocellulose structures from raw cotton linter. *Carbohydr. Polym.* **2013**, *91*, 229–235. [CrossRef]
157. Digaitis, R.; Thybring, E.E.; Thygesen, L.G. Investigating the role of mechanics in lignocellulosic biomass degradation during hydrolysis: Part II. *Biotechnol. Prog.* **2021**, *37*, 3083. [CrossRef]
158. Jordan, J.H.; Easson, M.W.; Thompson, S.; Wu, Q.; Condon, B.D. Lignin-containing cellulose nanofibers with gradient lignin content obtained from cotton gin motes and cotton gin trash. *Cellulose* **2021**, *28*, 757–773. [CrossRef]
159. Johar, N.; Ahmad, I.; Dufresne, A. Extraction, preparation and characterization of cellulose fibres and nanocrystals from rice husk. *Ind. Crop. Prod.* **2012**, *37*, 93–99. [CrossRef]
160. Sunasee, R.; Hemraz, U.D.; Ckless, K. Cellulose nanocrystals: A versatile nanoplatform for emerging biomedical applications. *Expert Opin. Drug Deliv.* **2016**, *13*, 1243–1256. [CrossRef] [PubMed]
161. Hu, H.; Yuan, W.; Liu, F.-S.; Cheng, G.; Xu, F.-J.; Ma, J. Redox-Responsive Polycation-Functionalized Cotton Cellulose Nanocrystals for Effective Cancer Treatment. *ACS Appl. Mater. Interfaces* **2015**, *7*, 8942–8951. [CrossRef] [PubMed]
162. Grishkewich, N.; Mohammed, N.; Tang, J.; Tam, K.C. Recent advances in the application of cellulose nanocrystals. *Curr. Opin. Colloid Interface Sci.* **2017**, *29*, 32–45. [CrossRef]
163. Bhowmik, S.; Islam, J.M.M.; Debnath, T.; Miah, M.Y.; Bhattacharjee, S.; Khan, M.A. Reinforcement of Gelatin-Based Nanofilled Polymer Biocomposite by Crystalline Cellulose from Cotton for Advanced Wound Dressing Applications. *Polymers* **2017**, *9*, 222. [CrossRef] [PubMed]
164. Mascheroni, E.; Rampazzo, R.; Ortenzi, M.A.; Piva, G.; Bonetti, S.; Piergiovanni, L. Comparison of cellulose nanocrystals obtained by sulfuric acid hydrolysis and ammonium persulfate, to be used as coating on flexible food-packaging materials. *Cellulose* **2016**, *23*, 779–793. [CrossRef]
165. Li, F.; Biagioni, P.; Bollani, M.; Maccagnan, A.; Piergiovanni, L. Multi-functional coating of cellulose nanocrystals for flexible packaging applications. *Cellulose* **2013**, *20*, 2491–2504. [CrossRef]
166. Claramunt, J.; Ventura, H.; Fernández-Carrasco, L.J.; Ardanuy, M. Tensile and flexural properties of cement composites reinforced with flax nonwoven fabrics. *Materials* **2017**, *10*, 215. [CrossRef]
167. Masmoudi, F.; Bessadok, A.; Dammak, M.; Jaziri, M.; Ammar, E. Biodegradable packaging materials conception based on starch and polylactic acid (PLA) reinforced with cellulose. *Environ. Sci. Pollut. Res.* **2016**, *23*, 20904–20914. [CrossRef]

Article

Enhanced Water Resistance of Recycled Newspaper/High Density Polyethylene Composite Laminates via Hydrophobic Modification of Newspaper Laminas

Binwei Zheng, Weiwei Zhang *, Litao Guan, Jin Gu, Dengyun Tu and Chuanshuang Hu *

College of Materials and Energy, South China Agricultural University, Guangzhou 510642, China; hnzbw@stu.scau.edu.cn (B.Z.); ltguan@scau.edu.cn (L.G.); gujin@scau.edu.cn (J.G.); tudengyun@scau.edu.cn (D.T.)
* Correspondence: zhangww@scau.edu.cn (W.Z.); cshu@scau.edu.cn (C.H.)

Abstract: A high strength recycled newspaper (NP)/high density polyethylene (HDPE) laminated composite was developed using NP laminas as reinforcement and HDPE film as matrix. Herein, NP fiber was modified with stearic acid (SA) to enhance the water resistance of the NP laminas and NP/HDPE composite. The effects of heat treatment and SA concentration on the water resistance and tensile property of NP and composite samples were investigated. The chemical structure of the NP was characterized with X-ray diffractometer, X-ray photoelectron spectroscopy and attenuated total reflectance Fourier transform infrared spectra techniques. The surface and microstructure of the NP sheets were observed by scanning electron microscopy. An expected high-water resistance of NP sheets was achieved due to a chemical bonding that low surface energy SA were grafted onto the modified NP fibers. Results showed that the hydrophobicity of NP increased with increasing the stearic acid concentration. The water resistance of the composite laminates was depended on the hydrophobicity of the NP sheets. The lowest value of 2 h water absorption rate (3.3% ± 0.3%) and thickness swelling rate (2.2% ± 0.4%) of composite were obtained when the SA concentration was 0.15 M. In addition, the introduction of SA can not only enhance the water resistance of the composite laminates, but also reduce the loss of tensile strength in wet conditions, which shows potential in outdoor applications.

Keywords: recycled newspaper; composite laminates; water resistance; high strength

1. Introduction

Recycled paper has been widely used to fabricate bio-based fuels, fertilizer and high-valued cellulose products in the past decades [1–4]. As a readily available sheet product with desirable mechanical properties, recycled newspaper has been used to fabricate paper-based composite laminates in recent years [5–7]. Excellent mechanical performances of recycled newspaper composite laminates were achieved by forming a compact structure, which could enhance the fiber-fiber bond of paper sheet [8,9]. However, recycled newspaper has poor water resistance due to its hydrophilic nature, resulting in unexpected water uptake behavior of its laminated composite [10]. The strength of paper sheets can be reduced significantly due to the dis-bonding of paper fibers after the absorption of water molecules, leading to a negative effect on mechanical performances of the composites [11]. The poor water resistance greatly limits the applications of this sort of composites, especially for composites with high paper content [12]. Although the composite can be protected laterally, it will inevitably produce scratches or even cracks when they applied outdoors. Water molecules can pass through these defects. Therefore, it is essential to endow hydrophobicity for paper sheets in preparing recycled paper-based composite laminates.

Some efforts have been made in developing a hydrophobic/superhydrophobic paper sheets to improve the water resistance [13]. It is well known that both high surface roughness with micro/nanoscale structures and low surface energy are beneficial to create a hydrophobic/superhydrophobic coating on paper sheet [14–19]. However, some disadvantages still should be considered. On one hand, paper is easily discomposed through physical or chemical damage [20]. On the other hand, superhydrophobic coating are usually weakly adhesive and poorly durable, so the maintenance of their superhydrophobicity is inconvenient and expensive [21]. An alternative method is chemical modification at fiber surface instead of depositing micro-scale coating layer on paper surface. Paper and paperboard are mainly composed of natural cellulosic fibers, which are composed of β-D-glucose polymer with many hydroxyl groups. Hydrophilic surface of cellulosic fibers could be transferred into hydrophobic surface through chemical methods, such as silanization, etherification, and esterification [22].

Stearic acid (SA), a saturated fatty derived from animal and vegetable fats and oils, possesses a long surface energy hydrocarbon chain and active carboxyl group that can form an ester linkage with cellulose hydroxyl groups. With the advantages of low cost, environmental friendliness, and biocompatibility with cellulose fibers, SA has been widely used as a hydrophobic agent for cellulose products in recent years [23,24]. In these cases, cellulose films or paper are usually dipped in organic solvents and stirred vigorously in high temperature to obtain the hydrophobicity, which would damage to the fiber network and even results in dis-bonding of cellulose fibers. The solvent reaction system is regarded as an inefficient method for intact paper sheets modification.

The objective of this work is to improve the water resistance of recycled newspaper enhanced high-density polyethylene composite laminates. Herein, low surface energy stearic acid was firstly deposited on the paper fiber using a solvent evaporation method and then heated to form an ester bonding with a gentle solid-phase reaction method so that the integrity of paper sheets can be maintained. The SA modified paper sheet was stacked as reinforcing layer by using the HDPE film as matrix layer to fabricate composite laminates via hot-pressing. Effects of the modified process and stearic acid concentrations on water resistance of recycled newspaper and the composite laminate were investigated.

2. Experimental

2.1. Materials

Recycled newspapers (NP), with grammage of 45 g/m^2 and thickness of 50 μm from Shanghai Securities News, were cut to 200 mm × 200 mm. High-density polyethylene films (HDPE, density 0.95 g/cm^3, MFR 0.24g/10min at 190 C/2.16 kg, T_m 119 °C, tensile strength 17 MPa), food packaging film in daily life with the thickness of 12.8 μm, were purchased from local company (Charoen Pokphand Group, Guangzhou, China). Stearic acid (SA) and absolute ethanol (analytical grade) were obtained from Guangzhou Chemical Reagent Factory.

2.2. Preparation of Hydrophobic NP

The modification process of NP sheets was shown in Scheme 1. The dried SA powder was dissolved in absolute ethanol and then stirred vigorously at 70 °C for 15 min, forming a homogeneous solution. NP sheets were dipped into SA/absolute ethanol solution (0.01, 0.05, 0.1, 0.15, 0.2 M) for 15 min and then dried in air at room temperature, which was named as deposited NP. The SA-deposited NP sheets were further heated in an oven at 105 °C for 2 h and then cooled down to room temperature, which was called modified NP.

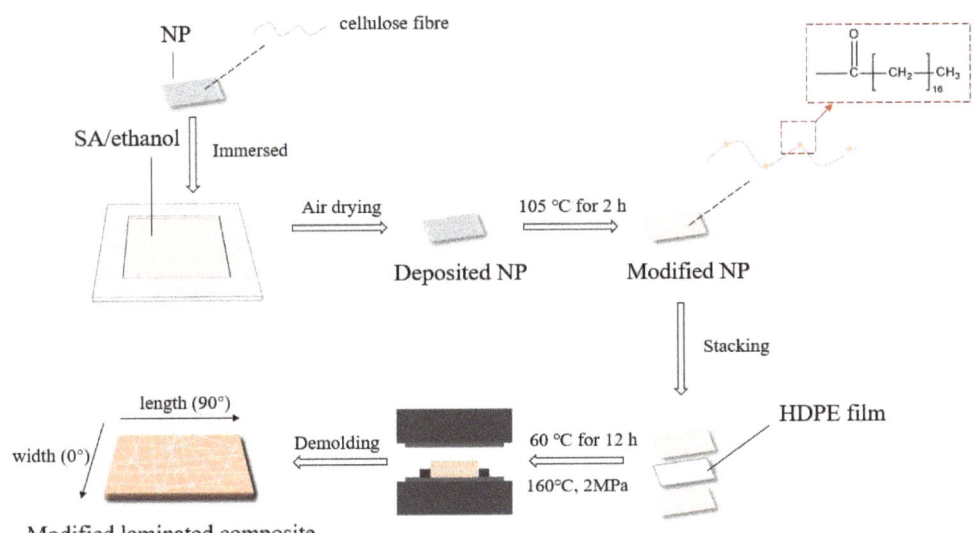

Scheme 1. Formation illustration of the hydrophobic NP and composite fabrication process.

2.3. Composite Fabrication

Laminates were prepared by stacking NP sheets and HDPE films alternately with the same direction. Herein, each NP sheet was stacked with one layer of the HDPE film to control paper content of the laminates at 78.5 wt%. In total, 20 layers of newspaper and 20 layers of HDPE film were used. The prepared laminates were oven-dried at 50 °C for 24 h and then transferred to the pre-heated hot press machine (BY302X2/2 150T, Suzhou New Cooperative, Suzhou, China). A steel bar frame as a thickness gauge was used to control the thickness of the composites at 1 mm. Poly-tetrafluoroethylene (PTFE) films were used as demolding layer to facilitate the demolding process. Composites were fabricated by setting the pressing pressure and the heating temperature at 2.0 MPa and 160 °C, respectively, holding for 25 min to ensure a sufficient compaction and infiltration. After the hot-pressing procedure was completed, the hot panel was quickly cooled down to room temperature by using circulation water cooling system. Finally, the pressure was released, and the composite laminates (NP/HDPE composite) was removed. As NP is a typical anisotropic material with different mechanical properties at fiber direction (0°, wide direction of newspaper, 52.0 MPa of tensile strength) and perpendicular fiber direction (90°, length direction of newspaper, 12.9 MPa of tensile strength). Specimens for property testing were cut from the composites along parallel directions by using a mold cutter (GT7016HA, Gotech testing machine Co., Ltd, Dongguan, China).

2.4. Characterizations

The phase structure of the NP samples was characterized by X-ray diffractometer (XRD) (D8 Advance, Bruker, Germany). The patterns were recorded in the region of 2θ from 4° to 70° with a scanning speed of 10°/min at 40 kV and 40 mA (copper Kα radiation λ = 0.154 nm). Attenuated Total Reflectance Fourier transform infrared spectra (ATR-FTIR) (TENSOR27, Bruker, Germany) was used to analyze functional groups of the NP and composites samples in the range of 4000–600 cm^{-1} with 32 scans and a resolution of 4 cm^{-1}. Chemical elements of the NP samples surface were obtained by using X-ray photoelectron spectroscopy (XPS) (Thermo EscaLAB 250Xi, Shanghai, China), with an Al X-radiation (Kα, hν = 1486.8 eV). All of the binding energies were corrected based on the C1s peak at 284.8 eV. The microstructure of pristine, deposited and modified NP samples were observed by scanning electron microscopy (SEM) (EVO-18, Zeiss, Jena, Germany) with an acceleration voltage of 10 kV. All of the samples were coated with a thin carbon conductive coating via sputter deposition before testing.

2.5. Physical and Mechanical Properties Testing

The hydrophobicity of the NP sheets was assessed by water contact angle (WCA) and water absorption rate according to ASTM D 724 and GB/T 461.3, respectively. WCA was performed with an optical Contact Angle Meter (DSA100, Kruss, Germany). A water droplet size of 5 µL was placed on NP sample surface. Water absorption rate of the NP sheets with the dimension of 100 mm × 100 mm was calculated by measuring the mass of the sample before and after immersing in a tank full of deionized water. Ten repeats of WCA and water absorption rate of each NP sample were carried out.

2 h-water absorption and thickness swelling tests were conducted to evaluate the water uptake behavior of composite laminates in accordance with ASTM D 570. The dimension of the sample was 76.2 × 25.4 mm^2. Samples were weighed and thickness measured before and after stored in distilled water. Three test bars were conducted for every composite sample with different hydrophobic modification conditions.

Tensile property of the pristine and hydrophobic modification NP sheets in dry and wet conditions were measured by using an electromechanical universal testing machine (CMT5504, Shenzhen Rethink Cooperation, China) according to GB/T 12914. The dimension of the sample was 15 mm × 200 mm and the loading speed during the testing was 25 mm/min. All NP samples with 0° direction were tested ten times.

The mechanical performances of composite laminates with 0° direction were also evaluated by investigating the tensile property in dry and wet condition, using an electromechanical universal testing machine (CMT5504, Shenzhen Rethink Cooperation, Shenzhen, China) according to ASTM D 638. For the tensile test, dumbbell specimen size was 165 × 13 mm^2 and the loading speed was 5 mm/min. Nine replicates of the tests were conducted.

3. Results and Discussion

3.1. Morphology and Microstructure

The appearance and microstructures of the pristine, the deposited NP and the modified NP samples are shown in Figure 1. Water drop spread on pristine NP surface instantaneously, leaving a large wetting spot. For the deposited NP, it formed a hemisphere firstly but penetrated into NP gradually. For the modified NP, the water beads were hold on the surface steadily. The pristine and deposited NP exhibited opaque while the modified NP turned to semi-transparent. It is known that the opacity of conventional paper is generally caused by a large amount of light scattering due to the different refractive index of cellulose (approximately 1.5) and air (1.0) when light transmits through fiber-air interfaces [25]. The completely exposure of hydrophilic fibers and the porous structure (Figure 1d) in pristine NP result in highly wetting and large amount of light scattering. As can be seen in Figure 1d–f, flaky SA occupied in gaps and pores of fiber network as well as at fiber surface for deposited NP. For modified NP, crystalloid SA disappeared which probably permeated into fiber network due to heat treatment. Li et al. suggested that the transparency of paper sheet can be improved by using transparent agents (with similar refractive index to cellulose) to fill the voids inside paper [26]. It could be inferred that, after heat treatment, SA (refractive index is 1.455) had melted and filled in part of voids inside the NP sheet, resulting in reduced light scattering and semi-transparent appearance of the modified NP.

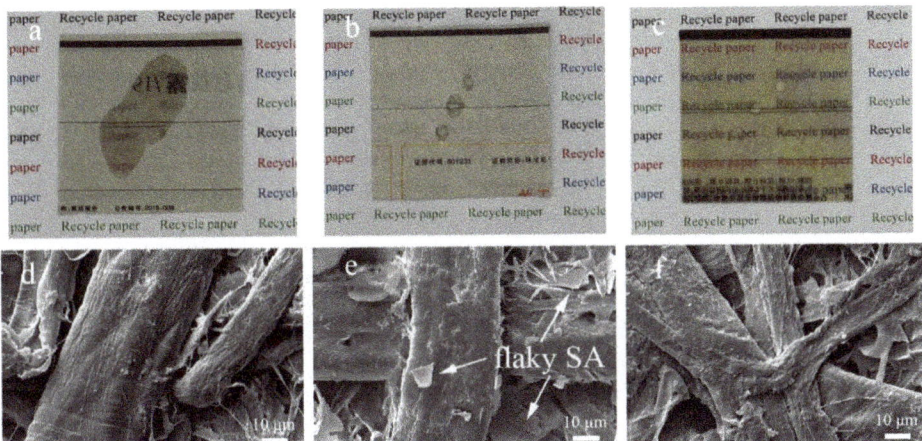

Figure 1. Photographs (a–c) and SEM (d–f) images of pristine NP (a,d), deposited NP (b,e) and modified NP (c,f).

The microstructures of unmodified and SA modified composite samples are presented in Figure 2. The multilayer structure of composite samples was obvious that NP sheets and HDPE films were stacked alternately. It was found that HDPE permeated into the NP sheets and occupied the voids, forming a dense composite structure. However, pores and voids could be found clearly because of lacking enough HDPE (only 21.5% HDPE content of the composite) to fully fill. In comparison to unmodified composite samples, the modified composite samples showed intact cross section (Figure 2b,c). It was considered that adding SA could improve the interface bonding between NP fiber and HDPE. Therefore, there was no clear delamination and transverse cracks in modified composite. Since the penetration of HDPE was promoted, the fiber network was broken in modified composite and even dis-bonding in composite modified with high SA concentration (Figure 2c).

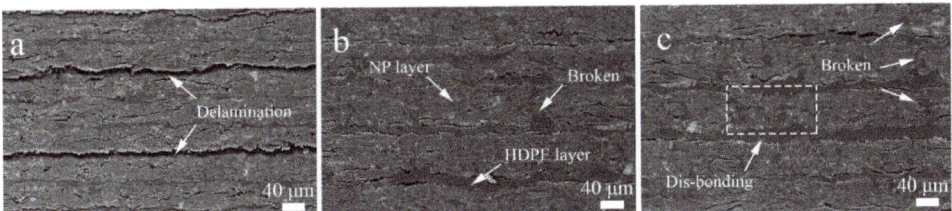

Figure 2. BSEM images of (a) unmodified composite; (b) 0.05M SA modified composite and (c) 0.15M SA modified composite.

3.2. Chemical Structure Characterization of NP

The XRD patterns of pristine NP, SA, deposited NP and modified NP are presented in Figure 3. For all NP samples, three strong diffraction peaks were observed at around $2\theta = 12°$, $16°$ and $22.2°$ which could be assigned to the typical crystalline plane (11$\bar{}$0), (110) and (020) of cellulose, respectively [27]. The diffraction patterns of NP samples were not completely consistent with those of pure cellulose because some non-fiber components, such as calcium carbonate filler, ink etc. were found in common waste newspaper [11]. For pure SA sample, the "short spacing" peaks at $2\theta = 21.4°$ and $23.7°$ were assigned to the (110) and (021) crystallographic planes of the SA respectively due to the presence of the hydrocarbon chain lateral packing order [28–30]. It can also be seen that the "short spacing" peak at around $2\theta = 21°$ in deposited NP sample indicates the SA existed as crystal in deposited NP [31,32]. However, the "short spacing" peak was indistinctively found in modified NP. The weak crystalline intensity of SA in modified NP was attributed to its uniform dispersion on the fiber surface. The uniform SA coating contributed to the restriction of hydrocarbon chain mobility so that SA crystal could not be clearly characterized. What's more, it provided a larger contact area for stearic acid and cellulose fiber to promote interaction.

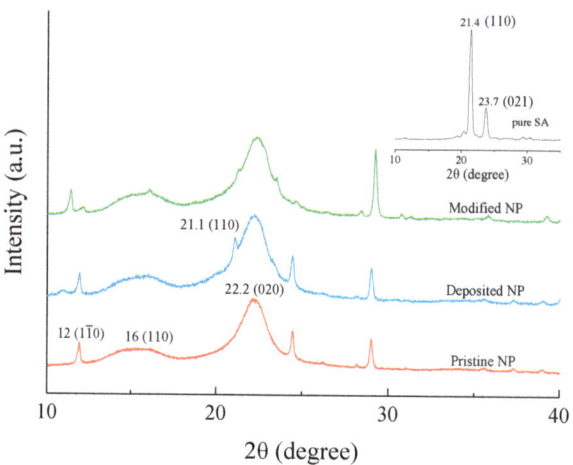

Figure 3. XRD pattern of pristine NP, deposited NP, modified NP and pure SA.

The XPS spectra of pristine NP, deposited NP and modified NP samples were presented in Figure 4. All of the NP samples showed the clear C1s (284 eV) and O1s (532 eV) signals (Figure 4a). Three characteristic functional groups C1 (C–O), C2 (C–C), and C3 (C=O) of carbon atoms for cellulose fibers can be found in pristine, deposited, and modified NP samples [16,33]. Compared with pristine NP, both deposited and modified NP samples showed an increase in C/O ratio and C2 (C–C, C–H) ratio (as shown in Table 1), since the coating layer of long carbon chain of SA increased the carbon content [18]. Moreover, the modified NP samples showed additional peaks C4 at 288.9 eV, which was attributed to O–C=O bonds signals from ester linkages. Therefore, it further confirmed that SA was grafted onto the modified NP fiber surface.

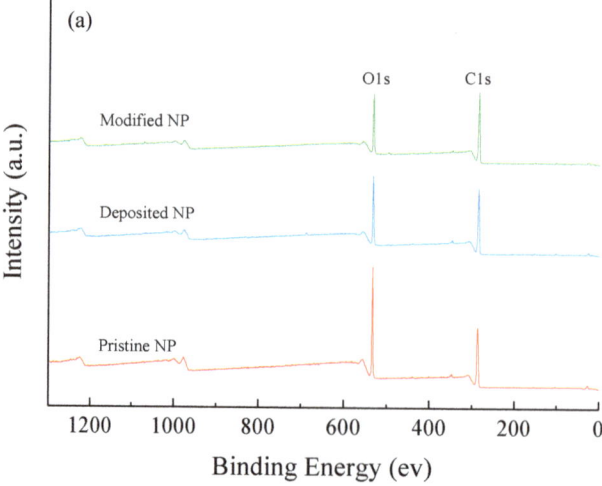

Figure 4. *Cont.*

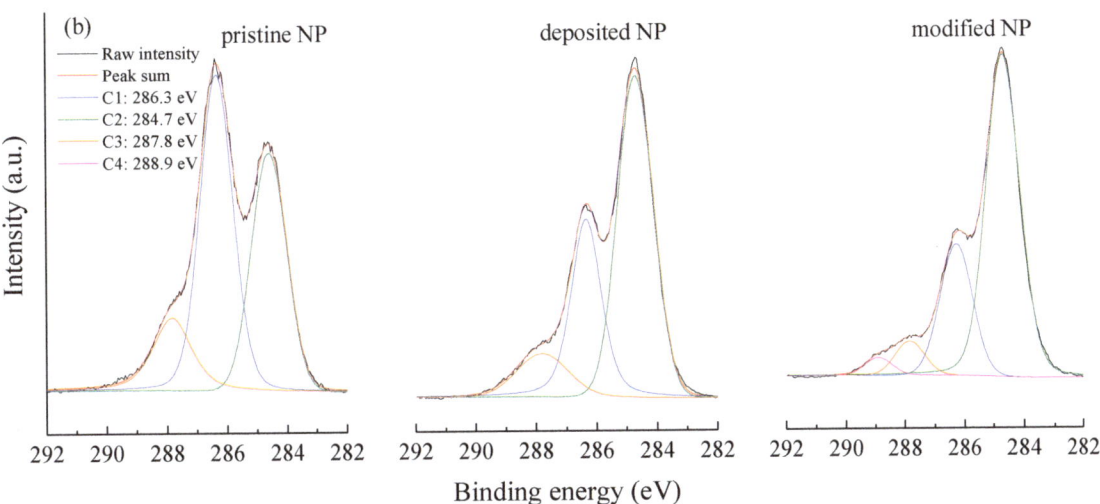

Figure 4. XPS spectra of pristine, deposited and modified NP samples: (**a**) survey spectra and (**b**) deconvolution of the C1s spectra.

Table 1. Assignment of the XPS C1s peaks.

Samples	Element Concentration (%)		Atomic Ratio	Ratios of C1s Deconvolution			
				C1	C2	C3	C4
	C1s	O1s	C/O	C–O	C–C/C–H	C=O	O–C=O
Pristine NP	65.7	34.3	1.9	48.6	36.6	14.9	0
Deposited NP	72.8	27.2	2.7	31.3	57.5	10.8	0
Modified NP	75.1	24.9	3.0	25.0	66.6	5.7	2.8

The ATR-FTIR spectrum was recorded to analyze the chemical bonds of the pristine, deposited and modified NP. As can be seen in Figure 5, the peaks at 2920 cm^{-1} and 2854 cm^{-1} corresponded to the stretching vibration of –CH$_3$ and –CH$_2$ respectively, which were strongly found in deposited NP due to in-phase stretching groups of long carbon chain of SA while relatively weak in modified NP. It was thought that the stretching vibration of –CH$_3$ and –CH$_2$ is difficult to be characterized in modified NP because of uniform coating of SA on the fiber surface. The bands at 1427 cm^{-1} and 1058 cm^{-1} were due to the stretching vibration of crystalline regions and C–O–C from cellulose macromolecule of NP fiber. The peak at 1645 cm^{-1}, assigned to the bending vibration of O–H from interlayer water molecules, was found obviously in pristine NP but not in the deposited and modified NP. New absorption band at 1703 cm^{-1} in the deposited and modified NP was attributed to the stretching vibration of C=O band after adding SA. Pristine and deposited NP showed broad bands in the region of 3600–3000 cm^{-1} that can be assigned to the stretching vibration of –OH from cellulose. After the heat treatment, both the stretching vibration of –OH from cellulose and C=O from SA were no longer strong, implying that –OH of the NP fiber reacted with –COOH of SA.

3.3. Water Resistance and Tensile Property of Modified NP

The water contact angle and 2 h-water uptake test were carried out to investigate the water resistance of pristine and modified NP. As shown in Figure 6, all of the SA modified NP sample showed better water resistance than the pristine NP. The water contact angles (WCA) and 2 h water absorption rates of pristine NP samples are 0° and 127.8% ± 2.3%, respectively. High water uptake of the pristine NP could be attributed to the presence of a large number of hydrophilic hydroxyl groups [34]. With increasing SA concentration, the WCA of modified NP samples increased gradually, and the 2 h water absorption rate decreased, since more SA was deposited to react with hydrophilic hydroxyl groups to cause more long carbon chains covering at modified NP surface. When the SA concentration was higher than 0.15 M, the water contact angle and water absorption curve became flat, indicating the SA solution reached the saturate level. After dissolving the SA again (NP sheets were dipped into pure ethanol solution at 70 °C for 15min), the modified NP still exhibited better water resistance (38.1% ± 5.1% of 2 h water absorption rates) than that of the pristine NP. Although the water resistance has been improved, the 2 h water absorption rate was still as high as 30%. As inks and dyes are usually organic compounds, which have a good polarity with SA, they are able to support the SA modification. Therefore, the high-water absorption rate of the modified NP might be attributed to the high amount of inorganic fillers in NP that was added during the paper making. Moreover, the lumen of natural cellulosic fibers could absorb water due to the capillary action.

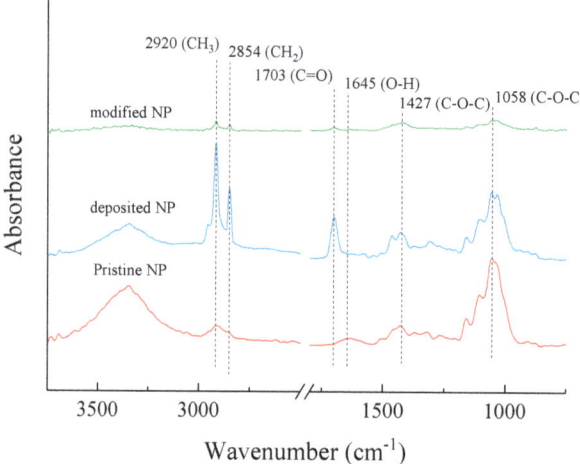

Figure 5. ATR-FTIR spectra of pristine, deposited NP and modified NP.

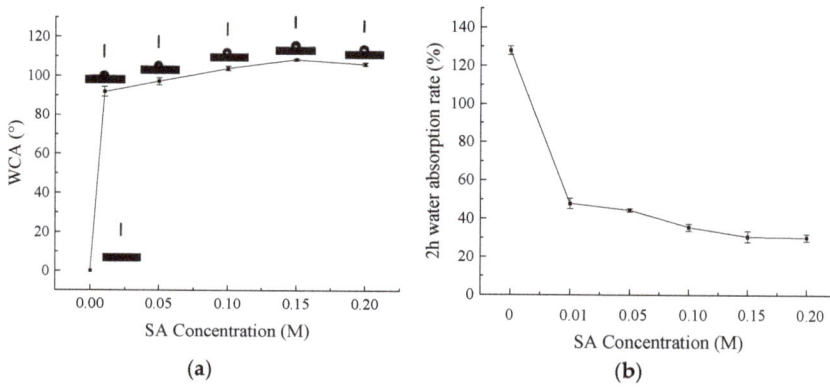

Figure 6. The water contact angle (**a**) and 2 h water absorption rate (**b**) of modified NP with different SA concentrations.

Figure 7 shows the effect of SA concentration on tensile properties of the NP samples in the dry and wet (2 h water immersion) states. Pristine NP exhibited desirable tensile property because of tight fiber network and hydrogen bonds among cellulose fibers. As in Figure 7a, modified NP samples showed higher tensile strength than the pristine NP while the tensile strength of NP increased with increasing the SA concentration. It can be observed in Figure 1c that a certain part of voids inside the modified NP sheets was occupied by SA. It was considered that the filling of SA inside NP sheets can increase the friction between fibers. The tensile strength of all NP samples declined when they were in wet condition because water molecular opened hydrogen bond between fibers [11]. For the pristine NP, it reduced significantly from 54.1 ± 1.4 MPa to 12.2 ± 2.5 MPa. As expected, the modified NP exhibited desirable water resistance due to the uniform coating of hydrophobic carbon chain. So, the tensile strength (in wet) increased from 12.2 ± 2.5 MPa to 47.9 ± 3.4 MPa when the SA concentration increased from 0 to 0.20 M, corresponding to strength loss rate decreased from 78.2% to 25.8%. In Figure 7b, the wet tensile modulus of the modified NP samples showed nearly three times as much as pristine NP while it could not change significantly with increasing the SA concentration.

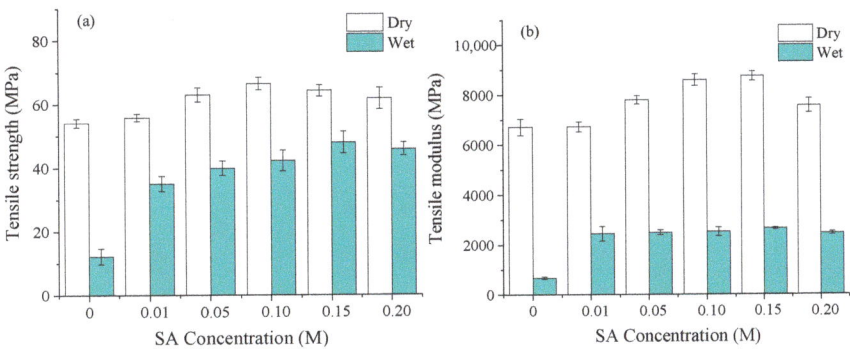

Figure 7. The effect of SA concentration on (**a**) tensile strength and (**b**) tensile modulus of modified NP in dry and wet conditions.

3.4. Water Resistance and Tensile Property of NP/HDPE Composite

Pristine and modified NP sheets were used to fabricate the NP/HDPE laminated composite via hot-pressing process. The 2 h water absorption rate and thickness swelling rate of composite samples at different SA concentrations were presented in Figure 8. The modified composite samples showed better water resistance than other paper-based composite laminates that 49.3% (with 78.5% paper content) and 10.0% (with 30.0% paper content) water absorption rate were found in literature [10,12]. The 2-h water absorption rate of the composite samples was remarkably lower than that of pristine NP samples, since visible pores and voids in the NP sheets can be filled by HDPE, result in low water uptake [12]. Compared with the unmodified composite, the modified composite (0.01–0.20 M) showed lower water absorption and thickness swelling. As the water adsorption in HDPE could be neglected, the water resistance of the composite laminates is mainly depended on the wettability of NP layer and its porosity. Due to the presence of non-polar long carbon chain of SA, the hydrophobic of the NP sheet was improved. Moreover, a compatible interface between the NP fiber and HDPE was obtained, so paper pores and voids could be fully filled by melted HDPE to block the water transmission. Therefore, with an increasing SA concentration, the water absorption rate fluctuated at around 3%, while the thickness swelling further decreased to 2.2% ± 0.4%.

The tensile properties of dry and wet NP/HDPE composite samples at different SA concentrations were presented in Figure 9. The dry composite samples showed comparable high tensile strength (from 47–89 MPa) to other paper-based composite laminates (in a range of 58–84 MPa) [7,10,11,35]. The relatively high tensile property was attributed to the dense structure of the composites which can bond the NP fibers tightly and enhance the mechanical interlocking. Results showed that the dry tensile strength and modulus of composites samples fabricated by modified NP decreased with increasing of SA concentration. It was thought that the non-polar long carbon chain coating (grafted from SA) of NP fiber could promote the penetration of non-polar HDPE matrix, leading to the excessive penetration of HDPE in the modified NP layer while 25 min hot-pressing time was suitable for pristine NP layer. As a result, the mechanical interlock of the NP fiber was weakened (Figure 2b,c). Research shows that the mechanical properties of the NP/HDPE composite decreased if the integrity of the NP sheets was destroyed due to excessive penetration of HDPE, confirmed via observing microstructure [36]. Therefore, with an increasing SA concentration, the tensile strength of the composite decreased gradually.

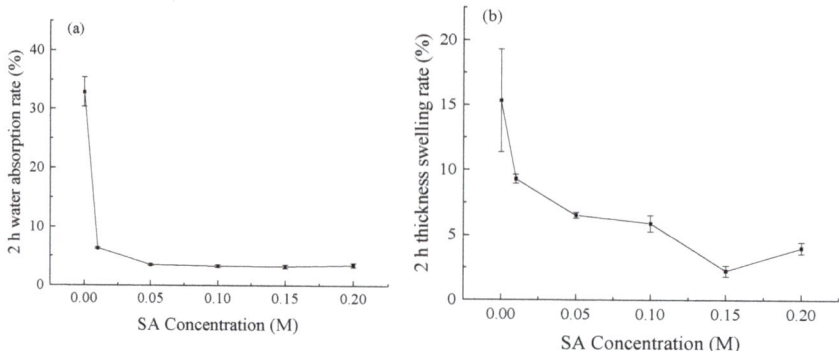

Figure 8. The 2 h water absorption rate (**a**) and 2 h thickness swelling rate (**b**) of NP/HDPE composite with different SA concentrations.

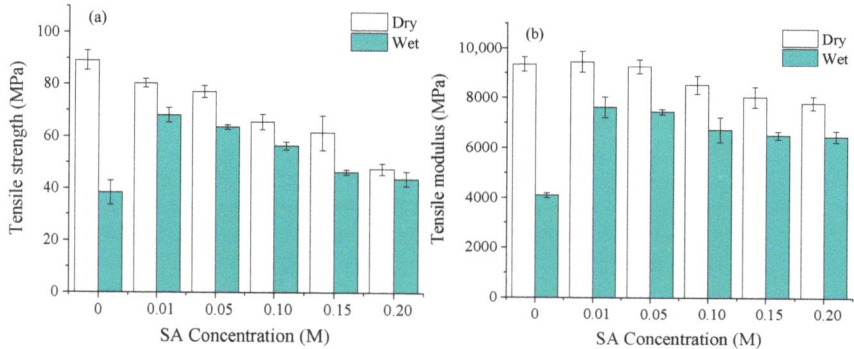

Figure 9. The effect of SA concentration on (**a**) tensile strength and (**b**) tensile modulus of NP/HDPE composite in dry and wet conditions.

As expected, the modified composite samples exhibited a better water resistance than the unmodified composites samples. Correspondingly, the modified NP showed clearly higher values of tensile strength and modulus than the unmodified NP in wet conditions. Therefore, the introduction of SA can not only improve the water resistance of the composite laminates, but also reduce the loss of tensile strength in wet.

4. Conclusions

In this paper, recycled newspaper composite laminates with good water resistance was fabricated using eco-friendly method. NP sheet was modified with stearic acid as reinforcement. The effects of heat treatment and stearic acid concentration on microstructure, water uptake behavior and tensile performance of NP and composite laminates were investigated. The main findings are as follows:

(1) The modified NP showed visible transparency and higher hydrophobicity with increasing the stearic acid concentration in comparison to the pristine and deposited NP.
(2) The water resistance of the composite laminates was significantly improved due to the hydrophobic modification of the NP sheets since lower water uptake and higher wet tensile strength were found in modified composite samples.
(3) The excellent water resistance of NP and its composite laminates were attributed to the esterification reaction between hydrophilic hydroxyl group of NP fiber and carboxyl group of stearic acid.
(4) The introduction of stearic acid can not only improve the water resistance of the composite laminates, but also reduce the loss of tensile strength in wet conditions, which shows potential in outdoor applications. As adding SA can promote the penetration of non-polar HDPE matrix, it was thought that the mechanical property can be improved via altering the holding time.

Author Contributions: Investigation, B.Z.; writing—original draft preparation, B.Z.; writing—review and editing, C.H. and W.Z. and L.G. and J.G. and D.T.; supervision, C.H. and W.Z.; project administration, C.H.; funding acquisition, C.H. and W.Z. All authors have read and agreed to the published version of the manuscript.

Funding: This research was funded by Department of science and technology of Guangdong Province, grant number 2017B020238003 and 2018A030313233.

Institutional Review Board Statement: Not applicable.

Informed Consent Statement: Not applicable.

Data Availability Statement: The data presented in this study are available on request from the corresponding author. The data are not publicly available due to our future research needs.

Conflicts of Interest: The authors declare no conflict of interest.

References

1. Abbasi, S.A.; Hussain, N.; Tauseef, S.M.; Abbasi, T. A novel FLippable Units Vermireactor Train System—FLUVTS—for rapidly vermicomposting paper waste to an organic fertilizer. *J. Clean. Prod.* **2018**, *198*, 917–930. [CrossRef]
2. Bourtsalas, A.C.T.; Zhang, J.; Castaldi, M.J.; Themelis, N.J. Use of non-recycled plastics and paper as alternative fuel in cement production. *J. Clean. Prod.* **2018**, *181*, 8–16. [CrossRef]
3. Lei, W.; Zhou, X.; Fang, C.; Li, Y.; Song, Y.; Wang, C.; Huang, Z.J. New approach to recycle office waste paper: Reinforcement for polyurethane with nano cellulose crystals extracted from waste paper. *Waste Manag.* **2019**, *95*, 59. [CrossRef] [PubMed]
4. Hietala, M.; Varrio, K.; Berglund, L.; Soini, J.; Oksman, K. Potential of municipal solid waste paper as raw material for production of cellulose nanofibres. *Waste Manag.* **2018**, *80*, 319–326. [CrossRef] [PubMed]
5. Kim, Y.M.; Han, T.U.; Watanabe, C.; Teramae, N.; Park, Y.K.; Kim, S.; Hwang, B.; Pyrolysis, A. Analytical pyrolysis of waste paper laminated phenolic-printed circuit board (PLP-PCB). *J. Anal. Appl. Pyrolysis* **2015**, *115*, 87–95. [CrossRef]
6. Bayatkashkoli, A.; Ramazani, O.; Keyani, S.; Mansouri, H.R.; Madahi, N.K. Investigation on the production possibilities of high pressure laminate from borax and recycled papers as a cleaner product. *J. Clean. Prod.* **2018**, *192*, 775–781. [CrossRef]
7. Guan, N.; Hu, C.S.; Guan, L.T.; Zhang, W.W.; Yun, H.; Hu, X.J. A Process Optimization and Performance Study of Environmentally Friendly Waste Newspaper/Polypropylene Film Layered Composites. *Materials* **2020**, *13*, 413. [CrossRef]
8. Du, Y.; Wu, T.; Yan, N.; Kortschot, M.T.; Farnood, R. Fabrication and characterization of fully biodegradable natural fiber-reinforced poly(lactic acid) composites. *Compos. Part B Eng.* **2014**, *56*, 717–723. [CrossRef]
9. Prambauer, M.; Paulik, C.; Burgstaller, C. Evaluation of the interfacial properties of polypropylene composite laminates, reinforced with paper sheets. *Compos. Part A Appl. Sci. Manuf.* **2016**, *88*, 59–66. [CrossRef]
10. Prambauer, M.; Paulik, C.; Burgstaller, C. The influence of paper type on the properties of structural paper—Polypropylene composites. *Compos. Part A Appl. Sci. Manuf.* **2015**, *74*, 107–113. [CrossRef]
11. Das, S. Mechanical and water swelling properties of waste paper reinforced unsaturated polyester composites. *Constr. Build. Mater.* **2017**, *138*, 469–478. [CrossRef]
12. Zheng, B.W.; Hu, C.S.; Guan, L.T.; Gu, J.; Guo, H.Z.; Zhang, W.W. Structural Characterization and Analysis of High-Strength Laminated Composites from Recycled Newspaper and HDPE. *Polymers* **2019**, *11*, 1311. [CrossRef] [PubMed]
13. Li, L.; Breedveld, V.; Hess, D.W. Design and fabrication of superamphiphobic paper surfaces. *ACS Appl. Mater. Interfaces* **2013**, *5*, 5381–5386. [CrossRef] [PubMed]
14. Tian, X.; Li, Y.; Wan, S.; Wu, Z.; Wang, Z. Functional Surface Coating on Cellulosic Flexible Substrates with Improved Water-Resistant and Antimicrobial Properties by Use of ZnO Nanoparticles. *J. Nanomater.* **2017**, *2017*, 1–9. [CrossRef]

15. Ogihara, H.; Xie, J.; Okagaki, J.; Saji, T. Simple method for preparing superhydrophobic paper: Spray-deposited hydrophobic silica nanoparticle coatings exhibit high water-repellency and transparency. *Langmuir* **2012**, *28*, 4605–4608. [CrossRef] [PubMed]
16. Liu, Z.; Yu, J.; Lin, W.; Yang, W.; Li, R.; Chen, H.; Zhang, X. Facile method for the hydrophobic modification of filter paper for applications in water-oil separation. *Surf. Coat. Technol.* **2018**, *352*, 313–319. [CrossRef]
17. Thakur, S.; Misra, M.; Mohanty, A.K. Sustainable Hydrophobic and Moisture-Resistant Coating Derived from Downstream Corn Oil. *ACS Sustain. Chem. Eng.* **2019**, *7*, 8766–8774. [CrossRef]
18. Fu, J.; Yang, F.; Guo, Z. Facile fabrication of superhydrophobic filter paper with high water adhesion. *Mater. Lett.* **2019**, *236*, 732–735. [CrossRef]
19. Gao, Z.; Zhai, X.; Liu, F.; Zhang, M.; Zang, D.; Wang, C. Fabrication of TiO2/EP super-hydrophobic thin film on filter paper surface. *Carbohydr. Polym.* **2015**, *128*, 24–31. [CrossRef]
20. Wang, Y.; Liu, Y.; Zhang, L.; Zhang, M.; He, G.; Sun, Z. Facile fabrication of a low adhesion, stable and superhydrophobic filter paper modified with ZnO microclusters. *Appl. Surf. Sci.* **2019**, *496*, 143743. [CrossRef]
21. Li, Y.; Li, L.; Sun, J. Bioinspired self-healing superhydrophobic coatings. *Angew. Chem. Int. Ed. Engl.* **2010**, *49*, 6129–6133. [CrossRef] [PubMed]
22. Gurunathan, T.; Mohanty, S.; Nayak, S.K. A review of the recent developments in biocomposites based on natural fibres and their application perspectives. *Compos. Part A Appl. Sci. Manuf.* **2015**, *77*, 1–25. [CrossRef]
23. He, M.; Xu, M.; Zhang, L. Controllable stearic acid crystal induced high hydrophobicity on cellulose film surface. *ACS Appl. Mater. Interfaces* **2013**, *5*, 585–591. [CrossRef] [PubMed]
24. Chen, Q.; Shi, Y.; Chen, G.; Cai, M. Enhanced mechanical and hydrophobic properties of composite cassava starch films with stearic acid modified MCC (microcrystalline cellulose)/NCC (nanocellulose) as strength agent. *Int. J. Biol. Macromol.* **2020**, *142*, 846–854. [CrossRef] [PubMed]
25. Guan, F.; Song, Z.; Xin, F.; Wang, H.; Yu, D.; Li, G.; Liu, W. Preparation of hydrophobic transparent paper via using polydimethylsiloxane as transparent agent. *J. Bioresour. Bioprod.* **2020**, *5*, 37–43. [CrossRef]
26. Li, G.; Yu, D.; Song, Z.; Wang, H.; Liu, W. Facile Fabrication of Transparent Paper with Tunable Wettability for Use in Biodegradable Substrate. *ACS Sustain. Chem. Eng.* **2020**, *8*, 2176–2185. [CrossRef]
27. French, A.D. Idealized powder diffraction patterns for cellulose polymorphs. *Cellulose* **2014**, *21*, 885–896. [CrossRef]
28. Teixeira, A.C.T.; Garcia, A.R.; Ilharco, L.M.; Goncalves da Silva, A.M.P.S.; Fernandes, A.C. Phase behaviour of oleanolic acid, pure and mixed with stearic acid: Interactions and crystallinity. *Chem. Phys. Lipids* **2010**, *163*, 655–666. [CrossRef]
29. Zhu, C.; Chen, Y.; Cong, R.; Ran, F.; Fang, G. Improved thermal properties of stearic acid/high density polyethylene/carbon fiber composite heat storage materials. *Sol. Energy Mater. Sol. Cells* **2021**, *219*. [CrossRef]
30. Ensikat, H.J.; Boese, M.; Mader, W.; Barthlott, W.; Koch, K. Crystallinity of plant epicuticular waxes: Electron and X-ray diffraction studies. *Chem. Phys. Lipids* **2006**, *144*, 45–59. [CrossRef]
31. Wu, S.; Tang, Z.J.; Jiang, Z.F.; Yu, Z.L.; Wang, L.J. Preparation and characterization of hydrophobic cotton fibre for water/oil separation by electroless plating combined with chemical corrosion. *Int. J. Environ. Res. Publ. Health* **2015**, *2*, 144–150. [CrossRef]
32. Sobhana, S.S.L.; Zhang, X.; Kesavan, L.; Liias, P.; Fardim, P. Layered double hydroxide interfaced stearic acid—Cellulose fibres: A new class of super-hydrophobic hybrid materials. *Colloids Surf. A Physicochem. Eng. Asp.* **2017**, *522*, 416–424. [CrossRef]
33. Hao, J.; Li, Y.; Liao, R.; Liu, G.; Liao, Q.; Tang, C. Fabrication of Al(2)O(3) Nano-Structure Functional Film on a Cellulose Insulation Polymer Surface and Its Space Charge Suppression Effect. *Polymers (Basel)* **2017**, *9*, 502. [CrossRef] [PubMed]
34. Ashori, A.; Sheshmani, S. Hybrid composites made from recycled materials: Moisture absorption and thickness swelling behavior. *Bioresour. Technol.* **2010**, *101*, 4717–4720. [CrossRef] [PubMed]
35. Graupner, N.; Prambauer, M.; Fröhlking, T.; Graf, C.; Becker, J.M.; Meyer, K.; Weber, D.E.; Weddig, N.B.; Wunsch, T.; Burgstaller, C.; et al. Copy paper as a source of reinforcement for biodegradable composites—Influence of fibre loading, processing method and layer arrangement—An overview. *Compos. Part A Appl. Sci. Manuf.* **2019**, *120*, 161–171. [CrossRef]
36. Zheng, B.W.; Guan, L.T.; Zhang, W.W.; Gu, J.; Tu, D.Y.; Hu, C.S. Production and Characterization of Large-Scale Recycled Newspaper Enhanced HDPE Composite Laminates. *Polymers* **2020**, *12*, 851. [CrossRef]

Article

Isolation of Textile Waste Cellulose Nanofibrillated Fibre Reinforced in Polylactic Acid-Chitin Biodegradable Composite for Green Packaging Application

Samsul Rizal [1], Funmilayo G. Olaiya [2], N. I. Saharudin [2,*], C. K. Abdullah [2], Olaiya N. G. [3,*], M. K. Mohamad Haafiz [2], Esam Bashir Yahya [2], F. A. Sabaruddin [2], Ikramullah [1] and Abdul Khalil H. P. S. [2,*]

1. Department of Mechanical Engineering, Universitas Syiah Kuala, Banda Aceh 23111, Indonesia; samsul_r@yahoo.com (S.R.); ikramullah@mhs.unsyiah.ac.id (I.)
2. School of Industrial Technology, Universiti Sains Malaysia, Penang 11800, Malaysia; phunmieoseyemi@gmail.com (F.G.O.); ck_abdullah@usm.my (C.K.A.); mhaafiz@usm.my (M.K.M.H.); essam912013@gmail.com (E.B.Y.); atiyah88@gmail.com (F.A.S.)
3. Department of Industrial and Production Engineering, Federal University of Technology, P.M.B.740 Akure, Nigeria
* Correspondence: nurizzaati@usm.my (N.I.S.); ngolaiya@futa.edu.ng (O.N.G.); akhalilhps@gmail.com (A.K.H.P.S.)

Citation: Rizal, S.; Olaiya, F.G.; Saharudin, N.I.; Abdullah, C.K.; N. G., O.; Mohamad Haafiz, M.K.; Yahya, E.B.; Sabaruddin, F.A.; Ikramullah; Khalil H. P. S., A. Isolation of Textile Waste Cellulose Nanofibrillated Fibre Reinforced in Polylactic Acid-Chitin Biodegradable Composite for Green Packaging Application. *Polymers* **2021**, *13*, 325. https://doi.org/10.3390/polym13030325

Received: 23 December 2020
Accepted: 18 January 2021
Published: 20 January 2021

Publisher's Note: MDPI stays neutral with regard to jurisdictional claims in published maps and institutional affiliations.

Copyright: © 2021 by the authors. Licensee MDPI, Basel, Switzerland. This article is an open access article distributed under the terms and conditions of the Creative Commons Attribution (CC BY) license (https://creativecommons.org/licenses/by/4.0/).

Abstract: Textile waste cellulose nanofibrillated fibre has been reported with excellent strength reinforcement ability in other biopolymers. In this research cellulose nanofibrilated fibre (CNF) was isolated from the textile waste cotton fabrics with combined supercritical carbon dioxide and high-pressure homogenisation. The isolated CNF was used to enhance the polylactic acid/chitin (PLA/chitin) properties. The properties enhancement effect of the CNF was studied by characterising the PLA/chitin/CNF biocomposite for improved mechanical, thermal, and morphological properties. The tensile properties, impact strength, dynamic mechanical analysis, thermogravimetry analysis, scanning electron microscopy, and the PLA/chitin/CNF biocomposite wettability were studied. The result showed that the tensile strength, elongation, tensile modulus, and impact strength improved significantly with chitin and CNF compared with the neat PLA. Furthermore, the scanning electron microscopy SEM (Scanning Electron Microscopy) morphological images showed uniform distribution and dispersion of the three polymers in each other, which corroborate the improvement in mechanical properties. The biocomposite's water absorption increased more than the neat PLA, and the contact angle was reduced. The results of the ternary blend compared with PLA/chitin binary blend showed significant enhancement with CNF. This showed that the three polymers' combination resulted in a better material property than the binary blend.

Keywords: textile waste; cellulose nanofibre; green materials; biopolymers; environmental

1. Introduction

Textile waste and synthetic polymers are part of the causative agent of pollution in the environment. The use of synthetic polymer for several industrial applications has increased plastic waste pollution [1]. The use of synthetic polymers for packaging is enormous, and this has been reported to take a larger percentage of plastic waste in landfills and ocean pollution. Among the solution proposed is the development of alternative biodegradable materials for industrial applications. This challenge has spurred the ongoing researches on the development of biodegradable composites [2].

Biodegradable polymers, also called biopolymers, are sourced from both plant and animal origin [3]. The biodegradable properties of biopolymers are majorly due to their sources. Biopolymers are also called biodegradable polymers and are isolated from renewable sources. Biopolymers are used as a replacement for synthetic polymers because of their biodegradable properties. Biopolymers are thermally degradable to carbon and biodegradable by microorganisms to carbon dioxide and water [4]. The biodegradable properties of biopolymers remove the environmental pollution concern of synthetic polymers. The rate of degradation of biopolymers is dependent on the microbial action of the soil where it is disposed of [5]. However, previous research on biopolymers' properties showed that their mechanical properties are low for packaging applications [6–8].

Several researches have been done on biopolymers to enhance their mechanical properties [9–11]. Biopolymers such as polylactic acid (PLA), poly (butyl acrylate) (PBA), and chitin has been majorly reinforced for improved mechanical properties. Polylactic acid has been at the forefront of biopolymers used in biodegradable packaging [12]. PLA is studied because of its availability, biodegradability, and eco-friendly properties in the environment. There are two primary methods for producing polylactic acid; the chemical method (polymerisation of lactic acid) and the industrial method (fermentation) [12]. Polylactic acid has different forms such as Poly (l-lactic acid) (PLLA), Poly Lactic-co-Glycolic Acid (PLGA), etc., and this is dependent on the stereoisomers and comonomers. Furthermore, there are low and high molecular weights of polylactic acids. PLA has been reported to be brittle and of low mechanical properties like other biopolymers.

Textile Waste cellulose nanofibre has been frequently used to reinforce biopolymer for enhanced strength [3,13]. Cellulose is usually isolated from plant sources, and the demand for it is on the increase. Recently, a review of textile waste's possible use showed the possibility of cellulose isolation from cotton fabric waste [14,15]. Textile waste as a source of cellulosic materials is a significant breakthrough because it reduces pollution. However, to date, no comprehensive study has been done using textile waste as reinforcement or fillers in biopolymers. Textile waste accounts for the major percentage of municipal waste, which majorly is comprised of disposed of clothing. Globally, tons of fabric waste is generated and this forms part of the land and ocean pollutants [16].

Cellulose nanofibre has been in used as reinforcement in PLA for packaging applications. Several types of research have been done on the combination of PLA and CNF to produce biocomposite film. Previous studies on PLA/CNF biocomposite films [17–19]. However, the PLA/CNF blend has poor miscibility (agglomeration) due to the difference in nature. PLA is hydrophobic, while CNF is hydrophilic. The difference in their nature has resulted in poor miscibility and low mechanical properties. However, a material that has good miscibility (with PLA and CNF) can improve the miscibility between them.

Chitin has been blended individually with both PLA and CNF [19–21]. Previous studies on PLA/chitin and chitin/CNF showed good miscibility and dispersion of chitin in PLA and CNF. Chitin is generally hydrophobic, and it is normally expected to have good miscibility with PLA. However, due to the hydroxyl functional group's presence, it also has good miscibility with CNF with a similar functional group. The compatibilizer effect of chitin, due to its good miscibility with PLA and CNF, was used in this study. The properties of PLA/chitin/CNF have not been reported in the literature. The cellulose nanofibrilated fibre used in this study was prepared using combine hydrolysis and high-pressure homogenisation. In this study, a polylactic acid-chitin blend was reinforced with cellulose nanofibrillated fibre for enhanced mechanical properties. The effect of inclusion of chitin on the morphology of the composite was studied.

2. Materials and Methodology
2.1. Materials

Practical grade 3505D polylactic acid was purchased from Sigma Aldrich Pasir Panjang Rd, Queenstown, Singapore. The polylactic acid has a tensile and yield strength of 62 MPa and 65 MPa respectively. The specific gravity and melt extrusion temperature of the PLA grade is 1.24, and 55–60 °C respectively. The value of PLA molecular flow rate (MFR), g/10 min (210 °C, 2.16 kg) is 14. Chitosan from prawn was deacetylated until 90% and used in this study. The cellulose nanofibrillated fibre used in this study was prepared from textile fabric waste. The percentage composition of polylactic acid and chitin ratio was kept constant at 90% to 10% based on the work by Nasril et al. [21]. The percentage of textile waste cellulose nanofibrillated fibre varied between 1% and 5% by weight.

2.2. Textile Waste Cellulose Nanofibrillated Fibre Isolation and Characterisation

Cellulose was isolated from textile waste (100% cotton-based fabrics) with alkaline hydrolysis based on the modified method of Thambiraj [22]. The cotton fabrics were cut into small pieces of 2–3 cm with a milling saw and using mild alkaline hydrolysis (NaOH) was converted into cotton pulp fibre [23]. The fabrics were heated in an alkaline solution of 25 wt% concentration of sodium hydroxide (NaOH) and 0.2 wt% of anthraquinone (AQ) (all percentages based on the weight of the fibre) at 160 °C for 4 h [24]. The cotton pulp fibre was bleached using ozone (gas flow rate of 0.5 L/min at 30 °C for all experiments), a chlorine-free process to avoid the isolated cellulose toxicity [25]. The isolated cellulose was converted to cellulose nanofibrillated fibre using combined supercritical carbon dioxide explosion and high-pressure homogenisation. The bleached cellulose pulp was exposed to supercritical carbon dioxide extraction for 2 h at 500 bars and a temperature of 60 °C to obtain $SC-CO_2$ cellulose microfibrillated fibre. After that, a high-pressure homogenisation of 56 MPa and a 44-homogenisation cycle was used to obtain cellulose nanofibrillated fibre (CNF).

The cellulose nanofibrillated fibre obtained was characterised by particle size analysis, FT-IR analysis, and transmission electron microscope (TEM). The FT-IR analysis was used to confirm the isolation of CNF by studying the functional group present. The FT-IR analysis was conducted using FT-IR EFTEM Libra—Carl Zeiss, Selangor, Malaysia. The CNF was mixed with KBr, pressed into the film, and placed in the FT-IR machine for analysis. The nanosize distribution of the particles was studied with a particle size analyzer (Zetasizer Ver. 6.11, Malvern, UK) under dynamic laser light scattering. The cellulose nanofibrillated fibre was dispersed in water at 1 mg to 5 mL and observed under a laser light.

Furthermore, the nano-size of the obtained CNF was further confirmed with TEM. The freeze-dried samples were first dissolved in water, and a drop from the solution was dispersed in acetone on a copper grid. The CNF was observed at 100 nm and 40 kV potential with TEM Perkin-Elmer, PC1600, Winter Street Waltham, MA, USA. The XRD analysis of the isolated cellulose nanofibre powder was done using PANalytical X'Pert PRO X-ray Diffraction (Malvern Panalytical, Techlink, Singapore) at a diffraction angle range of $2\theta = 10°$ to $70°$, 1.540598 for K-alpha 1 and K-alpha 2 wavelength, 45 volts, and a tube current of 40 A.

2.3. Preparation of Textile Waste Cellulose Nanofibrillated Fibre Reinforced in PLA/Chitin Biocomposite

Polylactic acid, chitosan and cellulose nanofibrillated fibre were extruded to filament form using twin-screw Process 11 extruder, Thermo Scientific (Waltham, MA, USA) at 100/min and temperature profile range of 120 to 180 °C. The extruded filament was pelletised using Thermo Scientific Varicut Pelletizer 11M (Thermo Fisher Scientific, Waltham, MA, USA) and the pellets were hot-pressed using a Carver Press (model 3851-0) (Carver, Wabash, IN, USA) compression moulding machine at 170 °C. The compressed composites were cut to test samples and were stored in a zip lock bag. The percentage by weight between PLA and Chitin was kept constant at 90:10 (Table 1) from previous works [8,11,20,21]. Furthermore, the percentage of cellulose nanofibrillated fibre vary between 0 to 5% by weight of the composite. The percentage of CNF added to the composite was limited to 5% based on previous studies on PLA/CNF [19,26,27].

Table 1. Composition variation.

Sample Name	Polylactic Acid (Wt%)	Chitosan (Wt%)	CNF (Wt%)
PLA	100	0	0
PCC0	90	10	0
PCC1	90	10	1
PCC3	90	10	3
PCC5	90	10	5

2.4. Characterisation of Textile Waste Cellulose Nanofibrillated Fibre Reinforced in PLA/Chitin Biocomposite

The tensile properties of the neat PLA, PLA/chitin, and the PLA/chitin/CNF were measured with MT1175 (Dia-Stron Instruments, Andover, UK) Instron Universal Testing machine at a force of 50 KN and ASTM 638. The test samples were cut into a dumbbell shape of a standard dimension 165 mm × 19 mm × 3 mm, and the values of five (5) replicates of the samples were documented. The tensile strength, elongation, and tensile modulus were obtained from the test. Furthermore, the fractured surface of the tensile samples was observed to study the miscibility of the polymers.

The impact properties of the neat PLA, PLA/chitin, and the PLA/chitin/CNF biocomposite were tested with Ceast Resil 7181 Impactor (Corporate Consulting, Service and Instruments (CCSI), Akron, OH, USA). The samples were prepared based on D256 standard sizes for impact testing. The value of the impact was obtained in Joule per meter square for five (5) replicates of each sample.

The morphology of the tensile fractured surface of neat PLA, PLA/chitin, and the PLA/chitin/CNF biocomposite was observed with scanning electron microscopy. The scanning electron microscopy (SEM) samples were coated with gold for improved conductivity. The fractured surface images at a magnification of 100 μm were observed with scanning electron microscope EVO MA 10, Carl-ZEISS SMT, Oberkochen, Germany. The XRD was conducted with the powdered form of the neat PLA and PLA/chitin/CNF biocomposite. PANalytical X'Pert PRO X-ray Diffraction (Malvern Panalytical, Techlink, Kaki Bukit Rd, Bedok, Singapore) was used at a diffraction angle range of $2\theta = 10°$ to $50°$, 1.540598 for K-alpha 1 and K-alpha 2 wavelength, 45 volts, and a tube current of 40 A.

The FT-IR analysis was conducted using FT-IR EFTEM Libra—Carl Zeiss, Selangor, Malaysia. The CNF was mixed with KBr, pressed into the film, and placed in the FT-IR machine for analysis. The nanosize distribution of the particles was studied with a particle size analyzer (Zetasizer Ver. 6.11, Malvern, UK) under dynamic laser light scattering.

The DMA properties of the neat PLA, PLA/chitin, and PLA/chitin/CNF were conducted at ASTM D4065 standard. The dynamic mechanical analysis of the samples was analysed to evaluate the thermomechanical properties of the material. The test was done with DMA analyser PerkinElmer Dynamic Mechanical Analyzer (DMA 8000) (PerkinElmer Inc., Akron, OH, USA) and the properties of the neat PLA and the PLA/chitin/CNF biocomposites were obtained with changing temperature. The value of storage modulus, loss modulus, and loss factor were obtained with temperature change for each sample.

The thermal properties of neat PLA, PLA/chitin, and the PLA/chitin/CNF biocomposites were studied with thermogravimetry analysis (TGA) and derivative thermogravimetry analysis (DTA). PerkinElmer TG-IR-GCMS Interface Q500, TA Instruments (PerkinElmer Inc., Akron, OH, USA) was used at a temperature range of 40 °C to 800 °C and a 20 °C/min temperature increase. A mass range of 5 mg to 10 mg of the neat PLA and Biocomposites was used as the TGA samples. The rate of degradation (mass decrease) of the biocomposite with temperature was recorded and analysed.

The biocomposite wettability properties were studied with contact angle and water absorption measurements. The samples' surface contact angle with water was observed with KSV CAM 10 (KSV Instruments Ltd., Espoo, Finland) machine, and the contact angle measurement for five replicates of each sample was obtained. The average contact angle of the neat PLA and the biocomposites were documented. The water absorption test was conducted with ASTM D570 standard for the neat PLA and the biocomposite samples to measure the rate of water intake. The amount of water absorbed by each sample within a 24 h interval was measured using Equation (1)

$$\text{water absorbed } (\%) = \frac{W_2 - W_1}{W_1} \times 100 \tag{1}$$

where W_1 and W_2 are the initial and final weight of the samples before and after 24 h.

3. Results and Discussion
3.1. Properties of Textile Waste Cellulose Nanofibrillated Fibre

Cellulose nanofibre isolation was confirmed with the FTIR functional group analysis and transmission electron microscopy (TEM). Figure 1 showed the result of the FT-IR analysis of the isolated cellulose nanofibrillated fibre from cotton. The FT-IR graph showed the absorption spectra of a typical bond present in cellulose nanofibrillated fibre. The stretched peak between 3200 to 3500 represents the OH bond, the 2500–3000 cm^{-1} band represented the C–C bond, and 1500 to 2000 cm^{-1} the C–H. The peaks at 1050 cm^{-1} is the C–O stretching vibrations. The band between 400 to 800 cm^{-1} is often referred to as the amorphous band of the nanocellulose. These peaks are typical of the chemical bonds present in nanocellulose, as reported in the literature [28]. The analysis showed that using the chemical in the isolation process of CNF in this method does not affect the final product's chemical structure. The FTIR graph showed similarity to those obtained from other methods as reported. The –OH stretch band is observed to reduce absorbance as the raw fibre approach the complete isolation of CNF. There is no indication of a new band after the isolation process of CNF, which indicates that no new bond was formed with the hydrolysis chemicals to produce cellulose (i.e., digestion and bleaching).

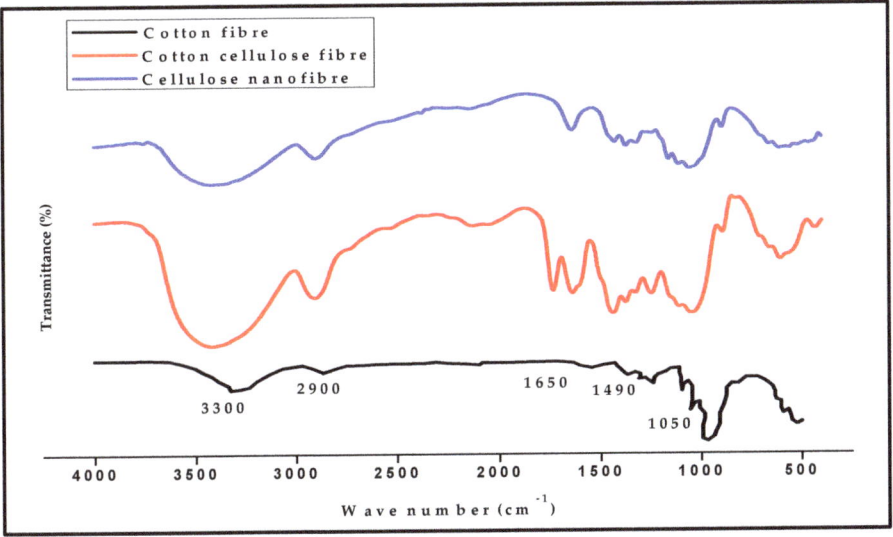

Figure 1. Fourier-transform infrared spectroscopy (FT-IR, Fourier-Transform Infrared Spectroscopy) analysis of Cellulose nanofibre extracted from textile waste fabrics.

Furthermore, the transmission electron microscopy and fibre size distribution of the isolated cotton cellulose nanofibre is shown in Figure 2a–c. The TEM images of cellulose nanofibre are presented in Figure 2a. The figure showed fibre diameter in the nanosize range between 10 nm and 30 nm, as measured using the TEM software fibre diameter measurement. Further confirmation of the fibre size diameter was done using Image J software. The nanofibre revealed internetworking with each other, as shown in the figure. The result of the particle size distribution of the isolated cellulose nanofibre is presented in Figure 2c. The particle sizes are measured on a percentage scale, and the intensity (%) of each particle size is plotted. The particle size analysis result showed a size range of 60 nm to 220 nm with a peak between 100 nm and 120 nm. The result showed a higher percentage of nanofibre between 100 nm and 120 nm, and this confirms the isolation of cellulose nanofibrillated fibre from textile waste [23,29,30].

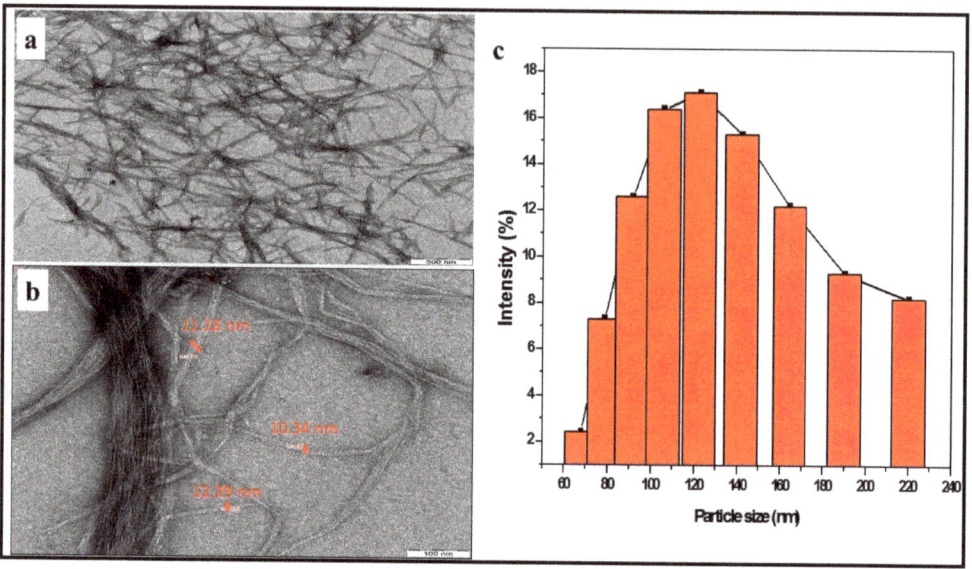

Figure 2. Isolated textile waste cellulose nanofibre transmission electron microscopy image (a,b) and percentage particle size analysis (c).

3.2. Characterisation of Textile Waste Cellulose nanofibre Reinforced in Polylactic Acid-Chitin Biocomposite

The result of the tensile and impact properties of the neat PLA, PLA/chitin, and PLA/chitin/CNF biocomposite is shown in Figure 3a. The tensile strength of the composite is seen to increase with the addition of cellulose nanofibrillated fibre. The highest tensile strength was obtained with PCC5, while the lowest was the neat PLA. The tensile strength values increased significantly from 47.5 MPa to 83.7 MPa. The increase in the tensile strength showed the effect of nanocellulose reinforcement on the tensile properties. A previous report explained that the increase in tensile strength is due to the addition of CNF with a larger surface area. The CNF increased the polymer mix's bonding, resulting in a well-arranged internal structure of the biocomposite [18].

The elongation (Figure 3a) of the biocomposite showed significant improvement compared to the neat PLA and PLA/chitin. This showed that the addition of CNF also affects the elongation of the biocomposite. However, the elongation is seen to reduce with more CNF though higher than the neat PLA. A similar result is noticed with the extension as they are directly proportional. The elongation trend observed has been reported in the literature with the PLA blend with CNF [17].

A similar trend was observed in the tensile modulus (Figure 3b) of the biocomposite. The tensile modulus increased from 4330.0 MPa to 7836.2 MPa. The tensile modulus value showed a significant effect of the CNF on the PLA/Chitin biocomposite, as observed in the plot. The value of the tensile properties obtained in this result seems to be far higher than those reported in the literature for PLA/chitin [8,21]. This result showed that the addition of CNF further enhanced PLA/chitin's modulus value compared to PCC1. This makes the ternary blend more suitable for packaging applications compared to PLA/chitin.

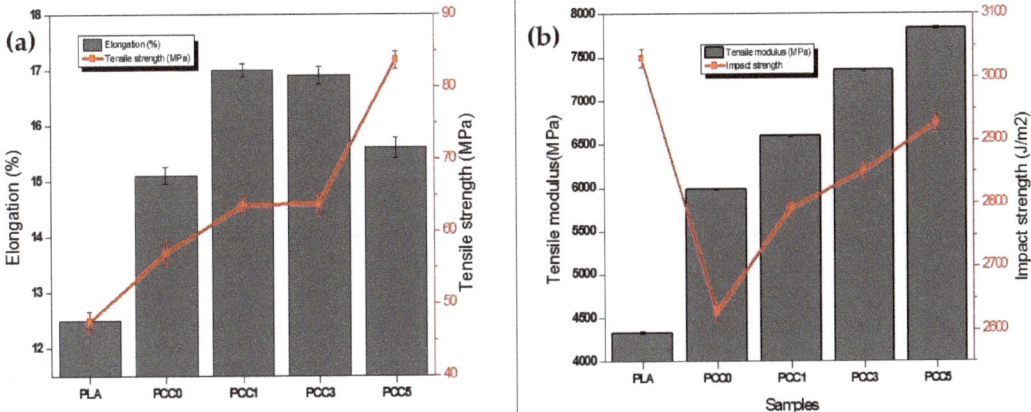

Figure 3. (a) Elongation and Tensile strength. (b) Tensile modulus and impact strength of neat PLA, PLA/chitin, and PLA/chitin/CNF biocomposite.

The impact test of the neat PLA, PLA/chitin, and the PLA/chitin/CNF biocomposite are shown in Figure 3b. The resilience energy per square metre result of the biocomposite is generally lower than the neat PLA. The addition of chitin resulted in a lower impact strength compared to the neat PLA. The impact energy of the biocomposite range from 2627.25 J/m^2 for PCC0 to 2927 J/m^2 for PCC5. The increase in resilience is likely due to the reinforcement effect of the cellulose nanofibre. The resilience of the material is seen to increase with the addition of CNF until PCC5. This shows that the material's ability to resist impact increased with the addition of CNF, which is needed for packaging applications. The result of the resilience obtained in this study is similar to those reported for PLA/chitin [8,19] and PLA/CNF [19,26]. Previous reports on nanofibre use as reinforcement stated that the fibre use's aspect ratio affects the resulting composite's strength. This differentiates the impact strength of composites from each other. Furthermore, the particle size, intermolecular adhesion, and intramolecular cohesion force also affect the polymer blend's impact strength [19]. Therefore, the increase in this study's impact strength can be attributed to one or more of these factors.

The result of the dynamic mechanical analysis (DMA) of the neat PLA, PLA/chitin, and PLA/chitin/CNF biocomposite is presented in Figure 4. Figure 4a showed the storage modulus result with a change in temperature for each biocomposite and neat PLA. The storage modulus is observed to increase significantly with the addition of CNF compared to neat PLA. A significant increase in the biocomposite storage modulus compared to the neat PLA indicates interfacial interaction between the matrix and the reinforcement. The storage modulus graph showed a distinct transition from the glassy region to the rubbery region [31]. The storage modulus reduction with temperature in all the samples was uniform with a single slope. The single slope in the rubbery region of the storage modulus graph is a characteristic of good miscibility [32]. The highest storage modulus value was obtained with 5% CNF composition, which showed enhancement of the inter or intra surface interaction with the addition of CNF. This means that the high storage modulus indicates less mobility of the polymer chain. More energy is required to stretch, which results in high tensile strength and modulus in the mechanical properties reported in Figure 3. Furthermore, a high storage modulus has a direct effect on the transition temperature of the material. This was observed in Figure 4c as a shift and widening in the tan delta peak temperature. The tan delta peak temperature value of the biocomposite increased with the addition of CNF. The shifted tan delta peak temperature is also termed the glass transition temperature of the material and indicated thermal properties enhancement of the biocomposite with the addition of CNF [8]. The loss modulus (Figure 4b) indicated the damping properties of the biocomposite. The effect of CNF addition on the biocomposite peak temperature is more obvious in the loss modulus peak. The loss modulus peak showed that the neat PLA peak was the lowest, while there was either a shift or widening peak of the biocomposite. The loss modulus peak is often referred to as the stiffness transition temperature, while the tan delta peak temperature is the glass transition temperature [33]. The value of the storage modulus is observed to be higher than the loss modulus, and this indicates that the biocomposite is mainly elastic.

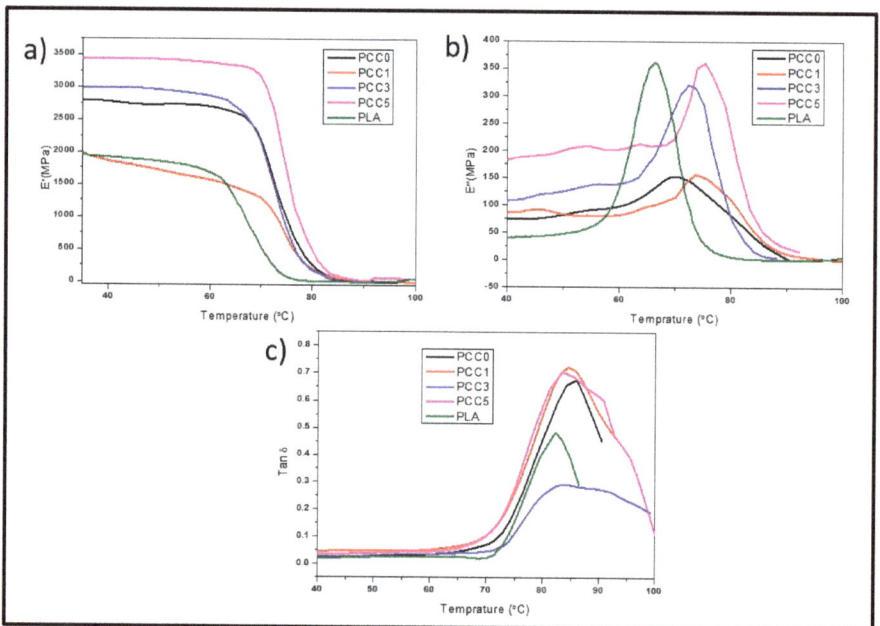

Figure 4. (a) Storage modulus (E'), (b) loss modulus (E"), (c) loss factor (tan δ) of dynamic mechanical analysis for neat PLA, PLA/CNF, and PLA/chitin/CNF biocomposite (PCC1, PCC2, PCC3).

The result of the tensile fractured surface morphology of the neat PLA, PLA/chitin (PCC0), and the PLA/chitin/CNF (PCC1, PCC3, and PCC5) biocomposite is shown in Figure 5. The morphological images showed good miscibility between biopolymers with no segregation. The images showed increased flakes in the biocomposite morphology with the addition of CNF. Additionally, the fractured surface of the biocomposite becomes rougher compared to the neat PLA. This observation means that the biopolymer blends are thoroughly compacted together with no void. This is probably responsible for the high tensile strength and modulus value observed in the mechanical analysis. The changes in the neat PLA's morphology when chitin and CNF were added can be significantly seen in Figure 5c to f. The SEM images with the addition of chitin only (PCC0) are characterised with wedges and flakes while the addition of CNF (PCC1, PCC3, PCC5) introduced a fibre network in the SEM images. The network of CNF fibres is shown to increase in the images from PCC1 to PCC5. Previous studies on PLA/CNF biocomposite reported agglomeration of CNF in PLA due to differences in nature [17,18], i.e., PLA is hydrophobic, and CNF is hydrophilic. The difference in the nature of PLA and CNF has resulted in poor mechanical properties, as reported in Reference [18].

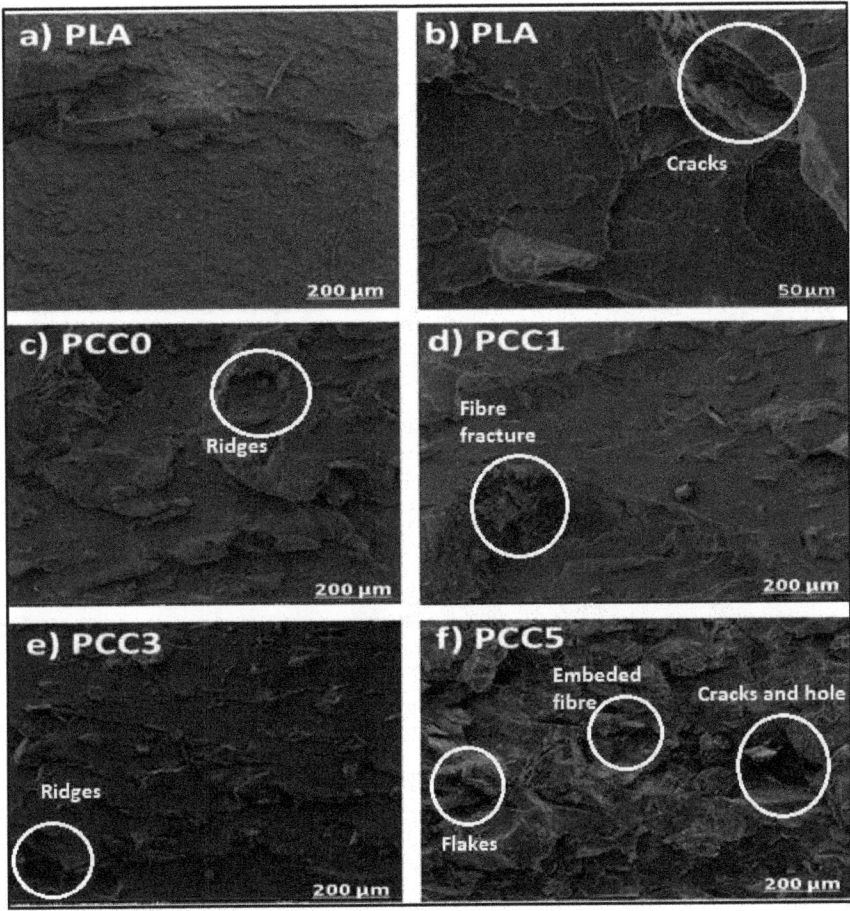

Figure 5. Fractured surface morphology of tensile sample for neat PLA (**a,b**), PLA/chitin (**c**), and PLA/chitin/CNF (**d–f**) biocomposite.

However, chitin has been reported to have good miscibility with PLA and CNF. Previous studies on PLA/Chitin and Chitin/CNF have shown good miscibility between these biopolymers [19,21]. In these studies, chitin addition is used as a compatibiliser to enhance the miscibility of the PLA and CNF biocomposite. This compatibiliser effect of chitin resulted in the uniform distribution of CNF in the composite. The miscibility between PLA and chitin is due to the similarity in their hydrophobic nature. In contrast, the miscibility between chitin and CNF is due to the presence of a hydroxyl group in its chemical structure [34]. The morphological images, as observed under scanning electron microscopy showed no void, agglomeration, and segregation. The morphological studies from the SEM corroborated the significant increase in mechanical properties reported in Figure 3. Additionally, the result of the impact strength can be justified from the morphological images. Miscibility and dispersion of the polymer blends can also influence the increase in impact strength [35,36]. The uniform distribution of the CNF in the polymer matrix enhanced the fibre network, which absorbed the impact energy from the impactor [11].

Figure 6 presents the result of the X-ray diffraction analysis of cellulose nanofibre extracted from textile waste fabrics, neat PLA, PLA/chitin, and PLA/chitin/CNF biocomposite. The crystalline properties of the cellulose nanofibre determine its strength reinforcement ability in the polymer matrix. Distinct peaks were found at two theta equals 15 and 23 degrees, respectively. From previous studies on cellulose nanofibre, similar peaks were obtained [37]. Cellulose nanofibre XRD pattern from the previous study has been reported as highly crystalline with a low amorphous part [38]. Cellulose nanofibre from soft and hardwood has been reported with a 40% to 80% crystalline part [38]. Similar crystallite properties were obtained with the CNF obtained in this study. The crystallinity index was 66.5% using Atiqah et al. [24] method. The crystal peak of the isolated CNF was observed between two theta 10 to 30 degrees, and no significant peak was seen between two theta 30 to 70 degrees, respectively. The XRD analysis confirms cellulose nanofibre's successful isolation, as established by the FT-IR analysis and previous reports [38].

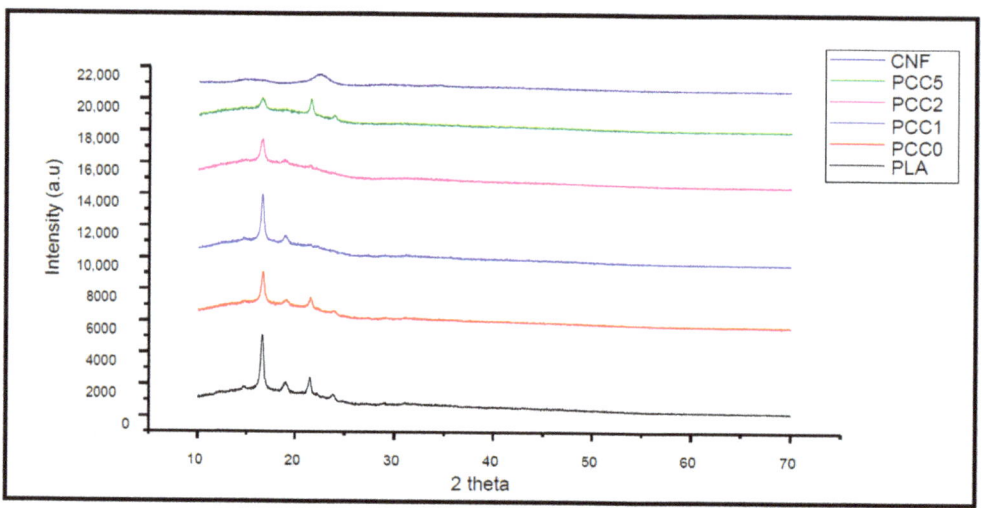

Figure 6. X-ray diffraction analysis of cellulose nanofibre (CNF) extracted from textile waste fabrics, neat PLA, PCC0, PCC1, PCC3, and PCC5.

The X-ray diffraction study of PLA and biocomposites showed a typical characteristic of an amorphous polymeric blend with few crystalline peaks between 15° and 23°. The formation of crystallite between material blends may be due to the PLA's semi-crystalline properties since the neat PLA showed similar peaks [39,40]. However, the XRD result showed lowered peaks with the addition of CNF, which showed that PLA became less crystalline and this resulted in less brittleness of the PLA [41]. The lowered peaks are probably possible because CNF has both crystalline and amorphous parts. The addition of CNF showed that the XRD peaks were further lowered, mainly due to fibre–matrix interaction. With the addition of chitin only (PCC0), the neat PLA peaks seem not to be significantly affected. This is probably reasonable considering the semi-crystalline of chitin. The brittle nature of PLA is one of its limitations and from a previous study, it has been reported that the addition of chitosan often has no lowered effect on the brittle nature of PLA [21]. However, the result of the XRD showed that biocomposite is more amorphous than crystalline [21].

The result of the FT-IR analysis of the neat PLA and the biocomposite (PCC0, PCC1, PCC3, PCC5) is shown in Figure 7a. The FT-IR plot of the neat PLA, PLA/chitin, and biocomposites showed wave numbers between 450 to 4000 cm^{-1}. Generally, the FT-IR of the graph showed a major stretch band between 3400–3600 cm^{-1}, and small peaks at 2350 cm^{-1}, 1650–1700 cm^{-1}, and 1150 cm^{-1} wave number. The neat PLA spectra between 3450–3550 cm^{-1}, 1700 cm^{-1}, and 1100 cm^{-1} are typical of the lactic terminal, ester group, and vibration of the ester unit, respectively [21,42].

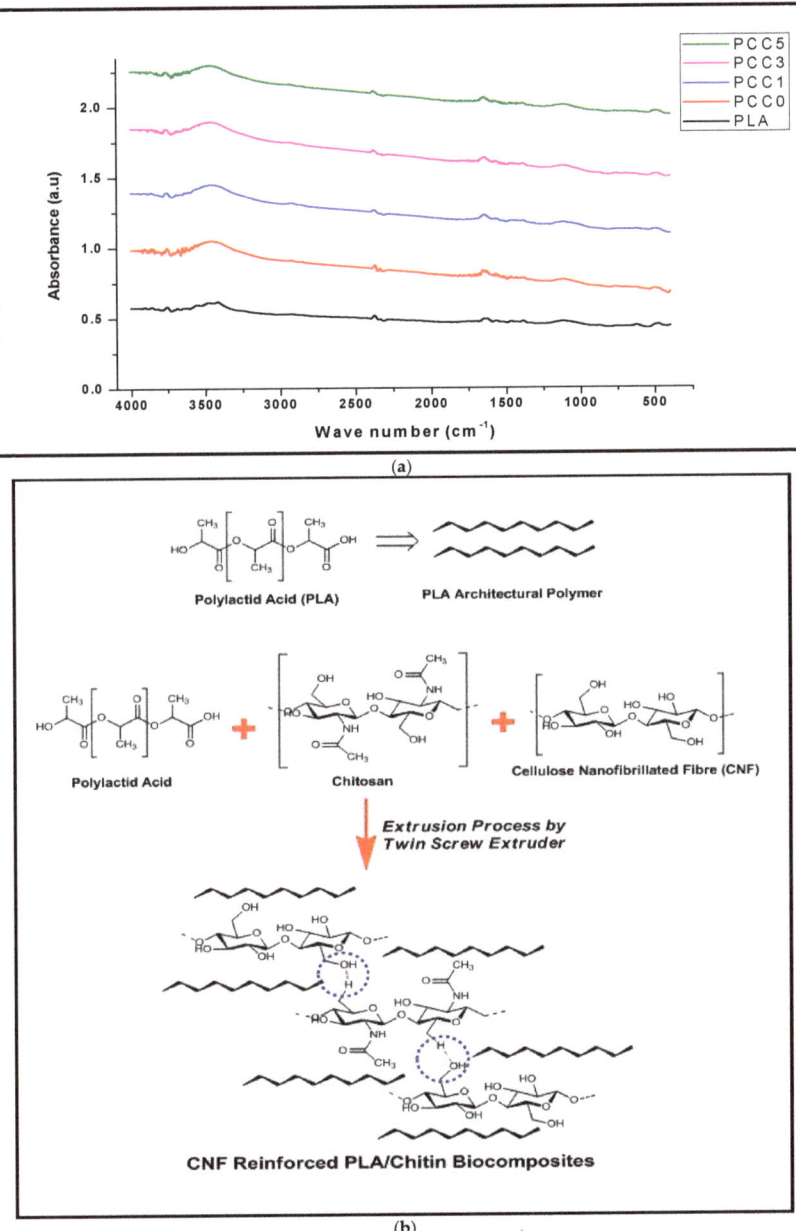

Figure 7. (a) FT-IR analysis of neat PLA, PLA/chitin (PCC0), and PLA/chitin/CNF (PCC1, PCC3, and PCC5). (b) Chemical reaction and bond formation between the PLA/chitin/CNF biocomposite.

The PLA/chitin (PCC0) spectra band between 3400–3550 cm^{-1} represents the lactic monomer's combined effect from PLA and the hydroxyl group in chitin. Additionally, the 1650 cm^{-1} in the PCC0 showed C=O and –NH$_2$ in PLA and chitin, respectively [21,40]. This is in regards to PLA/chitin/CNF spectra bands. The peak between 3400–3600 cm^{-1} is typical of the presence of –OH hydrogen bonding, 2350 cm^{-1} typical of –C–N, and

1650 cm^{-1} typical of –C=C– bond. The alcohol –OH stretch is typical of hydrogen bonds, resulting from chitin and CNF as both polymers, have the presence of OH in their chemical structure. C–N– bonding in the FT-IR graph is due to chitin's addition to the blend, while the C=C– is a typically in the presence of the formation of alkene bond between the polymers. The small wavenumber bands below 1000 cm^{-1} showed C–H– bond, common to the three polymeric materials. The presence of –C–N, –OH, and –C=C– is a clear indication of chemical interaction between the three polymeric materials [11,21]. The possibility of bond formation between the three polymeric materials is shown in Figure 7a.

This chemical bonding formation is similar to the one reported in the literature [43]. The FT-IR result showed that enhancement of the biocomposite's mechanical properties is due to the composite's physical interaction; the particle sizes of the CNF or the formation of a bond between the three (3) biopolymers. Furthermore, a previous report has shown possible interfacial bonding between PLA and CNF [44,45]. The mechanical properties' improvement indicates interfacial or intermolecular bonding between PLA, chitin, and CNF.

The thermal properties, as analysed with thermogravimetry analysis (TGA) and derivative thermogravimetry analysis (DTA), are shown in Figure 8 for neat PLA, PLA/chitin, and PLA/chitin/CNF biocomposite. The TGA result showed single degradation starting at 276 °C, 280 °C, 284 °C, 290 °C, 298 °C for PCC5, PCC3, PCC1, PCC0, and PLA, respectively.

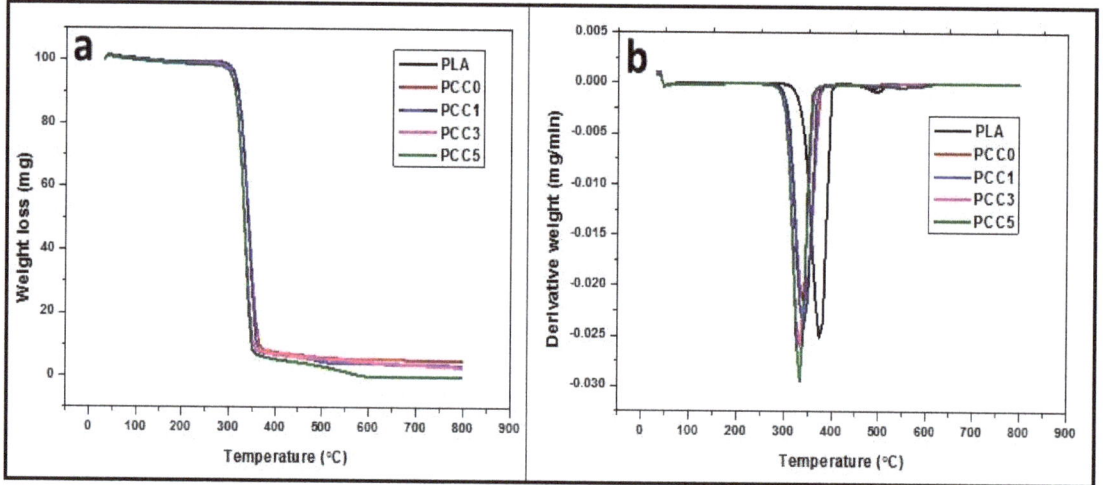

Figure 8. Thermal properties. (a) Thermogravimetry analysis (TGA). (b) Derivative thermogravimetry analysis (DTA) of neat PLA, PLA/chitin (PCC0), and PLA/chitin/CNF (PCC1, PCC3, and PCC5).

The TGA onset degradation temperature was observed to reduce with the addition of CNF, while the percentage weight loss of the composite is seen to increase with the addition of CNF. The percentage drop in the neat PLA's weight, PCC0, PCC1, PCC3, and PCC5 from the thermogravimetry analysis is 91%, 92%, 94%, 96%, and 99%, respectively. The result showed that the percentage decomposition of PLA increases with the addition of CNF, and less residue is found with samples that have a higher percentage of CNF. The TGA graph showed that PCC5 has the highest percentage decomposition and the neat PLA has the lowest. It can be inferred from the result that the addition of CNF to PLA and PLA/chitin improved the biodegradable properties of the biocomposite significantly. The PCC5 has about 8% less residue compared to the neat PLA. As reported in previous literature, chitin's addition to PLA improved the biodegradability [11,21]. This result showed that the biodegradability of PLA is further improved with the addition of CNF [19]. The degradation properties of the PLA/chitin/CNF biocomposite are a combined effect of chitin and CNF. This result is similar to previous literature on PLA/chitin and PLA/CNF [8,10,17,18].

The DTG result (Figure 8b) also corroborates the finding from the TGA result. The DTG showed that biocomposite thermal stability is lowered with the addition of CNF. This is observed with the peak value of the DTG result. The peak value on the temperature axis of the PCC5 is observed to be less than the neat PLA (Figure 8b). The DTA curve endothermic peak showed values of samples PLA, PCC0, PCC1, PCC3, and PCC5 as 375 °C, 351 °C, 348 °C, 345 °C, and 343 °C, respectively. The peak showed a drop in the biocomposite's thermal stability with the addition of chitin and CNF [19]. This explains the reason for the increase in the percentage of decomposition. This result showed that the biocomposite becomes less thermally stable at a temperature above the DTG peaks [21]. The peak of the DTG peaks is significant for industrial application, and it determines its thermal limit of usage. The result of the biocomposite's DTG peak corroborates the observation from the TGA graph, and similar trends have been reported in previous works from the literature [20,21].

The result of the water absorption and contact angle analysis of the neat PLA and the PLA/chitin/CNF biocomposites is presented in Figure 9 (line graph). The percentage water absorption of the neat PLA and the biocomposite (PCC0, PCC1, PCC3, and PCC5) increased significantly from 2% (neat PLA) to 12% (PCC5) with the addition of CNF and chitin. The increase in the biocomposite's water absorption is more significant with the addition of CNF than chitin. The increase in the percentage of the water absorbed can be explained by the nature of CNF being hydrophilic [18]. The addition of CNF increases the ability of neat PLA to form hydrogen bonding with water. This resulted in more water absorption of the biocomposite with the addition of CNF. The effect of the hydrophilic nature of CNF increased the amount of water intake of the neat PLA [11]. Though the water absorption ability of the neat PLA increased, the biocomposite is still highly hydrophobic, based on the value of the water absorbed. The result obtained conforms with previous studies on the PLA biocomposite with chitin and CNF [11,19,21].

Furthermore, the result of the contact angle analysis shown in Figure 9 (histogram) showed that the values are less than 90°. The contact angle values ranged from 83.8° for neat PLA to 66.3° for PCC5, respectively. The contact angle values showed an increase in the biocomposite surface's wettability with the addition of chitin and CNF. The wettability increase is more significant with the addition of CNF than the addition of chitin, probably because of the hydrophilic nature of CNF [17]. The result of the contact angle also corroborated the result obtained from the water absorption test. Generally, the biocomposite is still hydrophobic from the contact angle [46]. However, there is an increase in wettability. Wettability is an essential property of polymeric materials intended for packaging application. Biocomposite's hydrophobic nature is needed for the water-repelling properties of the packaged product [12]. The result of the wettability properties of the biocomposite showed that the water repellence of PLA is still retained despite the addition of CNF for mechanical strength enhancement.

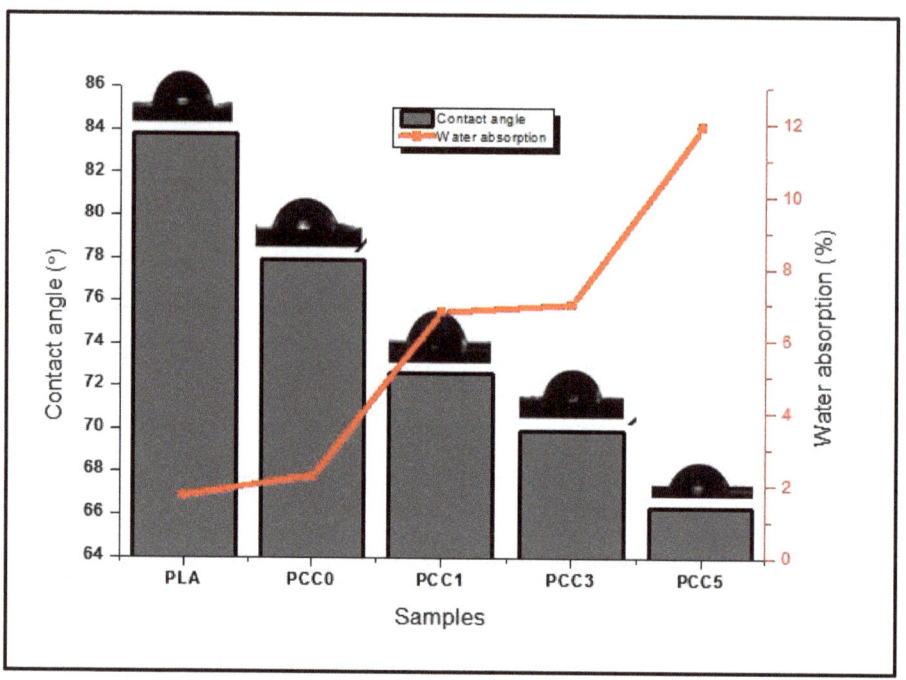

Figure 9. Water absorption and contact angle result of neat PLA, PLA/chitin (PCC0), and PLA/chitin/CNF (PCC1, PCC3, and PCC5).

4. Conclusions

The isolation of CNF was successfully done using combined supercritical carbon dioxide and high-pressure homogenisation. The CNF was used as reinforcement in PLA/chitin/CNF biocomposite, which was successfully prepared with combined melt extrusion and compression moulding techniques. The thermomechanical properties of the enhanced PLA/chitin/CNF biocomposite were analysed. The effect of the CNF on the obtained composite properties was studied and discussed in this study. The result showed improved mechanical, thermal, and wettability properties. The significant improvement in the properties was observed due to the combined effect of the chitin and CNF on the neat PLA. The PLA/chitin/CNF biocomposite result showed its potential use in packaging applications.

Author Contributions: Conceptualization, O.N.G., and A.K.H.P.S.; Data curation, F.G.O., C.K.A. and F.A.S.; Formal analysis, F.G.O. and O.N.G.; Funding acquisition, I.; Investigation, E.B.Y.; Project administration, N.I.S. and A.K.H.P.S.; Resources, S.R., N.I.S., E.B.Y., F.A.S. and I.; Supervision, A.K.H.P.S., M.K.M.H.; Writing—original draft, O.N.G.; Writing—review & editing, F.G.O., C.K.A. and O.N.G. All authors have read and agreed to the published version of the manuscript.

Funding: This research was funded by the Ministry of Education grant number FRGS-MRSA 203.PTEKIND.6711837. And The APC was funded by Ministry of Education, FRGS-MRSA 203.PTEKIND.6711837.

Institutional Review Board Statement: Not applicable.

Informed Consent Statement: Not Applicable.

Acknowledgments: The authors would like to acknowledge and express their gratitude towards the collaboration between Universitas Syiah Kuala, Banda Aceh, Indonesia, and Universiti Sains Malaysia, Penang, Malaysia.

Conflicts of Interest: The authors declare no conflict of interest.

References

1. Wei, R.; Zimmermann, W. Microbial enzymes for the recycling of recalcitrant petroleum-based plastics: How far are we? *Microb. Biotechnol.* **2017**, *10*, 1308–1322. [CrossRef] [PubMed]
2. Ahmed, J.; Varshney, S.K. Polylactides—Chemistry, properties and green packaging technology: A review. *Int. J. Food Prop.* **2011**, *14*, 37–58. [CrossRef]
3. Abdul Khalil, H.P.S.; Adnan, A.; Yahya, E.B.; Olaiya, N.; Safrida, S.; Hossain, M.; Balakrishnan, V.; Gopakumar, D.A.; Abdullah, C.; Oyekanmi, A. A Review on plant cellulose nanofibre-based aerogels for biomedical applications. *Polymers* **2020**, *12*, 1759. [CrossRef] [PubMed]
4. Alshabanat, M. Morphological, thermal, and biodegradation properties of LLDPE/treated date palm waste composite buried in a soil environment. *J. Saudi Chem. Soc.* **2019**, *23*, 355–364. [CrossRef]
5. Siracusa, V. Microbial degradation of synthetic biopolymers waste. *Polymers* **2019**, *11*, 1066. [CrossRef]
6. Sadasivuni, K.K.; Saha, P.; Adhikari, J.; Deshmukh, K.; Ahamed, M.B.; Cabibihan, J.J. Recent advances in mechanical properties of biopolymer composites: A review. *Polym. Compos.* **2020**, *41*, 32–59. [CrossRef]
7. Hermawan, D.; Lai, T.K.; Jafarzadeh, S.; Gopakumar, D.A.; Hasan, M.; Owolabi, F.T.; Aprilia, N.S.; Rizal, S.; Abdul Khalil, H.P.S. Development of seaweed-based bamboo microcrystalline cellulose films intended for sustainable food packaging applications. *BioResources* **2019**, *14*, 3389–3410.
8. Olaiya, N.; Surya, I.; Oke, P.; Rizal, S.; Sadiku, E.; Ray, S.S.; Farayibi, P.; Hossain, M.S.; Abdul Khalil, H.P.S. Properties and characterization of a PLA–chitin–starch biodegradable polymer composite. *Polymers* **2019**, *11*, 1656. [CrossRef]
9. Rizal, S.; Abdullah, C.; Olaiya, N.; Sri Aprilia, N.; Zein, I.; Surya, I.; Abdul Khalil, H.P.S. Preparation of Palm Oil Ash Nanoparticles: Taguchi Optimization Method by Particle Size Distribution and Morphological Studies. *Appl. Sci.* **2020**, *10*, 985. [CrossRef]
10. Olaiya, N.G.; Nuryawan, A.; Oke, P.K.; Abdul Khalil, H.P.S.; Rizal, S.; Mogaji, P.; Sadiku, E.; Suprakas, S.; Farayibi, P.K.; Ojijo, V. The role of two-step blending in the properties of starch/chitin/polylactic acid biodegradable composites for biomedical applications. *Polymers* **2020**, *12*, 592. [CrossRef]
11. Surya, I.; Olaiya, N.; Rizal, S.; Zein, I.; Sri Aprilia, N.; Hasan, M.; Yahya, E.B.; Sadasivuni, K.; Abdul Khalil, H.P.S. Plasticizer Enhancement on the Miscibility and Thermomechanical Properties of Polylactic Acid-Chitin-Starch Composites. *Polymers* **2020**, *12*, 115. [CrossRef] [PubMed]
12. Siakeng, R.; Jawaid, M.; Ariffin, H.; Sapuan, S.; Asim, M.; Saba, N. Natural fiber reinforced polylactic acid composites: A review. *Polym. Compos.* **2019**, *40*, 446–463. [CrossRef]
13. Sharma, A.; Thakur, M.; Bhattacharya, M.; Mandal, T.; Goswami, S. Commercial application of cellulose nano-composites—A review. *Biotechnol. Rep.* **2019**, *21*, e00316. [CrossRef] [PubMed]
14. Chuayjuljit, S.; Su-uthai, S.; Charuchinda, S. Poly (vinyl chloride) film filled with microcrystalline cellulose prepared from cotton fabric waste: Properties and biodegradability study. *Waste Manag. Res.* **2010**, *28*, 109–117. [CrossRef] [PubMed]
15. Wang, Z.; Yao, Z.; Zhou, J.; Zhang, Y. Reuse of waste cotton cloth for the extraction of cellulose nanocrystals. *Carbohydr. Polym.* **2017**, *157*, 945–952. [CrossRef] [PubMed]
16. Ma, Y.; Rosson, L.; Wang, X.; Byrne, N. Upcycling of waste textiles into regenerated cellulose fibres: Impact of pretreatments. *J. Text. Inst.* **2020**, *111*, 630–638. [CrossRef]
17. Clarkson, C.M.; El Awad Azrak, S.M.; Chowdhury, R.; Shuvo, S.N.; Snyder, J.; Schueneman, G.; Ortalan, V.; Youngblood, J.P. Melt spinning of cellulose nanofibril/polylactic acid (CNF/PLA) composite fibers for high stiffness. *ACS Appl. Polym. Mater.* **2018**, *1*, 160–168. [CrossRef]
18. Yang, Z.; Li, X.; Si, J.; Cui, Z.; Peng, K. Morphological, mechanical and thermal properties of poly (lactic acid)(PLA)/cellulose nanofibrils (CNF) composites nanofiber for tissue engineering. *J. Wuhan Univ. Technol. Mater. Sci. Ed.* **2019**, *34*, 207–215. [CrossRef]
19. Li, J.; Li, J.; Feng, D.; Zhao, J.; Sun, J.; Li, D. Comparative Study on Properties of Polylactic Acid Nanocomposites with Cellulose and Chitin Nanofibers Extracted from Different Raw Materials. *J. Nanomater.* **2017**, *2017*, 7193263. [CrossRef]
20. Hassan, M.M.; Koyama, K. Thermomechanical and viscoelastic properties of green composites of PLA using chitin micro-particles as fillers. *J. Polym. Res.* **2020**, *27*, 27. [CrossRef]
21. Nasrin, R.; Biswas, S.; Rashid, T.U.; Afrin, S.; Jahan, R.A.; Haque, P.; Rahman, M.M. Preparation of Chitin-PLA laminated composite for implantable application. *Bioact. Mater.* **2017**, *2*, 199–207. [CrossRef] [PubMed]
22. Thambiraj, S.; Shankaran, D.R. Preparation and physicochemical characterization of cellulose nanocrystals from industrial waste cotton. *Appl. Surf. Sci.* **2017**, *412*, 405–416. [CrossRef]
23. Wei, D.W.; Wei, H.; Gauthier, A.C.; Song, J.; Jin, Y.; Xiao, H. Superhydrophobic modification of cellulose and cotton textiles: Methodologies and applications. *J. Bioresour. Bioprod.* **2020**, *5*, 1–15. [CrossRef]
24. Atiqah, M.; Gopakumar, D.A.; FAT, O.; Pottathara, Y.B.; Rizal, S.; Aprilia, N.S.; Hermawan, D.; Paridah, M.; Thomas, S.; Abdul Khalil, H.P.S. Extraction of Cellulose Nanofibers via Eco-friendly Supercritical Carbon Dioxide Treatment Followed by Mild Acid Hydrolysis and the Fabrication of Cellulose Nanopapers. *Polymers* **2019**, *11*, 1813. [CrossRef]
25. Prabaharan, M.; Rao, J.V. Study on ozone bleaching of cotton fabric–process optimisation, dyeing and finishing properties. *Coloration Technol.* **2001**, *117*, 98–103. [CrossRef]

26. Ghasemi, S.; Behrooz, R.; Ghasemi, I.; Yassar, R.S.; Long, F. Development of nanocellulose-reinforced PLA nanocomposite by using maleated PLA (PLA-g-MA). *J. Thermoplast. Compos. Mater.* **2018**, *31*, 1090–1101. [CrossRef]
27. Safdari, F.; Bagheriasl, D.; Carreau, P.J.; Heuzey, M.C.; Kamal, M.R. Rheological, mechanical, and thermal properties of polylactide/cellulose nanofiber biocomposites. *Polym. Compos.* **2018**, *39*, 1752–1762. [CrossRef]
28. Luo, X.; Wang, X. Preparation and characterization of nanocellulose fibers from NaOH/Urea pretreatment of oil palm fibers. *BioResources* **2017**, *12*, 5826–5837. [CrossRef]
29. Du, H.; Liu, C.; Zhang, Y.; Yu, G.; Si, C.; Li, B. Sustainable preparation and characterization of thermally stable and functional cellulose nanocrystals and nanofibrils via formic acid hydrolysis. *J. Bioresour. Bioprod.* **2017**, *2*, 10–15.
30. Wang, Q.; Yuan, T.; Liu, S.; Yang, G.; Li, W.; Yang, R. Enzymatic activation of dissolving pulp with cationic polyacrylamide to enhance cellulase adsorption. *J. Bioresour. Bioprod.* **2017**, *2*, 16–19.
31. García-Campo, M.J.; Boronat, T.; Quiles-Carrillo, L.; Balart, R.; Montanes, N. Manufacturing and characterization of toughened poly (lactic acid)(PLA) formulations by ternary blends with biopolyesters. *Polymers* **2018**, *10*, 3. [CrossRef] [PubMed]
32. George, K.; Komalan, C.; Kumar, P.; Varughese, K.; Thomas, S. Dynamic Mechanical Analysis of Binary and Ternary Polymer Blends Based on Nylon Copolymer/EPDM Rubber and EPM Grafted Maleic Anhydride Compatibilize. *eXPRESS Polym. Lett.* **2007**, *1*, 641–653.
33. Goertzen, W.K.; Kessler, M. Dynamic mechanical analysis of carbon/epoxy composites for structural pipeline repair. *Compos. Part B Eng.* **2007**, *38*, 1–9. [CrossRef]
34. Wijesena, R.N.; Tissera, N.D.; Abeyratne, C.; Bangamuwa, O.M.; Ludowyke, N.; Dahanayake, D.; Gunasekara, S.; de Silva, N.; de Silva, R.M.; de Silva, K.N. In-situ formation of supramolecular aggregates between chitin nanofibers and silver nanoparticles. *Carbohydr. Polym.* **2017**, *173*, 295–304. [CrossRef] [PubMed]
35. Stark, N.M.; Rowlands, R.E. Effects of wood fiber characteristics on mechanical properties of wood/polypropylene composites. *Wood Fiber Sci.* **2003**, *35*, 167–174.
36. Rizal, S.; Gopakumar, D.A.; Thalib, S.; Huzni, S.; Abdul Khalil, H.P.S. Interfacial compatibility evaluation on the fiber treatment in the Typha fiber reinforced epoxy composites and their effect on the chemical and mechanical properties. *Polymers* **2018**, *10*, 1316. [CrossRef]
37. Jin, E.; Guo, J.; Yang, F.; Zhu, Y.; Song, J.; Jin, Y.; Rojas, O.J. On the polymorphic and morphological changes of cellulose nanocrystals (CNC-I) upon mercerization and conversion to CNC-II. *Carbohydr. Polym.* **2016**, *143*, 327–335. [CrossRef]
38. Martelli-Tosi, M.; Torricillas, M.d.S.; Martins, M.A.; Assis, O.B.G.d.; Tapia-Blácido, D.R. Using commercial enzymes to produce cellulose nanofibers from soybean straw. *J. Nanomater.* **2016**, *2016*, 1–9. [CrossRef]
39. Liao, Y.; Liu, C.; Coppola, B.; Barra, G.; Di Maio, L.; Incarnato, L.; Lafdi, K. Effect of porosity and crystallinity on 3D printed PLA properties. *Polymers* **2019**, *11*, 1487. [CrossRef]
40. Mihai, M.; Huneault, M.A.; Favis, B.D.; Li, H. Extrusion foaming of semi-crystalline PLA and PLA/thermoplastic starch blends. *Macromol. Biosci.* **2007**, *7*, 907–920. [CrossRef]
41. Kulachenko, A.; Denoyelle, T.; Galland, S.; Lindström, S.B. Elastic properties of cellulose nanopaper. *Cellulose* **2012**, *19*, 793–807. [CrossRef]
42. Tham, C.; Hamid, Z.A.A.; Ahmad, Z.; Ismail, H. *Surface engineered Poly (Lactic Acid)(PLA) Microspheres by Chemical Treatment for Drug Delivery System*; Trans Tech Publ.: Zurich, Switzerland, 2014; Volume 594.
43. Mokhena, T.; Sefadi, J.; Sadiku, E.; John, M.; Mochane, M.; Mtibe, A. Thermoplastic processing of PLA/cellulose nanomaterials composites. *Polymers* **2018**, *10*, 1363. [CrossRef] [PubMed]
44. Gazzotti, S.; Rampazzo, R.; Hakkarainen, M.; Bussini, D.; Ortenzi, M.A.; Farina, H.; Lesma, G.; Silvani, A. Cellulose nanofibrils as reinforcing agents for PLA-based nanocomposites: An in situ approach. *Compos. Sci. Technol.* **2019**, *171*, 94–102. [CrossRef]
45. Trifol, J.; Plackett, D.; Sillard, C.; Hassager, O.; Daugaard, A.E.; Bras, J.; Szabo, P. A comparison of partially acetylated nanocellulose, nanocrystalline cellulose, and nanoclay as fillers for high-performance polylactide nanocomposites. *J. Appl. Polym. Sci.* **2016**, *133*. [CrossRef]
46. Pang, X.; Zhuang, X.; Tang, Z.; Chen, X. Polylactic acid (PLA): Research, development and industrialization. *Biotechnol. J.* **2010**, *5*, 1125–1136. [CrossRef] [PubMed]

Article

Removal of Cadmium and Chromium by Mixture of Silver Nanoparticles and Nano-Fibrillated Cellulose Isolated from Waste Peels of Citrus Sinensis

Neha Tavker [1], Virendra Kumar Yadav [2], Krishna Kumar Yadav [3], Marina MS Cabral-Pinto [4,*], Javed Alam [5,*], Arun Kumar Shukla [5], Fekri Abdulraqeb Ahmed Ali [6] and Mansour Alhoshan [5,6]

1   School of Nano Sciences, Central University of Gujarat, Gandhinagar 382030, India; tavker.gini@gmail.com
2   School of Lifesciences, Jaipur National University, Jaipur 302017, India; yadava94@gmail.com
3   Institute of Environment and Development Studies, Bundelkhand University, Kanpur Road, Jhansi 284128, India; envirokrishna@gmail.com
4   Geobiotec Research Centre, Department of Geosciences, University of Aveiro, 3810-193 Aveiro, Portugal
5   King Abdullah Institute for Nanotechnology, King Saud University, P.O. Box-2455, Riyadh 11451, Saudi Arabia; ashukla@ksu.edu.sa (A.K.S.); mhoshan@ksu.edu.sa (M.A.)
6   Chemical Engineering Department, College of Engineering, King Saud University, P.O. Box-2455, Riyadh 11451, Saudi Arabia; falhulidy@ksu.edu.sa
*   Correspondence: marinacp@ua.pt (M.M.C.-P.); javaalam@ksu.edu.sa (J.A.)

Citation: Tavker, N.; Yadav, V.K.; Yadav, K.K.; Cabral-Pinto, M.M.; Alam, J.; Shukla, A.K.; Ali, F.A.A.; Alhoshan, M. Removal of Cadmium and Chromium by Mixture of Silver Nanoparticles and Nano-Fibrillated Cellulose Isolated from Waste Peels of Citrus Sinensis. *Polymers* **2021**, *13*, 234. https://doi.org/10.3390/polym13020234

Received: 13 December 2020
Accepted: 3 January 2021
Published: 12 January 2021

Publisher's Note: MDPI stays neutral with regard to jurisdictional claims in published maps and institutional affiliations.

Copyright: © 2021 by the authors. Licensee MDPI, Basel, Switzerland. This article is an open access article distributed under the terms and conditions of the Creative Commons Attribution (CC BY) license (https://creativecommons.org/licenses/by/4.0/).

Abstract: Nano-fibrillated cellulose (NFC) was extracted by a chemical method involving alkali and acid hydrolysis. The characterisation of the citrus sinensis fruit peel bran and nano-fibrillated cellulose was performed by XRD, FTIR, TEM, and FESEM. XRD confirmed the phase of NFC which showed monoclinic crystal with spherical to rod shape morphology with a size of 44–50 nm. The crystallinity index of treated NFC increased from 39% to 75%. FTIR showed the removal of lignin and hemicellulose from waste peels due to the alkaline treatment. Silver nanoparticles were also synthesised by utilizing extract of citrus sinensis skins as a reducing agent. Pharmaceutical effluent samples from an industrial area were tested by Atomic Absorption Spectrometry. Out of the four metals obtained, cadmium and chromium were remediated by silver nanoparticles with nano-fibrillated cellulose via simulated method in 100 mg/L metal-salt concentrations over a time period of 160 min. The highest removal efficiency was found for cadmium, i.e., 83%, by using silver and NFC together as adsorbents. The second highest was for chromium, i.e., 47%, but by using only NFC. The Langmuir and Freundlich isotherms were well fitted for the sorption of Cd (II) and Cr (II) with suitable high R^2 values during kinetic simulation. Thus, the isolation of NFC and synthesis of silver nanoparticles proved efficient for heavy metal sorption by the reuse of waste skins.

Keywords: citrus sinensis; nano-fibrillated cellulose; silver nanoparticles; acid hydrolysis; heavy metal sorption

1. Introduction

The fruit of the citrus species Citrus × sinensis is sweet orange. Orange peel is a waste by-product from fruit juice factories across the world. All the plant matter constitutes about 30% cellulose; cotton and wood being the highest, i.e., 90% and 50%, respectively [1]. The polysaccharide available in abundance having the repetitive unit $(C_6H_{10}O_5)_n$ and comprising of D-glucose is cellulose. This cellulose, being a fibrillary constituent of a plant cell, can be extracted from numerous natural sources such as wood and a few lignocellulosic fibres. There are chains that exist in cellulose which are stacked in an ordered format to make up a compact microfibril, which can be stabilized by hydrogen bonding whether it is inter-nuclear or intra-nuclear. Materials with cellulosic content have been investigated by researchers for decades due to their easy surface modifications and wide applications [2,3]. The existence of cellulose that is found as a common material in plant cell walls was first acknowledged by Anselm Payen in 1838, and this cellulose occurs in almost its purest form in the fibres of cotton. Yet, in wood, stalks and plant leaves, it is found in combination with lignin and hemicelluloses [4,5]. As a function of plant species along with its growth function, crystalline and amorphous domains are obtained in native cellulose fibres in variable ratios. Thus, it makes the properties of cellulose nanocrystal widely dependent on the sources of cellulose. The extraction, finding a suitable application, and its characterisation have given rise to varied terms, crystallites, whiskers, nanocrystals, nanofibers and nanofibrils, that have generated much activity globally: (i) NCC (NanoCrystalline Cellulose) and (ii) NFC (Nano-fibrillated Cellulose). Some of the novel methods for cellulose production include top-down methods that involve physical/enzymatic/chemical techniques for its isolation from agricultural/forest residue and wood, while bottom-up methods involve glucose bacteria to develop nano-fibrillated cellulose [6,7]. These cellulosic materials with one of their dimensions in the nanometre range are termed generically as nanocelluloses. Nano-fibrillated Cellulose (NFC) pertains to fibres that have been fibrillated to accomplish agglomerates of cellulose microfibril units; they are less than 100 nm in diameter with a length of several micrometres. Several terminologies exist for relating this material, but most often Nano/MicroFibrillated Cellulose (NFC/MFC) is used [8]. Numerous approaches can deliver cellulose nanofiber extraction, leading to diverse kinds of nanofibrillar materials, which depends on the raw material, pre-treatment, and disintegration of cellulose chains [9–13]. Since the citrus sinensis consists of a considerable amount of cellulose (14%), this material is potentially appropriate as a reinforcing component in high-performance composites. NFC can be produced by chemical or mechanical treatments such as acid hydrolysis. During the acid hydrolysis, chemistry for the hydrolytic cleavage of the glycosidic bonds takes place chiefly in the amorphous sections of the cellulose, which releases individual crystallites [14].

Bai and co-workers described a technique for the production of nanocrystalline cellulose that makes use of a narrow size distribution. The isolation and characterisation of cellulose obtained from sugar cane bagasse was reported by Sun et al. [15]. Wood definitely comprises of cellulose, but non-wood sources such as stems and leaves [16], cotton [17], sea animals [18], and sugar beet [19] have been used recently as raw materials to separate cellulose nanofibrils by chemical methods. Coconut husk fibres were used to prepare cellulose nanofibres by Imam et al. Gas-phase surface esterification of cellulose microfibrils and whiskers has been reported by Berlioz et al. Microfibrillated cellulose from the skins of prickly pear fruit was developed by Habibi et al. [20].

Metal nanoparticles constitute an important part of nanotechnology where numerous applications, depending on their tuneable chemical and physical properties, have been studied. Various capping agents have been utilized for synthesizing silver from silver nitrate, including chemical substrates and plant origins via top-down and bottom-up approaches. Researchers have made use of stem extracts, medicinal plants and alcoholic extracts and achieved different morphology and size [21,22].

Pharmaceutical industries generate compounds of waste containing organic and heavy metals which contaminate soil and water, and this is a major problem faced by the world [23]. Nanomaterials have been displaying faster rate kinetics in water treatment and also higher efficiency due to their high specific surface area and huge number of unsaturated atoms on their surfaces [24]. These advantages lead to an enhancement in adsorption capacity for the removal of organic and inorganic pollutants. The cleansing of toxic metals from wastewater by utilizing agricultural waste, based on the sorption phenomenon, has been considered a favourable technology. However, current research indicates that the use of agricultural waste can have certain drawbacks, such as the inclusion of colour, odour, lower sorption capacity, etc., restricting their commercial use. Hence, the search for cost-effective biomaterials with efficient and enhanced sorption along with their stability to use a minimal biomass dose is in great demand. Assorted inorganic nano-structured materials have been explored for remediating metallic ions but were recently associated with toxicity issues. One way to address such issues related to sustainability is to incorporate renewable materials of miniaturized elements of plant origin [25]. The skill to control, design, and manipulate organic materials on the nano scale to minimize contaminants and simultaneously avoid environmental risk is a major challenge for our 21st century.

In the previous literature reports, various types of cellulosic resources were used as precursors for the generation of nanocellulose. There are rarely any reports where metal nanoparticles have been used in combination with isolated polymers from discards and applied in simulated heavy metal removal. The removal of any heavy metals or any pollutant is executed by expensive clays, synthetic polymers and catalysts containing metals, which have costly precursors. Thus, as a novelty, we chose seasonal citrus sinensis peels grown in West India as the starting material for natural polymer extraction and also as an extract for reducing the nitrate of silver to obtain silver nanoparticles. Thus, waste as a discard can treat the effluent waste if applied at a commercial level with biodegradable polymers along with an economic route. The main objective was value addition to this agricultural waste, which is at present thrown away after use in juice shops or factories. This will open new avenues for using such abundant, renewable, and inexpensive agrowaste materials for developing value-added products.

2. Experimental Section

2.1. Materials

Analytical grades of sodium hydroxide, sodium hypochlorite, glacial acetic acid, hydrochloric acid, sulphuric acid and silver nitrate from Sigma Aldrich, India were used. The orange waste peels were obtained from juice vendors residing and selling by street side at Gandhinagar, Gujarat, India.

2.2. Extraction of Nano-Fibrillated Cellulose (NFC)

2.2.1. Bran Preparation

Waste peels of oranges were washed and immediately rinsed in 1% w/v potassium meta bisulphite solution for 24 h to inhibit oxidation. The rinsed peels were dried in a hot air oven at 60 °C for 24 h. After drying, the peels were crushed in a grinder and the finest particles were taken and sieved through 70 mesh screen. The bran was stored at 4 °C in sealed containers. The total yield of bran obtained was found to be 52.10%.

2.2.2. Chemical Treatment

The method for isolation was similar to the one described by Pelissari M. and co-workers [23]. In 5% KOH solution, the bran (8 gm) was kept under mechanical stirring at room temperature for 16 h. After the alkaline treatment, the insoluble residue was bleached with $NaClO_2$ solution for 1 h at 70 °C, pH 5.0 which was adjusted with 10% acetic acid. The residue was neutralised, washed and centrifuged at 6000 rpm for 20 min at 25 °C. Again, the second alkaline treatment was repeated with 5% KOH. The insoluble residue was subjected to acid hydrolysis for an hour with 1% H_2SO_4 at 80 °C. The final residue was neutralised, washed and centrifuged as performed in the earlier step and the diluted suspension was kept in a sealed container at 4 °C. The suspension of cellulose was dried by lyophilising and stored at 4 °C and was denoted as NFC. The protocol was similar to that followed by our previous reports [26,27]. The images captured at specific steps are depicted in Figure 1.

Figure 1. Steps involved in isolation of nano-fibrillated cellulose (NFC).

2.3. Synthesis of Silver Nanoparticles (Ag-NP)

A total of 10 g of waste citrus sinensis peels was washed and boiled in 100 mL distilled water. Further, the peels were crushed and the extract was filtered through Whatman filter paper 1. An aqueous solution (2 mM) of silver nitrate ($AgNO_3$) was prepared and used for the synthesis of silver nanoparticles. A total of 6 mL of extract was added to 80 mL of 2 mM $AgNO_3$ solution at room temperature at acidic pH 4. The reduction of silver nitrate to silver ions was confirmed by the colour change from colourless to brown when kept in the dark for 12 h. The solution that was reduced underwent centrifugation at 6000 rpm for 20 min. The supernatant liquid was discarded but the pellet obtained was re-dispersed in de-ionized water. This was repeated several times to rinse and remove the absorbed substances on the surface of the silver nanoparticles. The suspension was dried at 60 °C for 3 h to attain black powdered nanoparticles and used further for spectroscopic determination to assure its formation. The sample was noted by AgNPs.

2.4. Sorption Studies

Samples of effluent were taken from pharmaceutical industry at Vatva, GIDC phase—IV. The levels of heavy metals present in them were determined by Atomic Absorption Spectroscopy (AAS). Three inlet sites (E1 E2 E3) and 1 outlet (E4) were chosen.

Batch experiments were performed by opting for a simulated technique rather than extracting the inorganic metals from the effluents directly. Higher amounts of cadmium (Cd) and chromium (Cr) were detected amongst copper and lead as shown in Figure 2. So, Cd (II) and Cr (II) metal solutions (10 mg/L) were taken separately and adsorption experiments were carried out further. Three sets of each were prepared for biosorption, one having synthesized Ag nanoparticles, the other having NFC, and the third containing both. The pH of the solutions was kept 6.5 using 0.1 M HCl and 0.1 M NaOH, and 0.3 g NFC was taken in each single set up. Metal-loaded biomaterial was transferred to glass bottles with lid and shaken with 100 mL of each desorption reagent as a function of time (10, 20, 40, 80, 160) at room temperature. At each interval, suspensions were stirred and filtered using Whatman 42 filter paper and an estimation of metal ion concentration was carried out by AAS. The metal concentration retained in the solution was computed using Equation (1) and the sorption efficiency for the metal was analysed by Equation (2).

$$Q_e = (C_i - C_f) \times V/W \quad (1)$$

$$\% \text{ Metal sorption} = (C_i - C_f)/C_i \times 100 \quad (2)$$

where concentration at the beginning and at equilibrium is denoted by C_i and C_f, respectively, and the mass (g) of adsorbent noted by W and V is the volume (L) contained in the flask. The sorption efficiencies of silver nanoparticles (Ag), cellulose derived from discard (NFC), and the combination of silver nanoparticles with NFC have been compared for the removal of Cd (II) and Cr (II).

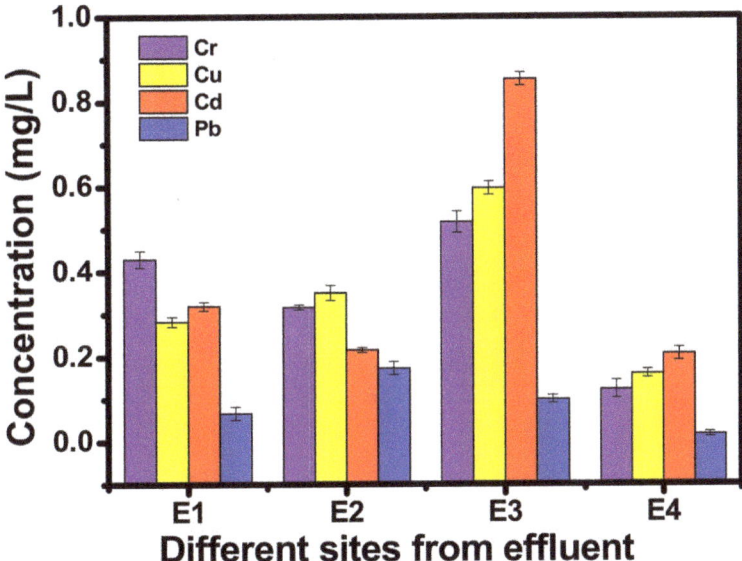

Figure 2. Graphical representation of the estimated heavy metals from pharmaceutical effluent.

Adsorption Isotherms

To demonstrate the adsorption by AgNPs, NFC, and Ag + NFC, Freundlich isotherm was utilized for cadmium and chromium heavy metals, for which the equation is given as:

$$q = K_f C_e^{1/n} \quad (3)$$

The respective data were fitted into the logarithmic equation form:

$$\text{Log } Q = \text{Log } K_f + \frac{1}{n} \text{Log } C_e \quad (4)$$

Here, q is the metal sorption taking place per unit mass of the adsorbent, C_e is the residual concentration of the simulated solution of metal ions, and K_f and n are constants. K_f is the biosorption capacity while $(1/n)$ is the biosorption intensity, and these were calculated from the slope and intercept plotted from Equations (3) and (4).

The general form of Langmuir adsorption equation is given by:

$$\frac{C_e}{q_e} = \frac{1}{Q_0 b} + \frac{C_e}{Q_0} \quad (5)$$

Equilibrium concentration is denoted by C_e, while the amount adsorbed by chromium and cadmium is q_e. Two Langmuir constants, b and Q_0, are in relation to the energy of adsorption and adsorption capacity, respectively. The linear plot of C_e/q_e vs. C_e infers that the adsorption of heavy metals obeyed the Langmuir model [28,29].

2.5. Characterisation of Nano-Fibrillated Cellulose and Silver Nanoparticles

2.5.1. Ultraviolet Spectroscopy (UV)

The silver nanoparticles synthesised were characterised after 24 h with the help of a UV–VIS spectrometer (Dynamica Halo DB-30). The silver nanoparticle solution was diluted and sonicated for 20 min and thus absorbance spectra were obtained. A resolution of 1 nm between 300 and 700 nm possessing a scanning speed of 300 nm/min was used.

2.5.2. Field Emission Scanning Electron Microscopy (FESEM)

The morphology and structure of synthesised NFC can be analysed under a Field emission scanning electron microscope. The Nova Nano FEG-SEM 450 was used throughout the experiment at an accelerating voltage of 15 KV. The lyophilised sample was mounted with carbon tape on the sample holder. Each sample was coated with gold using a sputtering technique. Similarly, the silver nanoparticles were also analysed by FESEM Hitachi SU9000 UHR.

2.5.3. Fourier-Transform Infrared Spectroscopy (FTIR)

FTIR spectra were recorded on a spectrophotometer (IR affinity 1S, Shimadzu). Samples of orange peel bran and cellulose nanoparticles were finely ground and mixed with potassium bromide. The mixture was compressed to pellet form and analysis was performed in the range of 400–4000 cm^{-1}.

2.5.4. Transmission Electron Microscopy (TEM)

The size of NFC was analysed using Transmission Electron Microscopy (TEM). Ultrathin units of isolated nanocellulose were cut by microtone at about $-100\ °C$ to allow electrons to pass, and the images were taken by JEM 2010 with an acceleration voltage of 200 kV. A total of 1 mg NFC was suspended into ethanol and ultrasonicated for 5 min.

2.5.5. X-ray Diffraction (XRD)

X-ray diffraction was used to determine the crystallinity of the cellulose based material obtained. Each material was placed on a sample holder and levelled to obtain uniform X-ray exposure. Samples were analysed using an X-ray diffractometer (Panalytical Xpert Pro MPD) at room temperature with a monochromatic Cu kα radiation source ($\lambda = 0.154$ nm) with 2θ angle ranging from $10°$ to $50°$, operated at a voltage of 45 kV and a current of 40 mA.

2.5.6. Atomic Absorption Spectroscopy (AAS)

The initial amount of chromium and cadmium in the metal-ion solution was determined by acid digestion and subsequent AAS analysis. Firstly, the pharmaceutical effluent was subjected to AAS, which showed 4 heavy metals in it. Later, the amount of metal salt taken was 100 mg/kg, and after the remediation for 160 min the samples at specific intervals were sent for AAS analysis.

3. Results and Discussion

3.1. UV

The reaction mixtures containing AgNO$_3$ with orange peel extract turned yellow to brown after 12 h of incubation in the dark. Control silver nitrate solutions without the extract of orange peels did not develop a brown colour. This shows that the waste orange peels successfully acted as a reducing and capping agent. Plasmon resonance was displayed by silver nanoparticles. Figure 3 shows the absorption spectrum of silver nanoparticles at 419 nm, which confirmed its presence [30].

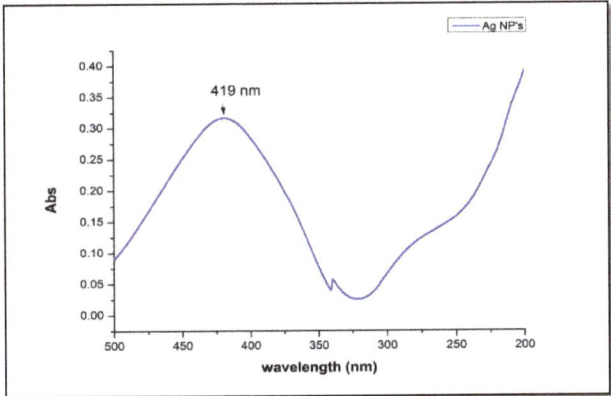

Figure 3. Absorption spectrum of silver nanoparticles synthesized from waste peel extract.

3.2. XRD

Orange bran and nano-fibrillated cellulose were analysed by XRD for their crystallinity index. Figure 4 shows X-ray diffractogram of isolated NFC and fruit bran. Nano-fibrillated cellulose showed a characteristic peak at $2\theta = 29°$ and $2\theta = 30°$. Mostly, the diffraction angle lies between 25–35° for nanocellulose, which can be indexed as (100), (002), and (004) inferring to the monoclinic phase. This result is in accordance with the findings of Morais J and group [25]. Using X-ray diffraction data and Scherrers equation $t = k\lambda/\beta Cos\theta$, the average crystallite size was estimated where t is crystallite size of the sample, β is full width half maxima (FWHM), and k is the wavelength of X-ray used, i.e., 1.548. The average grain size calculated was 6 nm. With the help of the intensity of amorphous peaks, the crystallinity index of orange peel bran and NFC was calculated. The crystallinity index is related to the strength and stiffness of fibres [31]. The crystallinity index of orange bran was 39% and that of NFC was 75%. Hence, the crystallinity index was increased by 36% after the chemical and mechanical treatments. It seems to be that the acid hydrolysis had much less effect on the morphology of these particles.

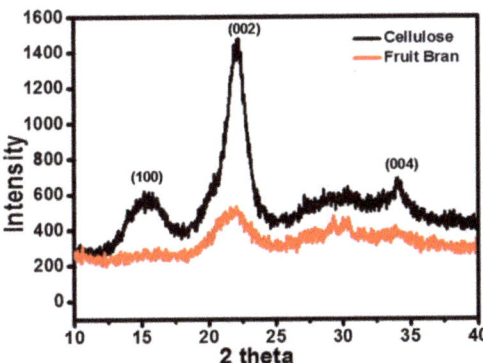

Figure 4. X-ray diffractogram of NFC and orange bran.

3.3. FTIR

The FTIR spectra of orange peel bran and nano-fibrillated cellulose are shown in Figure 5. Both spectra are dominated by peaks at 720 cm^{-1} and 3025 cm^{-1}, which correspond to -C-H and = C-H stretching vibrations of hemicelluloses. The sharp band between 3000 and 3050 cm^{-1} shows -OH groups present and reflects the hydrophilic character of orange bran and nano-fibrillated cellulose. Aliphatic saturated C-H stretching vibration in NFC is seen in the peak at 2970 cm^{-1}. The band at 1550 cm^{-1} demonstrates the bleaching step, which removed most of the lignin from nano-fibrillated cellulose. The peak at 760 cm^{-1} of C-H disappeared after NaClO$_2$ treatment. The band near 1050 cm^{-1} is related to xylans, which was significantly less intense for the NFC sample. These features proved the β-glycosidic linkages between anhydro-glucose units in cellulose.

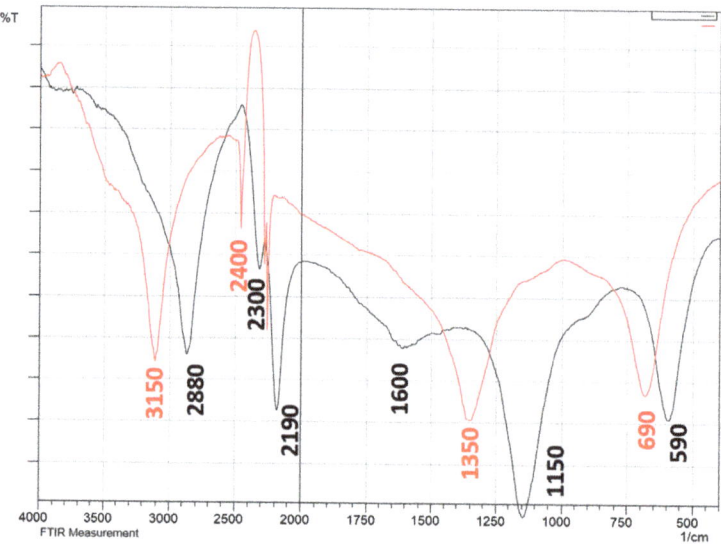

Figure 5. Comparative FTIR spectra of NFC and waste peel bran. —FTIR Spectra of waste peel Bran; —FTIR Spectra of lyophilised Cellulose.

3.4. FESEM

Figure 6a,b shows scanning electron microscopy micrographs of waste orange peel bran and NFC. The bran had an irregular surface and few grinding residues. Structural and chemical changes were depicted after the chemical treatment and morphological changes occurred in fibres with these steps. Amorphous components such as pectin and hemicelluloses were gradually removed.

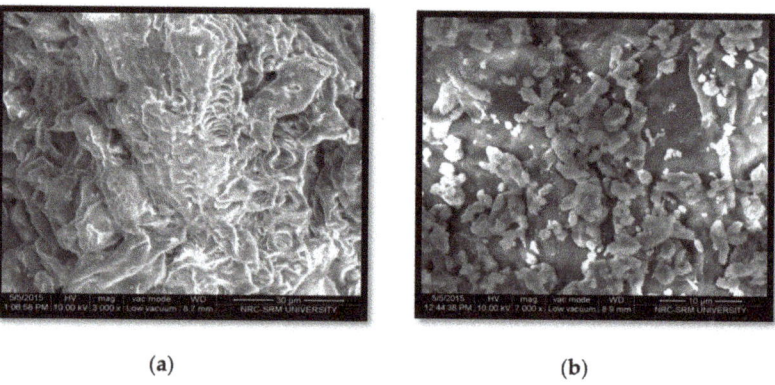

(a)   (b)

Figure 6. FE-SEM micrographs of waste peel bran (**a**) and NFC (**b**).

The mineral acid employed in the hydrolysis step had a major influence on the surface properties of NFC. Additionally, the source of cellulose plays a wider role in the morphology and dimensions of NFC. Algal and tunicate cellulose produces nanocrystals of several microns, whereas wood fibres liberate shorter ones.

3.5. TEM

3.5.1. Nano-Fibrillated Cellulose

Figure 7a shows TEM micrographs of a diluted suspension of nano-fibrillated cellulose from waste orange peels. It can be seen that nano-fibrillated cellulose is mostly spherical and rarely oval. The majority of the overall size of particles lies in the nanometric range, i.e., 44–50 nm. Cellulose particles in amorphous regions are randomly overlapping each other and are spherical in shape. They have no irregular surface as they consist of amorphous and crystalline regions. Incomplete removal of hemicelluloses during chemical treatment and the formation of inter-fibrillar hydrogen bonds account for the presence of residual nanoparticle bundles [32].

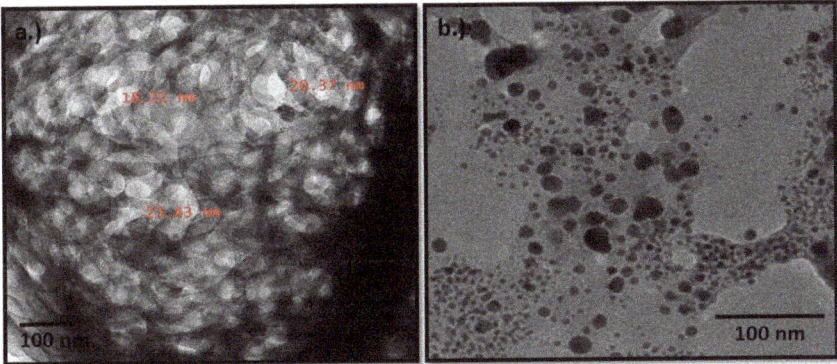

Figure 7. (a) TEM micrographs of nano-fibrillated cellulose (b) TEM micrograph of silver nanoparticles.

3.5.2. Ag Nanoparticles

Figure 7a of TEM confirmed the presence of silver nanoparticles. This study showed that the silver nanoparticles are spherical in shape and are polydispersed. The size of particles obtained was 30–34 at 25 °C. The synthesized silver nanoparticles from waste orange peels were stable in solution over a period of 3 months at room temperature [33].

3.6. AAS

The removal efficiency at each interval was calculated by the formula $(C_i - C_f)/C_i \times 100$, where C_i and C_f are the initial and final metal concentrations, respectively. Figure 8a–c shows the comparative chart for the removal efficacy of Cr (II) and Cd (II). The highest removal efficiency was found for cadmium, i.e., 83.49%, by using silver and NFC together as a bio sorbent. The cellulose acted as an absorber, where it makes an interface with the silver nanoparticles and enhances the efficacy. The second highest was for cadmium, i.e., 47.21%, but by using only nano-fibrillated cellulose as a bio sorbent. The maximum sorption for both cadmium and chromium was observed with silver along with NFC, i.e., 83.49% and 32.20%, respectively. In one set with cadmium containing silver as a bio-sorbent, after 40 min the metal-ion estimation which reduced showed an increase of 10 mg/lit at 80 min. This was because the adsorbate and adsorbent attained equilibrium and were incapable of further sorption. Thus, at one point after equilibrium, concentration was lowered in the metal-ion concentration solution and, again, the substance undergoing the sorption changed to its bulk state [34,35].

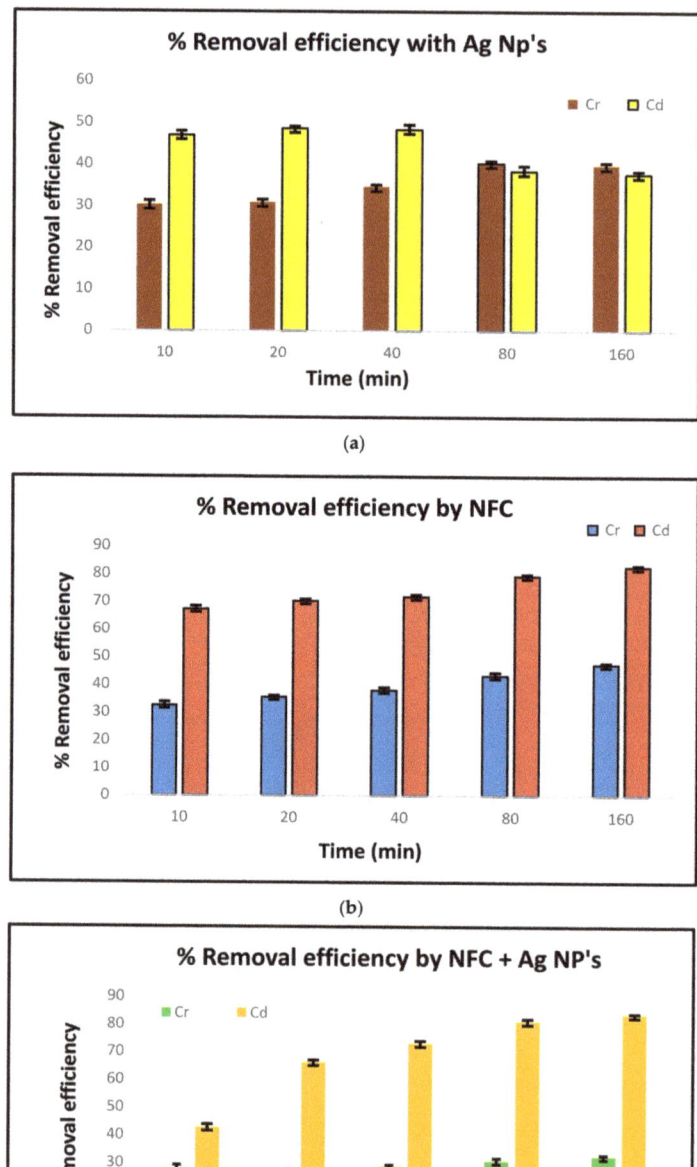

Figure 8. (**a**) Comparison chart of % removal efficiency in cadmium and chromium containing metal solution with silver nanoparticles at specific intervals. (**b**) Comparison chart of % removal efficiency in cadmium and chromium containing metal solution with NFC at specific intervals. (**c**) Comparison chart of % removal efficiency in cadmium and chromium containing metal solution with silver nanoparticles and NFC at specific intervals.

3.7. Sorption Isotherms

The maximum extent to which chromium and cadmium can perform sorption by NFC has been measured through the Langmuir and Freundlich isotherm. Figure 9a–f shows the kinetic simulation plots for Freundlich and Langmuir isotherms with three sets of adsorbent within 160 min. Due to wider gaps in the data obtained for Cd (II) and Cr (II), the linear plots have been separately visible on graph of each isotherm. The concentration was measured by taking sample for atomic absorption spectrometry at specific interval by doubling the time. So within 160 min, the concentration reduced up to ~17 ppm from 100 ppm in both the heavy metals. The kinetic plots for both isotherms with three adsorbents were fitted by using Origin 8.5. The linearity in the plots of Freundlich and Langmuir isotherms suggested the reaction mechanism followed first order rate kinetics.

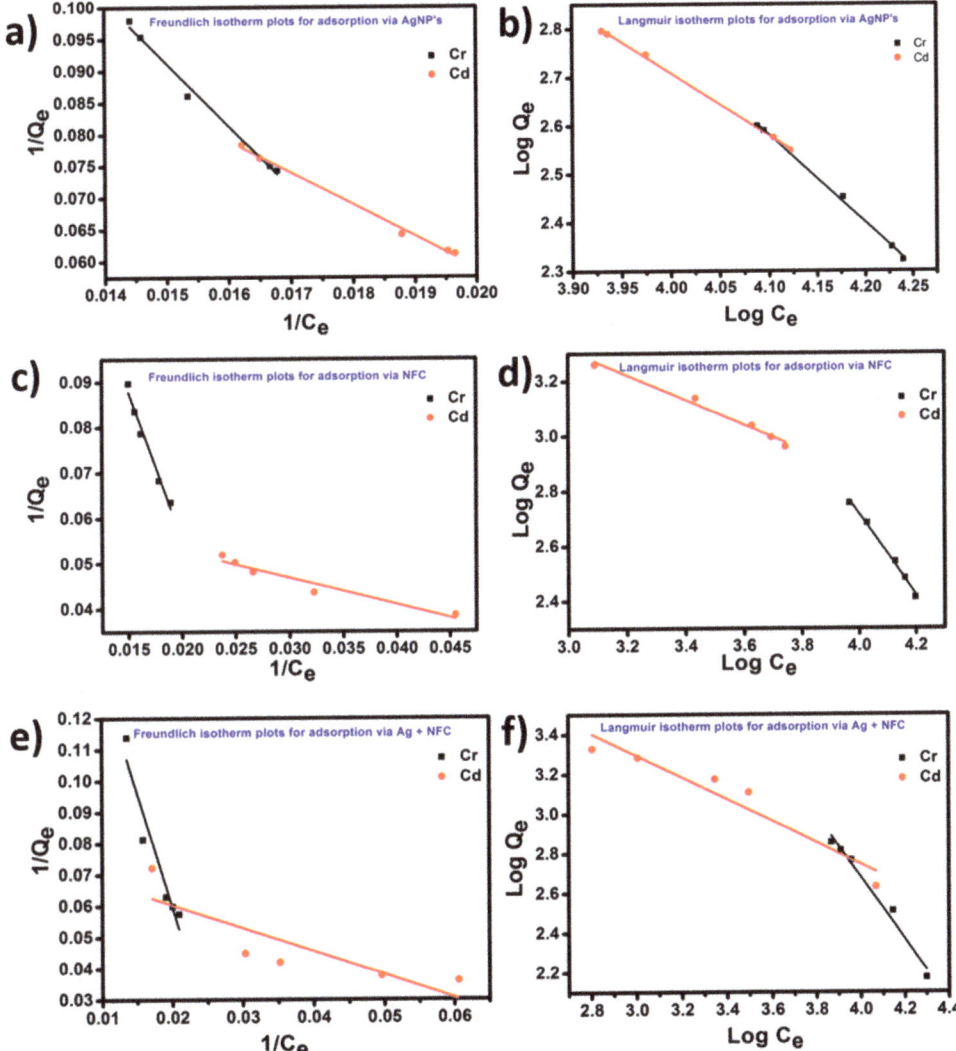

Figure 9. Freundlich and Langmuir isotherm kinetic simulation plots for Cd and Cr with three different adsorbents.

As per the model designed by Freundlich, the overall optimum biosorption capacity (K_f) was obtained for Cd (II): -0.74, while its intensity ($1/n$) was optimum for Cr (II): 0.21. K_f and ($1/n$) were analyzed for the adsorbent mixture of AgNPs and NFC. This denotes higher efficiency for the sorption of divalent ions when combined together rather than utilizing bare AgNPs or bare NFC. Similarly, biosorption capacity (Q_0) and energy (b) from Equation (5) of the Langmuir isotherm were found highest in energy for Cr (II): 8.92 while the biosorption capacity for Cd (II) was found to be -0.54. The appropriate high values of R^2 inferred that both Freundlich and Langmuir isotherm models were fitted well to the ions of adsorbent Ag + NFC. The linear graphs pertained to the formation of a monolayer of Cd (II) and Cr (II) that may be homogenous [36].

4. Conclusions

The present study offers a simple and convenient method of synthesis of nano-fibrillated cellulose from agricultural residues of fruit. Experimental results showed that produced nanocellulose had a diameter within the range of 44–50 nm. FTIR showed the removal of lignin and hemicellulose due to the alkaline treatment. TEM confirmed the presence of nano-fibrillated cellulose, which showed a spherical shape of 44 nm. Silver nanoparticles were also synthesised and characterised using orange peel extract as a reducing agent. Their size was found to be 15–16 nm at 25 °C. The heavy metal ion removal efficiency was successfully analysed for cadmium and chromium by adding silver and nano-fibrillated cellulose at specific intervals. Thus, the present work showed the successful sorption activity of cadmium and chromium by nano-fibrillated cellulose with a removal efficiency of 83.49% and 32.20%, respectively. These can also be a possible commercial use for minimizing or removing toxic metal ions from water bodies.

Author Contributions: Conceptualization, Data Interpretation, writing original draft: N.T.; Methodology: V.K.Y., N.T.; Investigation: K.K.Y., M.M.C.-P.; Visualisation: A.K.S., J.A., V.K.Y.; Writing, review and editing: A.K.S., F.A.A.A., M.A., K.K.Y., V.K.Y.; Project administration: J.A.; Funding acquisition: A.K.S., J.A. and M.M.C.-P. All authors have read and agreed to the published version of the manuscript.

Funding: The authors extend their appreciation to the Deputyship for Research and Innovation, "Ministry of Education" in Saudi Arabia for funding this research work through the project number IFKSURG-1439-085.

Institutional Review Board Statement: Not applicable.

Informed Consent Statement: Not applicable.

Data Availability Statement: The data presented in this study are available on request from the corresponding author.

Acknowledgments: All the authors are grateful to Si-Cart, Anand; SRM university Chennai; St. Xavier's' college, Ahmadabad, and Gujarat laboratories for the analysis of samples throughout the work.

Conflicts of Interest: The authors declare no conflict of interest.

References

1. Djenane, D. Chemical Profile, Antibacterial and Antioxidant Activity of Algerian Citrus Essential Oils and Their Application in Sardina pilchardus. *Foods* **2015**, *4*, 208–228. [CrossRef]
2. Salas, C.; Nypelö, T.; Rodriguez-abreu, C.; Carrillo, C.; Rojas, O.J. Nanocellulose properties and applications in colloids and interfaces. *Curr. Opin. Colloid Interface Sci.* **2014**, *19*, 383–396. [CrossRef]
3. Kalia, S.; Dufresne, A.; Cherian, B.M.; Kaith, B.S.; Avérous, L.; Njuguna, J.; Nassiopoulos, E. Cellulose-based bio- and nanocomposites: A review. *Int. J. Polym. Sci.* **2011**. [CrossRef]
4. Ahmadzadeh, S.; Keramat, J.; Nasirpour, A.; Hamdami, N.; Behzad, T.; Aranda, L.; Vilasi, M.; Desobry, S. Structural and mechanical properties of clay nanocomposite foams based on cellulose for the food-packaging industry. *J. Appl. Polym. Sci.* **2016**, *133*, 2079. [CrossRef]
5. Da Silva, A.E.; Rodrigues, H.; Salgado Gomes, M.C.; Eleamen, E.; Nagashima, T.; Tabosa Egito, E.S. Xylan, a Promising Hemicellulose for Pharmaceutical Use. In *Products and Applications of Biopolymers*; InTech: Rijeka, Croatia, 2012.

6. Kim, J.-H.; Shim, B.S.; Kim, H.S.; Lee, Y.-J.; Min, S.-K.; Jang, D.; Abas, Z.; Kim, J. Review of nanocellulose for sustainable future materials. *Int. J. Precis. Eng. Manuf. Technol.* **2015**, *2*, 197–213. [CrossRef]
7. Kargarzadeh, H.; Ioelovich, M.; Ahmad, I.; Thomas, S.; Dufresne, A. Methods for Extraction of Nanocellulose from Various Sources. *Handb. Nanocellulose Cellul. Nanocomposites* **2017**, 1–49. [CrossRef]
8. Rajinipriya, M.; Nagalakshmaiah, M.; Robert, M.; Elkoun, S. Importance of Agricultural and Industrial Waste in the Field of Nanocellulose and Recent Industrial Developments of Wood Based Nanocellulose: A Review. *ACS Sustain. Chem. Eng.* **2018**, *6*, 2807–2828. [CrossRef]
9. Purkait, B.S.; Ray, D.; Sengupta, S.; Kar, T.; Mohanty, A. Isolation of Cellulose Nanoparticles from Sesame Husk. *Ind. Eng. Chem. Res.* **2011**, *50*, 871–876. [CrossRef]
10. Tavker, N.; Gaur, U.K.; Sharma, M. Agro-waste extracted cellulose supported silver phosphate nanostructures as a green photocatalyst for improved photodegradation of RhB dye and industrial fertilizer effluents. *Nanoscale Adv.* **2020**. [CrossRef]
11. Tavker, N.; Sharma, M. Designing of waste fruit peels extracted cellulose supported molybdenum sulfide nanostructures for photocatalytic degradation of RhB dye and industrial effluent. *J. Environ. Manag.* **2020**, *255*. [CrossRef]
12. Trilokesh, C.; Uppuluri, K.B. Isolation and characterization of cellulose nanocrystals from jackfruit peel. *Sci. Rep.* **2019**, *9*, 16709. [CrossRef] [PubMed]
13. Yadav, C.; Saini, A.; Maji, P.K. Energy efficient facile extraction process of cellulose nanofibres and their dimensional characterization using light scattering techniques. *Carbohydr. Polym.* **2017**, *165*, 276–284. [CrossRef] [PubMed]
14. Pasquini, D.; de MoraisTeixeira, E.; da Silva Curvelo, A.A.; Belgacem, M.N.; Dufresne, A. Extraction of cellulose whiskers from cassava bagasse and their applications as reinforcing agent in natural rubber. *Ind. Crop. Prod.* **2010**, *32*, 486–490. [CrossRef]
15. Liu, Y. Homogeneous isolation of nanocellulose from sugarcane bagasse by high pressure homogenization. *Carbohydr. Polym.* **2012**, *90*, 1609–1613. [CrossRef]
16. Pennells, J.; Godwin, I.D.; Amiralian, N.; Martin, D.J. Trends in the production of cellulose nanofibers from non-wood sources. *Cellulose* **2020**, *27*, 575–593. [CrossRef]
17. Theivasanthi, T.; Anne Christma, F.L.; Toyin, A.J.; Gopinath, S.C.B.; Ravichandran, R. Synthesis and characterization of cotton fiber-based nanocellulose. *Int. J. Biol. Macromol.* **2018**, *109*, 832–836. [CrossRef]
18. DeNiro, M.J.; Epstein, S. Isotopic composition of cellulose from aquatic organisms. *Geochim. Cosmochim. Acta* **1981**, *45*, 1885–1894. [CrossRef]
19. Li, M.; Wang, L.; Li, D.; Cheng, Y.-L.; Adhikari, B. Preparation and characterization of cellulose nanofibers from de-pectinated sugar beet pulp. *Carbohydr. Polym.* **2014**, *102*, 136–143. [CrossRef]
20. Habibi, Y.; Mahrouz, M.; Vignon, M.R. Microfibrillated cellulose from the peel of prickly pear fruits. *Food Chem.* **2009**, *115*, 423–429. [CrossRef]
21. Pontaza-Licona, Y.S.; Ramos-Jacques, A.L.; Cervantes-Chavez, J.A.; López-Miranda, J.L.; de JesúsRuíz-Baltazar, Á.; Maya-Cornejo, J.; Rodríguez-Morales, A.L.; Esparza, R.; Estevez, M.; Pérez, R.; et al. Alcoholic extracts from Paulownia tomentosa leaves for silver nanoparticles synthesis. *Results Phys.* **2019**, *12*, 1670–1679. [CrossRef]
22. Aghdam, S.Z.; Karimi, M.S.; Tabatabaee, A.; Minaeian, S. The antibacterial effects of the mixture of silver nanoparticles with the shallot and nettle alcoholic extracts. *J. Appl. Biotechnol. Rep.* **2019**. [CrossRef]
23. Cabral Pinto, M.; da Silva, E.A.F. Heavy Metals of Santiago Island (Cape Verde) Alluvial Deposits: Baseline Value Maps and Human Health Risk Assessment. *Int. J. Environ. Res. Public Health* **2018**, *16*, 2. [CrossRef] [PubMed]
24. Feng, Z.; Odelius, K.; Rajarao, G.K.; Hakkarainen, M. Microwave carbonized cellulose for trace pharmaceutical adsorption. *Chem. Eng. J.* **2018**, *346*, 557–566. [CrossRef]
25. Cabral-Pinto, M.M.S.; Inácio, M.; Neves, O.; Almeida, A.A.; Pinto, E.; Oliveiros, B.; da Silva, E.A.F. Human Health Risk Assessment Due to Agricultural Activities and Crop Consumption in the Surroundings of an Industrial Area. *Expo. Health* **2020**, *12*, 629–640. [CrossRef]
26. Chen, W.; Yu, H.; Liu, Y.; Hai, Y.; Zhang, M.; Chen, P. Isolation and characterization of cellulose nanofibers from four plant cellulose fibers using a chemical-ultrasonic process. *Cellulose* **2011**, *18*, 433–442. [CrossRef]
27. Tavker, N.; Gaur, U.K.; Sharma, M. Highly Active Agro-Waste-Extracted Cellulose-Supported CuInS$_2$ Nanocomposite for Visible-Light-Induced Photocatalysis. *ACS Omega* **2019**, *4*, 11777–11784. [CrossRef]
28. Tavker, N.; Sharma, M. Enhanced photocatalytic activity of nanocellulose supported zinc oxide composite for RhB dye as well as ciprofloxacin drug under sunlight/visible light. In Proceedings of the AIP Conference Proceedings; AIP Publishing LLC: Melville, NY, USA, 2018; Volume 1961, p. 030013.
29. Desta, M.B. Batch sorption experiments: Langmuir and freundlich isotherm studies for the adsorption of textile metal ions onto teff straw (eragrostis tef) agricultural waste. *J. Thermodyn.* **2013**, *1*. [CrossRef]
30. Ahmed, R.; Yamin, T.; Ansari, M.S.; Hasany, S.M. Sorption Behaviour of Lead(II) Ions from Aqueous Solution onto Haro River Sand. *Adsorpt. Sci. Technol.* **2006**, *24*, 475–486. [CrossRef]
31. Ranoszek-Soliwoda, K.; Tomaszewskaa, E.; Małeka, K.; Celichowski, G.; Orlowski, P.; Krzyzowska, M.; Grobelny, J. *The Synthesis of Monodisperse Silver Nanoparticles with Plant Extracts*; Elsevier: Amsterdam, The Netherlands, 2019. [CrossRef]
32. Lefatshe, K.; Muiva, C.M.; Kebaabetswe, L.P. Extraction of nanocellulose and in-situ casting of ZnO/cellulose nanocomposite with enhanced photocatalytic and antibacterial activity. *Carbohydr. Polym.* **2017**, *164*, 301–308. [CrossRef]

33. Liu, S.; Tao, D.; Bai, H.; Liu, X. Cellulose-Nanowhisker-Templated Synthesis of Titanium Dioxide/Cellulose Nanomaterials with Promising Photocatalytic Abilities. *J. Appl. Polym. Sci.* **2011**. [CrossRef]
34. Mosaviniya, M.; Kikhavani, T.; Tanzifi, M.; Yarakibc, M.T.; Tajbakhsh, P.; Lajevardi, A. *Facile Green Synthesis of Silver Nanoparticles Using Crocus Haussknechtii Bois Bulb Extract: Catalytic Activity and Antibacterial Properties*; Elsevier: Amsterdam, The Netherlands, 2019. [CrossRef]
35. Pietrelli, L.; Francolini, I.; Piozzi, A.; Sighicelli, M.; Silvestro, I.; Vocciante, M. Chromium(III) Removal from Wastewater by Chitosan Flakes. *Appl. Sci.* **2020**, *10*, 1925. [CrossRef]
36. Vikrant, K.; Kumar, V.; Vellingiri, K.; Kim, K.H. Nanomaterials for the abatement of cadmium (II) ions from water/wastewater. *Nano Res.* **2019**. [CrossRef]

Article

Effect of Ink and Pretreatment Conditions on Bioethanol and Biomethane Yields from Waste Banknote Paper

Omid Yazdani Aghmashhadi [1,*,†], Lisandra Rocha-Meneses [2,*,†], Nemailla Bonturi [3], Kaja Orupõld [4], Ghasem Asadpour [1], Esmaeil Rasooly Garmaroody [5], Majid Zabihzadeh [1] and Timo Kikas [2]

1. Department of Wood and Paper Engineering, Sari University of Agricultural Sciences and Natural Resources, Km 9 Farah Abad Road, Sari 66996-48181, Mazandaran Province, Iran; asadpur2002@yahoo.com (G.A.); m.zabihzadeh@sanru.ac.ir (M.Z.)
2. Institute of Technology, Chair of Biosystems Engineering, Estonian University of Life Sciences, Kreutzwaldi 56, 51006 Tartu, Estonia; Timo.Kikas@emu.ee
3. Institute of Technology, University of Tartu, 50411 Tartu, Estonia; Nemailla.Bonturi@ut.ee
4. Institute of Agricultural and Environmental Sciences, Estonian University of Life Sciences, Kreutzwaldi 5, 51006 Tartu, Estonia; Kaja.Orupold@emu.ee
5. Department of Bio-refinery Engineering, Faculty of New Technologies Engineering, Shahid Beheshti University, Zirab P.O. Box 47815-168, Mazandaran, Iran; e_rasooly@sbu.ac.ir
* Correspondence: omidyazdani29@yahoo.com (O.Y.A.); Lisandra.Meneses@emu.ee (L.R.-M.)
† These authors contributed equally to this work.

Abstract: Waste banknote paper is a residue from the banking industry that cannot be recycled due to the presence of ink, microbial load and special coating that provides protection against humidity. As a result, waste banknote paper ends up being burned or buried, which brings environmental impacts, mainly caused by the presence of heavy metals in its composition. To minimize the environmental impacts that come from the disposal of waste banknote paper, this study proposes to produce value-added products (bioethanol and biogas) from waste banknote paper. For this, the effect of ink and pretreatment conditions on bioethanol and biomethane yields were analyzed. Waste banknote paper provided by the Central Bank of Iran was used. The raw material with ink (WPB) and without ink (WPD) was pretreated using sulfuric acid at different concentrations (1%, 2%, 3%, and 4%) and the nitrogen explosive decompression (NED) at different temperatures (150 °C, 170 °C, 190 °C, and 200 °C). The results show that the use of NED pretreatment in WPD resulted in the highest glucose concentration of all studies (13 ± 0.19 g/L). The acid pretreatment for WPB showed a correlation with the acid concentration. The highest ethanol concentration was obtained from the fermentation using WPD pretreated with NED (6.36 ± 0.72 g/L). The maximum methane yields varied between 136 ± 5 mol/kg TS (2% acid WPB) and 294 ± 4 mol/kg TS (3% acid WPD). Our results show that the presence of ink reduces bioethanol and biogas yields and that the chemical-free NED pretreatment is more advantageous for bioethanol and biogas production than the acid pretreatment method. Waste banknote paper without ink is a suitable feedstock for sustainable biorefinery processes.

Keywords: anaerobic digestion; biofuel; biomass; cotton-based waste; closed-loop; lignocellulose

1. Introduction

The population growth and economic development have led to an increase in the production and consumption of materials and resources to satisfy global demand. However, this has also led to an increase in the amount of waste produced, such as municipal solid waste, industrial waste, and wastepaper [1,2]. Instead of being discarded after primary use, these residues can be used for the production of value-added products e.g., production of low-cost biofuels [3,4]. Among wastepaper, great attention has been paid to waste banknotes since they cannot be recycled due to their properties, such as the presence of pathogenic bacteria, heavy ink, and presence of formaldehyde melamine resin that provides protection against humidity and makes the digestion process more challenging [5–8].

The most common handling options for banknote paper are incineration or landfilling [9]. However, these strategies increase air pollution, and at the same time, pollute soils, and groundwater [10]. As a result, there is a search for more suitable and environmentally friendly handling options that can be applied to waste banknote paper. Banknote paper is to a large extent made of cotton and thus, has a high percentage of cellulose and low amounts of lignin in its composition, which makes it a suitable feedstock for the production of biofuels [7]. As banknote paper has high amounts of ink in its composition, a deinking process is required before its further treatment. The deinking process is composed of three sequential steps: pulping, soaking, and screening [7]. In the pulping stage, chemical and mechanical methods are used to separate ink and other non-fibrous contaminants from the fibrous material [7,11,12]. In the soaking stage, sodium hydroxide is used at high temperatures, followed by sieving and rinsing to separate and remove ink particles from the pulp [13]. Finally, in the screening process, a centrifugal cleaning is used to select and separate contaminants from the pulp fibers. Banknote paper has high amounts of alpha-cellulose in its composition (88 to 96%) which makes it a suitable feedstock for bioethanol and biogas production. However, banknote paper has also a very crystalline structure that requires effective and harsh pretreatments methods to make the cellulose accessible for enzymatic hydrolysis [7,14].

Bioethanol production is composed of four sequential steps: pretreatment, hydrolysis, fermentation, and distillation [15]. A wide variety of chemical, physical, biological, and physio-chemical pretreatment methods have been reported in the literature as effective in breaking down the plant cell wall and in making cellulose accessible to the enzymes [16]. Chemical pretreatment methods include acid, alkali, ionic liquids, and organosolv. Although chemical pretreatment methods have been widely used and are effective in dissolving lignin and hemicellulose, they are still expensive because of the high cost of chemicals [17]. Besides, several inhibitors, such as aliphatic carboxylic acids, phenolic compounds, and furans, are formed during chemical pretreatment, which may disrupt subsequent hydrolysis and fermentation processes [18,19]. Physical pretreatment methods include comminuting, irradiation, and freezing, while biological pretreatment methods use fungi, enzymes, or microorganisms to break down the lignin and hemicellulose bonds [14,16]. Combined physical and chemical pretreatment methods include ammonia fiber expansion, steam explosion (SE), carbon dioxide explosion (CO_2 explosion, and NED).

From all the pretreatment methods reported before, great attention has been paid to chemical-free pretreatment methods (e.g., SE, CO_2 explosion and NED) since these methods reduce environmental impacts and costs that incur when chemicals are used [20]. Steam explosion pretreatment method uses high pressure, saturated steam, and a rapid decompression to disrupt the plant cell wall and dissolve the hemicellulose [16]. CO_2 explosion uses supercritical CO_2 and low pretreatment temperatures to improve the digestibility of the biomass [14]. NED is one of the most effective physical and chemical pretreatments methods reported in the literature [21]. This pretreatment method used nitrogen and high temperatures to open the biomass structure more effectively. NED is economically and environmentally attractive because neither catalysts nor chemicals are used in the processes [20]. However, the principle of operation and the processes that take place during this pretreatment still need further research [19].

Waste paper, and particularly waste banknote paper can also be used as a raw material for biogas production since it is a cheap source of organic material [22]. Anaerobic digestion is a biological process that occurs in the absence of oxygen and uses micro-organisms to convert organic waste into high-quality nutrient-rich fertilizer and yield energy in the form of biogas [23]. Biogas is mainly composed of methane (50–70%) and carbon dioxide (30–50%). Minor compounds include vapor water, nitrogen, oxygen, hydrogen sulfide, and ammonia [24]. Biogas can be upgraded to produce biomethane.

This study aims to investigate the potential of waste banknote paper (with and without ink) for bioethanol and biomethane production. For this purpose, two different pretreatment methods were applied—nitrogen explosive decompression and sulfuric acid pretreatment methods.

2. Materials and Methods

The production pathway utilized in this study to evaluate the potential of waste banknote paper for bioethanol and biogas production is illustrated in Figure 1.

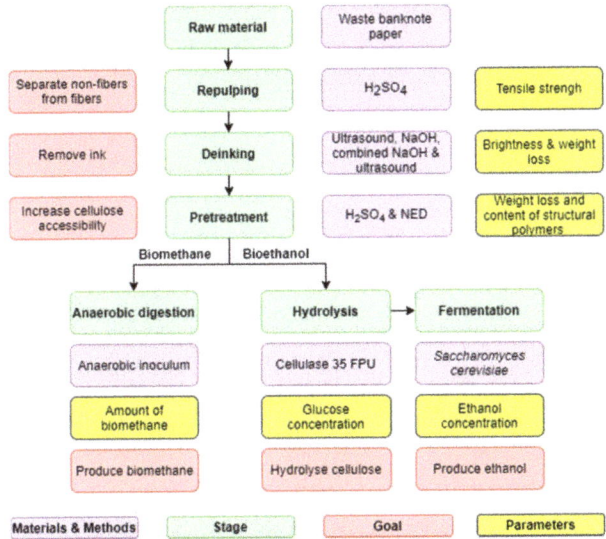

Figure 1. Production pathway utilized in this study to evaluate the potential of waste banknote paper for bioethanol and biomethane production.

2.1. Raw Material

Waste banknote paper obtained from the Central Bank of the Islamic Republic of Iran was used as a raw material in this study. After collection, the waste banknote paper was placed in the laboratory for 48 h to ensure equilibrium moisture content. After measuring the moisture content, the raw material was stored in bags until use.

2.2. Repulping

Wet strength is one of the main factors that should be taken into account when talking about waste banknote paper. This property of the paper is related to the presence of melamine formaldehyde, which is added to the banknote paper to increase its mechanical resistance and reduce water absorption. Therefore, the repulping process is required to break down melamine resin cross-links. In order to find the best conditions for the repulping process, the tensile strength was measured at different pHs (from pH = 1 to pH = 12). The lowest tensile strength indicates the best repulping conditions. The tensile strength was measured with a tensile tester, from FRANK-PTI GMBH company (Birkenau, Germany), following the standard method TAPPI T456 OM-15 [25]. For repulping, 0.5 mL of 98% sulfuric acid was added to 30 g of pulp, and distilled water was added to a 500 mL Erlenmeyer flask until a total working volume of 300 mL was reached. The samples were placed on a hot plate with a magnetic stirrer (M TOPS MS300HS, from Misung Scientific Co., Ltd., Gyeonggi-do, Korea) at 90 °C to 95 °C, and stirred at 500 rpm, for 90 min [26,27]. All the experiments were performed in triplicate.

2.3. Deinking

After repulping, the samples were deinked using three deinking methods: chemical (1%, 2%, 3%, and 4% sodium hydroxide), ultrasonic (30Am, for 5 min) and integrated ultrasonic and sodium hydroxide. In the ultrasonic deinking method, a Q700 Sonicator was used (from Qsonica, Newtown, USA). In the deinking process, 3g (dry weight) of pulp was put in 100 mL Erlenmeyer flasks and treated at a temperature of 90–95 °C. To increase the contact during the process, a mixer model FTDS 41 from Sci Finetech Co (Seoul, Korea) was used for 90 min, at 500 rpm. In order to separate and remove the ink from the fibrous materials, the samples were washed using distilled water and a 200 mesh filter. After finishing the deinking process, a new paper was produced, for the weight loss, tensile strength, and brightness measurements. The brightness properties of the papers were measured according to the standard T452 om 08 using a Zb-a Powders brightness colorimeter testing from Hangzhou Zhibang Automation Technology Co. Ltd. (Hangzhou, China) [7,26]. The sodium hydroxide with a purity of 100% was from Merck Group (Darmstadt, Germany).

2.4. Pretreatment

2.4.1. Chemical Pretreatment

H_2SO_4 with 98% purity was used for the chemical pretreatment of the banknote pulp samples. The acid was added in different concentrations (1%, 2%, 3%, and 4%) to Erlenmeyer flasks with 100g (dry weight) deinked waste banknote pulp (WPD) and non-de-inked waste banknote pulp (WPB). The samples were autoclaved for 30 min at 121 °C. After the pretreatment, solid and liquid fractions of the samples were separated by centrifugation (Thermo Scientific Heraeus megacentrifuge, Waltham, USA) at 10,000 rpm for 30 min. Samples from the solid fraction were left to dry at the atmospheric pressure and their weight loss was calculated.

2.4.2. Physio-Chemical Pretreatment

In NED pretreatment, 800 mL of distilled water was added to 100 g of dried WPD and WPB. The samples were mixed and pretreated at 150 °C, 170 °C, 190 °C, and 200 °C, at a pressure of 30 bars using compressed nitrogen gas. After reaching the desired temperatures, the samples were cooled down to 80 °C and the pressure was released from the vessel in an explosive manner [28]. After the pretreatment, solid and liquid fractions of the samples were separated by centrifugation (Thermo Scientific Heraeus megacentrifuge, Waltham, USA) at 10,000 rpm for 30 min.

2.5. Enzymatic Hydrolysis

Pretreated WPD and WPB samples from the solid fraction were used to make a broth with 2.5% dry matter content that was further used for the enzymatic hydrolysis. The enzyme complex Accellerase 1500 (from DuPont de Nemours) was used at a ratio of 0.3 mL per g of biomass. Distilled water was added to the flasks to obtain a total working volume of 200 mL. The hydrolysis took place for 72 h, at a temperature of 50 °C, under constant stirring in the orbital shaker (IKA®-Werke GmbH & Co. KG, Staufen im Breisgau, Germany) [18]. After the hydrolysis, the solid and liquid fractions of the samples were separated by centrifugation (Thermo Scientific Heraeus megacentrifuge, Waltham, USA) at 10,000 rpm for 30 min. All the experiments were performed in triplicate.

2.6. Fermentation

Liquid samples that were obtained from the hydrolysis stage were further fermented using the yeast *Saccharomyces cerevisiae*. This yeast was added to the samples from the liquid fraction at the ratio of 0.025 g/g. The fermentation process lasted 7 days and it was carried out at 25 °C [20].

2.7. Biomethane Potential

The biomethane potential (BMP) test used in this article is based on an adapted version of the protocol reported by Angelidaki et al. [29]. The inoculum was obtained from Tartu municipal wastewater treatment plant (Estonia), sieved through 2 mm mesh and pre-incubated for 4 days at 36 °C for degasification, before use [24]. The experiments were carried out in 575 mL plasma bottles with a working volume of 300 mL using the substrate to inoculum volatile solids (VS) ratio of 0.25. The blank test (just inoculum without substrate) was included to study the biogas and methane production of the inoculum, which later was subtracted from that of the samples with the substrate. The test bottles were flushed with N_2 to assure anaerobic conditions. The BMP tests were performed at 36 °C in the lab incubator (Memmert GmbH + Co. KG, Schwabach, Germany) for 32 days (up until the methane content was constant). The experiments were carried out in triplicate. During BMP test the pressure in the headspace of test bottles was measured with pressure meter BMP-Testsystem WAL (WAL Mess- und Regelsysteme GmbH).

2.8. Chemical Analysis

The neutral detergent fiber (NDF), acid detergent fiber (ADF) and acid detergent lignin (ADL) were determined using an ANKOM 2000 I, fiber Analyzer (ANKOM Technology Corporation, NY 14502, USA).

The dry matter content (TS) was analyzed with a moisture analyzer Ohaus MB 45. The volatile solids (VS) were analyzed as loss on ignition at 550 °C. The pH of the samples before and after pretreatment, hydrolysis, and fermentation was measured using a pH meter, model SevenCompact pH/Ion S220 from Mettler-Toledo AG (Schwerzenbach, Switzerland).

Glucose, glycerol, acetic acid, and ethanol were quantified by HPLC (LC-2030C Plus, Shimadzu, Kyoto, Japan) equipped with a refractive index detector (RID-20A, Shimadzu, Kyoto, Japan) using a Rezex ROA Organic Acid column (Phenomenex, Torrance, CA, USA) column at 45 °C, and isocratic elution at 0.6 mL/min of 5 mmol/L H_2SO_4.

The quantification of methane in the produced biogas was done by gas chromatography (CP-4900 Micro-GC, from Varian Inc., Palo Alto, USA).

2.9. Calculations

The hydrolysis and fermentation efficiency were calculated based on Equation 1 and Equation (2), respectively [30].

$$E_{HY} = \frac{m_{glc}}{m_{cel} \cdot 1.11} \cdot 100\% \qquad (1)$$

$$E_F = \frac{C_{eth}}{C_{glc} \cdot 0.51} \cdot 100\% \qquad (2)$$

where E_{HY} is the hydrolysis efficiency, m_{glc} the amount of glucose, m_{cel}, the amount of cellulose, 1.11 is the conversion factor of cellulose to glucose, E_F is the fermentation efficiency, C_{eth} is the concentration of ethanol, c_{glc} is the concentration of glucose, and 0.51 is the conversion factor of glucose to ethanol

The methane production was fitted and modelled in the software Graph-Pad Prism 5.0 with a non-linear regression model, using the equation Gompertz growth (Equation (3)).

$$Y = Y_M * \left(\frac{Y_0}{Y_M}\right)^{\exp(-K*X)} \qquad (3)$$

where Y is the cumulative methane produced (L/kg TS), Y_M is the maximum population (L/kg TS), Y_0 is the starting population (L/kg TS), K is the lag time (d^{-1}), X is time (days).

2.10. Statistical Analysis

The statistical analysis was performed in software Graph-Pad Prism 5.0. The normal distribution of the results was investigated using the Shapiro-Wilk normality test. For glucose and ethanol yields, the differences between the variables were studied using two-way ANOVA, followed by the post hoc test Tukey's multiple comparisons test. For biomethane yields, one-way ANOVA was used to investigate the statistically significant differences between the means of the different variables. The *post hoc* Dunn's multiple comparisons test. The results were statistically significant when $p \leq 0.05$ (confidence interval 95%).

3. Results

The results of the deinking and repulping process are reported in a previous paper published by the authors [26].

3.1. Chemical Composition

The results of the chemical composition, ash content, moisture, pH, and weight loss of WPB and WPD after being pretreated with NED and H_2SO_4 are reported in Table 1. The percentage of cellulose, hemicellulose, lignin, and ash for untreated WPB and WPD was 77–89%, 3–7%, 1–2%, and 0.6–1%, respectively. The cellulose content of WPB varied between 78% and 85%, while for WPD samples it varied between 84% and 89%. The percentage of hemicellulose differed from 1% and 6% for WPB samples, and from 1% to 8% for WPD samples. The amount of lignin and ash in all the samples was less than 2% and 1%, respectively. The moisture content in all the samples varied between 3% and 9%, and the pH between 5 and 6. The weight loss of all the samples was $\leq 10.2\%$. Statistically significant differences were found in the cellulose content of samples that were pretreated with acid 3% WPD and samples that were pretreated with acid 3% WPB (Table A1, Appendix A).

Table 1. Hemicellulose, cellulose, lignin, ash, and moisture content of the samples after the pretreatment stage.

Description	Cellulose (%)	Hemicellulose (%)	Lignin (%)	pH	Weight Loss (%)
NED 150 °C WPB	85 ± 3	6 ± 3	1 ± 0.7	5.5 ± 0.3	5.2
NED 170 °C WPB	84 ± 0.2	4 ± 2	1 ± 0.6	5.5 ± 0.2	6.1
NED 190 °C WPB	78 ± 5	4 ± 0.5	0.9 ± 0.0	5.7 ± 0.0	6.7
NED 200 °C WPB	83 ± 1	3 ± 0.5	0.6 ± 0.0	5.5 ± 0.0	6.3
Acid 1% WPB	85 ± 1	3 ± 2	0.7 ± 0.0	5.9 ± 0.1	6.2
Acid 2% WPB	83 ± 0.6	1 ± 3	0.6 ± 0.0	5.7 ± 0.0	9.4
Acid 3% WPB	70 ± 7	1 ± 1	0.7 ± 0.0	6.0 ± 0.0	10.2
Acid 4% WPB	79 ± 4	2 ± 2	0.6 ± 0.2	5.8 ± 0.0	9.7
NED 150 °C WPD	89 ± 0.6	8 ± 0.0	0.2 ± 0.1	5.6 ± 0.0	0.6
NED 170 °C WPD	88 ± 0.1	2 ± 1	0.3 ± 0.1	5.5 ± 0.3	3.1
NED 190 °C WPD	89 ± 2	2 ± 1.6	0.0 ± 0.0	5.5 ± 0.3	5.0
NED 200 °C WPD	87 ± 12	1 ± 0.3	0.0 ± 0.0	6.0 ± 0.0	9.3
Acid 1% WPD	88 ± 0.8	3 ± 1	0.5 ± 0.1	5.6 ± 0.4	2.4
Acid 2% WPD	88 ± 0.0	3 ± 0.4	0.3 ± 0.0	5.3 ± 0.1	1.7
Acid 3% WPD	84 ± 2	2 ± 1	0.2 ± 0.1	5.8 ± 0.1	2.1
Acid 4% WPD	86 ± 0.0	4 ± 0.0	0.2 ± 0.0	5.7 ± 0.4	2.6
Untreated WPB	77 ± 2	7 ± 3	2 ± 0.2	5.5 ± 0.5	-
Untreated WPD	89 ± 0.4	3 ± 0.4	1 ± 0.1	5.6 ± 0.2	-

The TS and VS contents of the samples used in the experiments are presented Table 2. The TS content of untreated banknote wastepaper varied between 915 g/kg (WPB) and 956 g/kg (WPD). For the remaining samples, the amount of dry matter differed from 918 g/kg and 970 g/kg. The amount of VS for untreated WPB was 955 g/kgTS and for untreated WPD 970 g/kgTS. The VS content was the lowest for WPB samples that were pretreated with acid 3% (903 g/kgTS), and highest for WPD samples that were pretreated with NED 200 °C (979 g/kgTS).

Table 2. Total Solids and volatile solids content.

Substrate	TS (g/kg)	VS (g/KgTS)
NED 150 °C WPB	960 ± 5	963 ± 9
NED 170 °C WPB	957 ± 9	944 ± 13
NED 190 °C WPB	934 ± 11	945 ± 1
NED 200 °C WPB	958 ± 6	940 ± 3
Acid 1% WPB	947 ± 8	956 ± 7
Acid 2% WPB	934 ± 4	930 ± 3
Acid 3% WPB	929 ± 7	903 ± 6
Acid 4% WPB	970 ± 1	917 ± 8
NED 150 °C WPD	959 ± 14	966 ± 2
NED 170 °C WPD	954 ± 0	971 ± 2
NED 190 °C WPD	962 ± 6	973 ± 0
NED 200 °C WPD	961 ± 0	979 ± 0
Acid 1% WPD	953 ± 6	972 ± 3
Acid 2% WPD	945 ± 2	950 ± 1
Acid 3% WPD	918 ± 5	937 ± 0
Acid 4% WPD	952 ± 1	916 ± 0
Untreated WPB	915 ± 7	955 ± 16
Untreated WPD	953 ± 3	970 ± 0

3.2. Glucose Content from Pretreatment and Ethanol Content after Fermentation

The increase in temperature used for the NED pretreatment of WPB resulted in a more efficient cellulose conversion to glucose as its concentration raised from 5.5 ± 0.71 g/L at 150 °C to 9.4 ± 0.44 g/L at 200 °C (Figure 2). The same trend was not observed for WPD as different temperatures used for the NED pretreatment did not result in statistically different ($p > 0.05$) glucose concentrations (Table A2, Appendix A). The use of NED pretreatment in WPD resulted in the highest glucose concentration of all studies (13 ± 0.19 g/L). The glucose yield of acid pretreated WPB showed a correlation with the acid concentration. The acid concentration of 2 and 3% resulted in 9.6 ± 0.18 and 9.8 ± 0.61 g/L of glucose, respectively. These concentrations were not different statistically and probably represent the range of optimum acid concentration for WPB pretreatment, as the performances of pretreatment using 1 and 4% of acid were inferior. The acid concentration on WPD pretreatment resulted in no significant statistical difference between the glucose concentrations obtained when using 1, 2, and 4% of acid (10.6 g/L of glucose on average). Unexpectedly, the pretreatment with 3% of acid pretreatment in WPD resulted in a lower glucose concentration (8.8 ± 0.60 g/L).

The glucose-containing liquid phases from the different pretreatments of WPB and WPD were used as a substrate for bioethanol production using *S. cerevisiae* (Figure 3). As expected, the ethanol concentration reflected the initial glucose concentration. The highest ethanol concentration was obtained from the fermentation using WPD pretreated with NED (6.4 ± 0.72 g/L), however there was no statistical difference amongst the different temperatures used (Table A3, Appendix A). The ethanol yield from glucose was also the highest, 0.5 g ethanol/g glucose (98% of theoretical yield). The second highest ethanol concentration and yield were also from WPD using acid pretreatment, 4.9 ± 0.20 g/L and 0.46 g ethanol/g glucose (90% of the theoretical yield). The best results with WPB were when using pretreatments with 2% of acid and NED at 200 °C, both about 4 g/L of ethanol and 0.43 g ethanol/g glucose (84% of the theoretical yield).

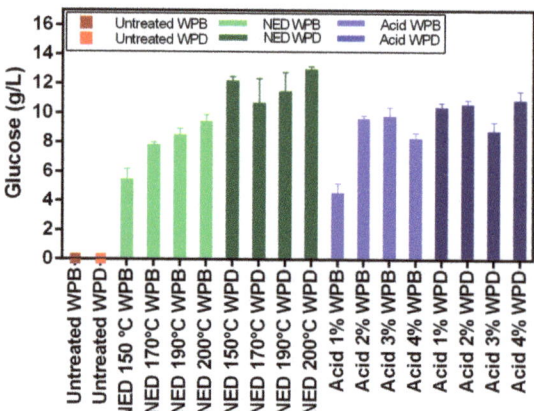

Figure 2. Glucose concentration for hydrolyzed WPB and WPD samples that were pretreated with NED at 150 °C, 170 °C, 190 °C, and 200 °C, and with acid 1%, 2%, 3%, and 4%.

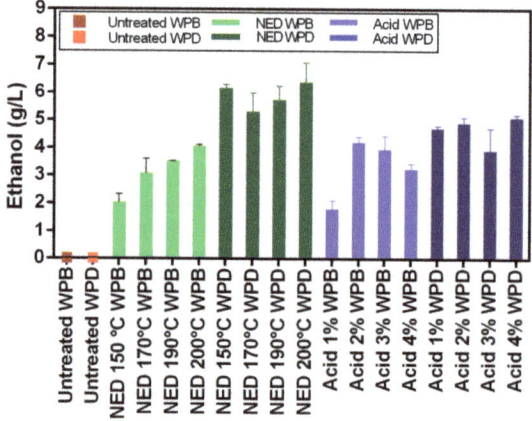

Figure 3. The concentrations of ethanol after the fermentation stage for WPB and WPD samples that were pretreated with NED at 150 °C, 170 °C, 190 °C, and 200 °C and with acid 1%, 2%, 3%, and 4%.

When it comes to the hydrolysis efficiency (Table 3), the results show relatively low values (between 5–15%). This can be attributed to the nature of the banknote paper. The cotton structure (main component of banknote paper) is highly crystalline in nature and thus, recalcitrant to enzymatic hydrolysis. Both pretreatments have only limited effect on reducing crystallinity of the cellulose in the paper. The highest hydrolysis efficiencies were obtained with acid pretreatment of WPB samples however, this is due to the fact that in the deinking process some of the amorphous cellulose is lost. On the other hand, the fermentation efficiencies of these same samples are lower due to the presence of inhibitory compounds originating from the ink. Fermentation efficiency varies between 73% and 99% and is clearly higher for deinked samples.

Table 3. Hydrolysis and fermentation efficiency.

	Hydrolysis Efficiency	Fermentation Efficiency
NED 150 °C WPB	6	73
NED 170 °C WPB	8	77
NED 190 °C WPB	10	81
NED 200 °C WPB	10	84
Acid 1% WPB	13	77
Acid 2% WPB	12	85
Acid 3% WPB	15	79
Acid 4% WPB	15	76
NED 150 °C WPD	5	99
NED 170 °C WPD	10	97
NED 190 °C WPD	10	98
NED 200 °C WPD	9	96
Acid 1% WPD	11	89
Acid 2% WPD	11	91
Acid 3% WPD	9	87
Acid 4% WPD	11	92

3.3. Potential of Waste Banknote Paper for Biomethane Production

3.3.1. Methane Yields

In Figure 4 the biomethane potentials for untreated sample and for WPB samples that were pretreated with NED at 150 °C, 170 °C, 190 °C, and 200 °C are presented. Samples that were pretreated at 150 °C had the highest biomethane yields (266 L/kg TS), followed by those that were pretreated at 190 °C (261 L /kg TS), untreated material (238 L/kg TS), 200 °C (208 L/kg TS), and 170 °C (202 L/kg TS).

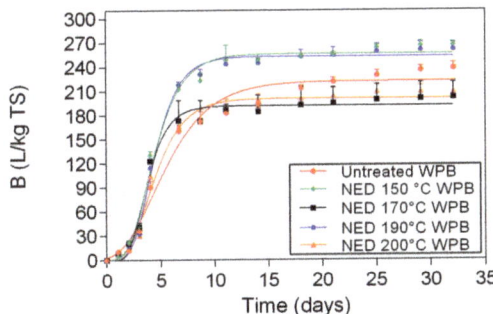

Figure 4. Biomethane potential measurement results and respective fitting curves for untreated samples and for WPB samples that were pretreated with NED at 150 °C, 170 °C, 190 °C, and 200 °C.

The biomethane results of untreated sample and WPB samples that were pretreated with acid at 1%, 2%, 3%, and 4% are shown in Figure 5. The biomethane yield varied between 150 L/kg TS (samples that were pretreated with acid 2%) and 263 L/kg TS (samples that were pretreated with acid 4%).

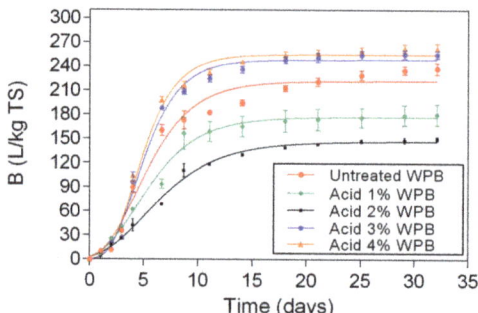

Figure 5. Biomethane potential measurement results and respective fitting curves for untreated samples and for WPB samples that were pretreated with acid at 1%, 2%, 3%, and 4%.

The biomethane potential of untreated WPD and of samples that were pretreated with NED (WPD) is illustrated in Figure 6. Untreated WPD had the lowest biomethane yields (202 L/kg TS), while samples that were pretreated with NED at 200 °C (WPD) had the highest biomethane yields (291 L/kg TS). The biomethane yields of samples that were pretreated with NED 170 °C (WPD) was 217 L/kg TS, followed by samples that were pretreated with NED 150 °C (WPD) (232 L/kg TS), and NED 190 °C (WPD) (236 L/kg TS). Statistically significant differences were found between untreated material and samples that were pretreated with NED at different temperatures (Table A4, Appendix A).

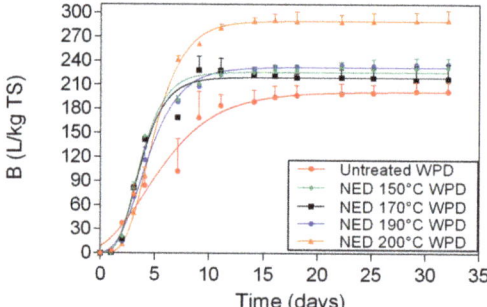

Figure 6. Biomethane potential measurement results and respective fitting curves for untreated samples and for WPD samples that were pretreated with NED at 150 °C, 170 °C, 190 °C, and 200 °C.

The biomethane potential of untreated WPD and samples that were pretreated with acid at 1%, 2%, 3%, and 4% (WPD) is reported in Figure 7. The biomethane yield of WPD was 202 L/kg TS, followed by samples that were pretreated with 1% acid WPD (218 L/kg TS), 4% acid WPD (260 L/kg TS), 2% acid WPD (289 L/kg TS), and 3% acid WPD (299 L/kg TS). Statistically significant differences were found between untreated WPD and samples pretreated with acid (Table A4, Appendix A).

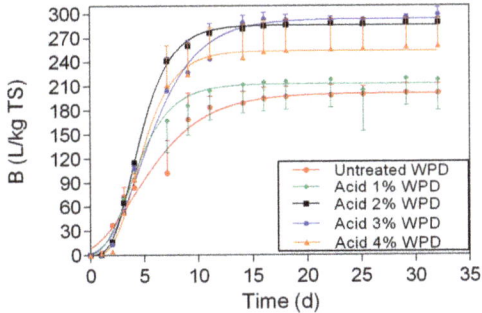

Figure 7. Biomethane potential measurement results and respective fitting curves for untreated samples and for WPD samples that were pretreated with acid at 1%, 2%, 3%, and 4%.

The maximum methane yield for WPB samples that were pretreated with NED at 150 °C, 170 °C, 190 °C, and 200 °C and with acid at 1%, 2%, 3%, and 4% shown in Figure 8. The maximum methane yield of untreated WPB was 222 L/kg TS. For samples (WPB) that were pretreated with NED, the maximum methane yields varied between 192 L/kg TS (samples that were pretreated with NED at 170 °C) and 256 L/kg TS (samples that were pretreated with NED at 150 °C). For samples (WPB) that were pretreated with acid the maximum methane yield was lower at 2% (136 L/kg TS) and higher at 4% (255 L/kg TS). The maximum methane yield of untreated WPD was 201 L/kg TS. For samples (WPD) that were pretreated with NED, the maximum methane yields ranged between 219 L/kg TS (NED 170 °C WPD) and 290 L/kg TS (NED 200 °C WPD), while for samples (WPD) that were pretreated with acid it ranged between 147 L/kg TS (acid 2% WPD) and 294 L/kg TS (acid 3% WPD). Statistically significant differences were found between NED 200 °C WPD vs. NED 200 °C WPB and Acid 2% WPD vs. Acid 2% WPB (Table A4, Appendix A).

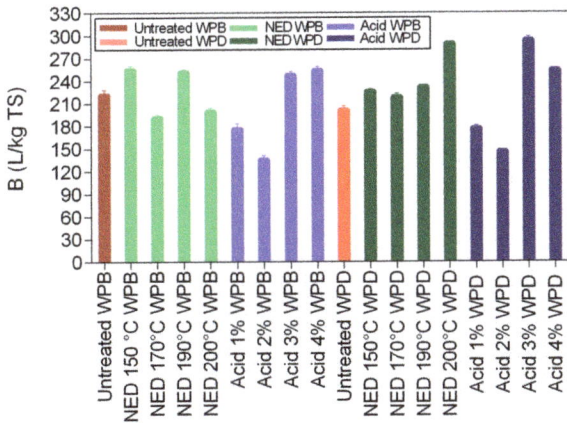

Figure 8. Maximum methane yields (B_{max}) of the fitting curves for WPB and WPD samples that were pretreated with NED at 150 °C, 170 °C, 190 °C, and 200 °C and with acid at 1%, 2%, 3%, and 4%.

3.3.2. Digestion Time

The digestion time to reach 85% (B_{85}) and 95% (B_{95}) of maximum biomethane potential and the Gompertz growth parameters for WPB and WPD samples that were pretreated with NED at 150 °C, 170 °C, 190 °C, and 200 °C and with acid at 1%, 2%, 3%, and 4% are shown in Table 4. Untreated WPB needed 14 days (201 L/kg TS) to achieve B85 and 21 days (225 L/kg TS) to reach B95, while untreated WPD required 7.4 days (172 L/kg TS) to achieve B85 and 11 days (193 L/kg TS) to reach B95. Overall, WPB samples that were pretreated with 1% acid and WPB samples that were pretreated with NED 150 °C and NED 170 °C require less time to achieve B85. WPD samples that were pretreated with 3% acid and WPB samples that were pretreated with 3% acid and 4% acid need more time to reach B85. A similar trend is followed for B95.

Table 4. Digestion time to achieve 85% (B_{85}) and 95% (B_{95}) of maximum biomethane potential and Gompertz growth parameters for WPB and WPD that were pretreated with NED at 150 °C, 170 °C, 190 °C, and 200 °C and with acid at 1%, 2%, 3%, and 4%.

	B_{85}		B_{95}		Gompertz Growth		
	L/kg TS	Days	L/kg TS	Days	Y_M	Y_0	K
NED 150 °C WPB	231	11	258	17	256 ± 4	0.06 ± 0.1	0.6 ± 0.07
NED 170 °C WPB	173	10	193	16	192 ± 4	0.0002 ± 0.001	0.8 ± 0.1
NED 190 °C WPB	228	11	255	17	253 ± 3	0.001 ± 0.003	0.6 ± 0.06
NED 200 °C WPB	181	12	203	18	200 ± 4	0.09 ± 0.2	0.5 ± 0.08
Acid 1% WPB	154	7.7	172	12	177 ± 3	3 ± 2	0.3 ± 0.03
Acid 2% WPB	126	11	140	17	147 ± 3	3 ± 1	0.3 ± 0.02
Acid 3% WPB	224	13	251	19	249 ± 4	0.3 ± 0.4	0.4 ± 0.04
Acid 4% WPB	230	13	257	19	255 ± 4	0.2 ± 0.3	0.5 ± 0.05
NED 150 °C WPD	202	9.7	226	15	226 ± 3	0.08 ± 0.1	0.7 ± 0.07
NED 170 °C WPD	195	8.0	218	12	219 ± 4	0.08 ± 0.2	0.6 ± 0.1
NED 190 °C WPD	207	11	232	17	232 ± 2	0.6 ± 0.5	0.5 ± 0.04
NED 200 °C WPD	260	12	291	18	290 ± 1	0.02 ± 0.01	0.5 ± 0.02
Acid 1% WPD	190	11	212	17	213 ± 3	1 ± 1	0.5 ± 0.05
Acid 2% WPD	256	12	286	17	286 ± 1	0.1 ± 0.06	0.5 ± 0.02
Acid 3% WPD	264	14	295	21	294 ± 4	2 ± 1	0.3 ± 0.03
Acid 4% WPD	229	12	256	18	254 ± 2	0.07 ± 0.07	0.5 ± 0.03
Untreated WPB	201	14	225	21	222 ± 6	3 ± 2	0.3 ± 0
Untreated WPD	172	7.4	193	11	201 ± 5	9 ± 4	0.3 ± 0.04

Untreated WPB needed 14 days (201 L/kg TS) to achieve B_{85} and 21 days (225 L/kg TS) to reach B_{95}, while untreated WPD required 7.4 days (172 L/kg TS) to achieve B_{85} and 11 days (193 L/kg TS) to reach B_{95}. Overall, WPB samples that were pretreated with 1% acid and WPB samples that were pretreated with NED 150 °C and NED 170 °C require less time to achieve B_{85}. WPD samples that were pretreated with 3% acid and WPB samples that were pretreated with 3% acid and 4% acid need more time to reach B_{85}. A similar trend is followed for B_{95}.

4. Discussion

4.1. Chemical Composition

The cellulose content reported in this study was 6% to 18% lower than the values reported in the literature [314 however, the overall percentage of cellulose was still very high (>77%). The high percentage of cellulose that was found in the waste banknote paper is due to its composition. Waste banknote paper has high amounts of cotton, which has mainly cellulose on its composition. The results obtained in this study were also compared with the chemical composition of cotton crops reported by Rocha-Meneses et al. [15]. The authors described the content of cellulose and hemicellulose present in cotton as 80–95% and 5–20%, respectively, which is in line with the results found in our study. These results are favorable for bioethanol and biogas production since research has shown that high amounts of cellulose contribute to high yields of bioethanol and biomethane. The hemicellulose and lignin contents are within the range of values reported in the literature [31]. Moreover, low lignin content is advantageous since the amount of energy that is required to break down the chemical bonds between hemicellulose and lignin and to make cellulose accessible for the enzymatic hydrolysis is low [20,32].

The amounts of TS and VS are fundamental when it comes to biogas production. High amounts of TS specify the amount of substrate that is accessible for the anaerobic digestion process, while high amounts of VS refer to the amount of substrate that can be transformed into biomethane. When compared with similar subtracts, higher amounts of VS lead to higher amounts of biogas [32,33].

4.2. Glucose Content after Hydrolysis and Ethanol Content after Fermentation

The use of WPD resulted in higher glucose and ethanol concentrations and yields when compared to WPB regardless of the pretreatment. A previous study also showed that the deinking process improved ethanol production of the pretreated waste banknote [26]. In this present work, when fermenting the NED-pretreated WPD, we were able to achieve 6.36 g/L of ethanol and 98% of the theoretical ethanol yield from glucose (0.50 g ethanol/g glucose). Although the ethanol concentration achieved is lower than the one obtained by Rocha-Meneses et al. [32] when using NED pretreatment to Napier grass as a fermentation substrate (10.3 g/L), a lower bioethanol yield (90% of the theoretical yield) was reported by the authors, probably by the presence of microbial growth inhibitors. No microbial inhibition effect was observed when using pretreated WPD as a fermentation substrate. Jeihanipour and Taherzadeha (2009) [34] reported a yield of 0.48 g ethanol/g textile from alkali pretreated cotton linter and waste jeans as a substrate for *S. cerevisiae* during a simultaneous saccharification and fermentation process (SSF). The yield reported by Jeihanipour and Taherzadeha (2009) [34] is higher than the one obtained in the present work (0.21 g ethanol/g waste banknote), but this yield can possibly be enhanced by optimizing the enzymatic hydrolysis step increasing glucose content, and, therefore, the ethanol production, as shown previously by Aghmashhadi et al. (2020) [26], widening the potential of this waste biomass utilization.

4.3. Potential of Waste Banknote Paper for Biomethane Production

From Figure 4 it is evident that samples of WPB that were pretreated with NED at 150 °C and 190 °C produce 12–13% more biomethane than untreated WPB, while samples that were pretreated at 170 °C and 200 °C have biomethane yields 9–15% lower than untreated WPB. At the moment, the mechanism behind these differences is not clear. As there is clearer trend in WPD, the hectic results of WPB might be due to the presence of ink and its interactions with the other compounds that are produced during the pretreatment at high temperature. Moreover, it might be due to the different amounts of ink present in different samples of WPB. Not all of the banknote is uniformly covered with the same kind of ink. Furthermore, these pretreatment methods are effective only under certain temperatures, which could require further studies in order to investigate the optimum pretreatment temperature for the anaerobic digestion process [35,36]. If the experiments are performed outside the optimum pretreatment temperatures it can lead to the production of inhibitory compounds and a decrease in the substrate biodegradability [37,38].

As it can be seen from Figure 5 only pretreatment with 3% and 4% acid increased biomethane production compared to untreated sample for WPB. Samples that were pretreated with high acid concentrations showed biomethane content 10–12% higher than untreated banknote paper. Similar results were reported by Sarto, Hildayati and Syaichurrozi (2019) [39]. On the other hand, lower concentrations of acid in the pretreatment seems to inhibit the biomethane production. This can be due to the composition of the samples. As the samples still have ink in their composition, probably harsher pretreatment methods are required to disrupt the structure of the banknote paper and access the cellulose, as well as longer retention times, since the degradation rate will be slower. Moreover, the utilization of acid to adjust the pH can have led to excessive sugar degradation and the production of inhibitory compounds [40]. A study reported by Venturin et al. [41] showed that H_2SO_4 pretreatment inhibited biogas production. In order to ascertain the exact mechanism, further studies should be performed, with different retention times in the acid pretreatment, and different reagents of pH adjustment.

The biomethane yields of samples that were pretreated with NED (WPD) (Figure 6) were 8-44% higher than the biomethane yields of untreated material. As it is shown in Figure 7, the biomethane content of samples that where pretreated with acid (WPD) was 8-48% higher than the biomethane content of untreated banknote paper. Overall, the biomethane yields of deinking samples (NED WPD and acid WPD) were highly improved, when compared with blank samples (NED WPB and acid WPB). For samples that were pretreated with NED, the biomethane yields of WPD were 8–35% higher than the biomethane yields of WPB. For samples that were pretreated with acid, the biomethane yields of WPD were 17–93% higher than the biomethane yields of WPB. These differences between WPB and WPD samples can be explained by the deinking process. The ink removal uses alkali as a solution. This means that deinking samples are subjected to two combined pretreatment methods alkali and NED. This can leave waste banknote paper with an increased amorphous structure, higher traits and quality, which will increase its biodegradability by the anaerobic microorganisms and as a result, increase biomethane yields.

As it is evident from Figure 8, samples that were pretreated with acid 2% WPB, acid 1% WPB, NED 170 °C WPB, NED 200 °C WPB, had the lowest biomethane yields (between 136 L/kg TS and 200 L/kg TS), while samples that were pretreated with acid 4% WPD, NED 200 °C WPD, and acid 3% WPD had the highest biomethane yields (between 254 mol/kg TS and 294 mol/kg TS). These results can be explained by the deinking process, by the effect of high acid concentrations, and by the effect of high pretreatment temperatures. Research has shown that the presence on ink in waste paper inhibits the enzymatic hydrolysis, and reduces the amount of sugars available for the fermentation stage, which results in inefficient bioethanol production [42]. Therefore, as expected, samples without ink resulted in better yields than samples with ink. On the other hand, high acid concentrations reduce the crystalline structure of cellulose more efficiently, making it more accessible for the enzymatic hydrolysis, which will lead to higher bioethanol and biogas yields [39,43,44]. The temperature also plays a role in biomethane yields. Research has shown that cellulose solubilization increases proportionally with the increment of the pretreatment temperature [39,45].

The differences in the digestion time between WPD and WPB can be explained by the presence of ink. Research has shown that the presence of ink can inhibit biorefinery processes [26].

These results are particularly important when it comes to the reduction of environmental impacts caused by the incineration or landfilling of waste banknote paper. At the moment, these are the most common handling options that are being applied to waste banknote paper. However, incineration and landfilling are a source of environmental pollution, contaminating the air, soil and water. Therefore, the Central Bank of Iran has decided to take action and reduce its environmental footprint caused by waste banknote paper and shift to waste management solutions that are more environmentally friendly. Therefore, biogas and bioethanol production are proposed as alternative handling options for these residues since it is known that they have a lower footprint when compared to incineration or landfilling [46]. Further research should be performed in order to make this solution competitive and to quantify the energy output from biogas, bioethanol, incineration and landfilling of waste banknote paper.

5. Conclusions

In this study, two different pretreatment methods (chemical and physio-chemical) were used to investigate the potential of waste banknote paper with and without ink for bioethanol and biomethane production. The results of this study show that glucose and ethanol concentrations are higher in WPD samples than for WPB. The glucose yields varied between 4.57 g/L (1% acid WPB) and 12.98 g/L (NED 200 °C WPD), while ethanol yields varied between 1.8 g/L (1% Acid WPB) and 6.36 g/L (NED 200 °C WPD). For WPB samples, the highest biomethane yields were reported in samples that were pretreated with NED 150 °C (266 L/kg TS) and 4% acid (263 L/Kg TS). For WPD samples, samples that were pretreated with 3% acid (299 L/Kg TS) and NED 200 °C (292 L/Kg TS) pretreatments insured the highest biomethane production. In general, the biomethane potential of WPD samples was higher than that of similarly pretreated WPB samples. NED pretreatment method gives higher glucose, ethanol, and biomethane yields than samples that were pretreated with sulfuric acid. The deinking process, the acid concentration, and the pretreatment temperature all influence glucose, bioethanol, and biomethane yields. More research should be done to find the most efficient ways to increase glucose, bioethanol, and biomethane yields.

Author Contributions: Conceptualization, O.Y.A., L.R.-M., and T.K.; methodology, O.Y.A., L.R.-M. and N.B.; software, L.R.-M.; validation, O.Y.A., L.R.-M. and N.B., and T.K.; formal analysis, L.R.-M.; investigation, O.Y.A., L.R.-M. and N.B.; resources, T.K.; data curation, L.R.-M. and N.B.; writing—original draft preparation, O.Y.A., L.R.-M., N.B., K.O. and T.K.; writing—review and editing, L.R.-M., N.B., K.O. and T.K.; visualization, L.R.-M.; supervision, L.R.-M., G.A., E.R.G., M.Z. and T.K.; project administration, T.K.; funding acquisition, T.K. All authors have read and agreed to the published version of the manuscript.

Funding: We gratefully acknowledge the financial support of the Doctoral School of Energy and Geotechnology III, supported by the European Union, European Regional Development Fund (Estonian University of Life Sciences ASTRA project "Value-chain based bio-economy"). The HPLC analyses were funded by European Union's Horizon 2020 research and innovation program under grant agreement No 668997 and the Estonian Research Council (grant PUT1488P).

Institutional Review Board Statement: Not applicable.

Informed Consent Statement: Not applicable.

Data Availability Statement: Not applicable.

Conflicts of Interest: The authors declare no conflict of interest.

Appendix A

Table A1. Two-way ANOVA, Tukey's multiple comparisons test, for cellulose content.

Dunn's Multiple Comparisons Test	Mean Rank Diff.	Significant?	Summary	Adjusted p Value
Acid 3% WPB vs. NED 150 °C WPB	−29.58	Yes	****	<0.0001
Acid 3% WPB vs. NED 170 °C WPB	−28.6	Yes	***	0.0001
Acid 3% WPB vs. NED 190 °C WPB	−22.89	Yes	**	0.0074
Acid 3% WPB vs. NED 200 °C WPB	−27.8	Yes	***	0.0002
Acid 3% WPB vs. NED 150 °C WPD	−33.52	Yes	****	<0.0001
Acid 3% WPB vs. NED 170 °C WPD	−32.84	Yes	****	<0.0001
Acid 3% WPB vs. NED 190 °C WPD	−32.67	Yes	****	<0.0001
Acid 3% WPB vs. NED 200 °C WPD	−32.33	Yes	****	<0.0001
Acid 3% WPB vs. Acid 1% WPB	−29.52	Yes	****	<0.0001
Acid 3% WPB vs. Acid 2% WPB	−27.38	Yes	***	0.0003
Acid 4% WPB vs. Acid 3% WPB	23.46	Yes	**	0.0052
Acid 1% WPD vs. Acid 3% WPB	33.26	Yes	****	<0.0001
Acid 2% WPD vs. Acid 3% WPB	32.5	Yes	****	<0.0001
Acid 3% WPD vs. Acid 3% WPB	29.23	Yes	****	<0.0001
Acid 4% WPD vs. Acid 3% WPB	30.37	Yes	****	<0.0001
Untreated WPB vs. Acid 3% WPB	22.15	Yes	*	0.0118
Untreated WPD vs. Acid 3% WPB	33.76	Yes	****	<0.0001

* $p \leq 0.05$, ** $p \leq 0.01$, *** $p \leq 0.001$, **** $p \leq 0.0001$.

Table A2. Two-way ANOVA, Tukey's multiple comparisons test, for glucose yields.

Dunn's Multiple Comparisons Test	Mean Rank Diff.	Significant?	Summary	Adjusted p Value
NED 150 °C WPB vs. NED 170 °C WPB	−2.36	Yes	***	0.0002
NED 150 °C WPB vs. NED 190 °C WPB	−3.02	Yes	****	<0.0001
NED 150 °C WPB vs. NED 200 °C WPB	−3.95	Yes	****	<0.0001
NED 150 °C WPB vs. NED 150 °C WPD	−6.72	Yes	****	<0.0001
NED 150 °C WPB vs. NED 170 °C WPD	−5.21	Yes	****	<0.0001
NED 150 °C WPB vs. NED 190 °C WPD	−6.02	Yes	****	<0.0001
NED 150 °C WPB vs. NED 200 °C WPD	−7.5	Yes	****	<0.0001
NED 150 °C WPB vs. Acid 2% WPB	−4.15	Yes	****	<0.0001
NED 150 °C WPB vs. Acid 3% WPB	−4.3	Yes	****	<0.0001
NED 150 °C WPB vs. Acid 4% WPB	−2.8	Yes	****	<0.0001
NED 150 °C WPB vs. Acid 1% WPD	−4.92	Yes	****	<0.0001
NED 150 °C WPB vs. Acid 2% WPD	−5.12	Yes	****	<0.0001
NED 150 °C WPB vs. Acid 3% WPD	−3.32	Yes	****	<0.0001
NED 150 °C WPB vs. Acid 4% WPD	−5.42	Yes	****	<0.0001
NED 170 °C WPB vs. NED 150 °C WPD	−4.36	Yes	****	<0.0001
NED 170 °C WPB vs. NED 170 °C WPD	−2.85	Yes	****	<0.0001
NED 170 °C WPB vs. NED 190 °C WPD	−3.66	Yes	****	<0.0001
NED 170 °C WPB vs. NED 200 °C WPD	−5.14	Yes	****	<0.0001
NED 170 °C WPB vs. Acid 1% WPB	3.27	Yes	****	<0.0001
NED 170 °C WPB vs. Acid 2% WPB	−1.79	Yes	*	0.0164
NED 170 °C WPB vs. Acid 3% WPB	−1.94	Yes	**	0.0058
NED 170 °C WPB vs. Acid 1% WPD	−2.56	Yes	****	<0.0001
NED 170 °C WPB vs. Acid 2% WPD	−2.76	Yes	****	<0.0001
NED 170 °C WPB vs. Acid 4% WPD	−3.06	Yes	****	<0.0001
NED 190 °C WPB vs. NED 150 °C WPD	−3.7	Yes	****	<0.0001
NED 190 °C WPB vs. NED 170 °C WPD	−2.19	Yes	***	0.0009
NED 190 °C WPB vs. NED 190 °C WPD	−3	Yes	****	<0.0001
NED 190 °C WPB vs. NED 200 °C WPD	−4.48	Yes	****	<0.0001
NED 190 °C WPB vs. Acid 1% WPB	3.93	Yes	****	<0.0001
NED 190 °C WPB vs. Acid 1% WPD	−1.9	Yes	**	0.0077
NED 190 °C WPB vs. Acid 2% WPD	−2.1	Yes	**	0.0018
NED 190 °C WPB vs. Acid 4% WPD	−2.4	Yes	***	0.0002
NED 200 °C WPB vs. NED 150 °C WPD	−2.77	Yes	****	<0.0001
NED 200 °C WPB vs. NED 190 °C WPD	−2.07	Yes	**	0.0022
NED 200 °C WPB vs. NED 200 °C WPD	−3.55	Yes	****	<0.0001
NED 200 °C WPB vs. Acid 1% WPB	4.86	Yes	****	<0.0001
NED 150 °C WPD vs. Acid 1% WPB	7.63	Yes	****	<0.0001
NED 150 °C WPD vs. Acid 2% WPB	2.57	Yes	****	<0.0001
NED 150 °C WPD vs. Acid 3% WPB	2.42	Yes	***	0.0001
NED 150 °C WPD vs. Acid 4% WPB	3.92	Yes	****	<0.0001
NED 150 °C WPD vs. Acid 1% WPD	1.8	Yes	*	0.0154
NED 150 °C WPD vs. Acid 3% WPD	3.4	Yes	****	<0.0001
NED 170 °C WPD vs. NED 200 °C WPD	−2.29	Yes	***	0.0004
NED 170 °C WPD vs. Acid 1% WPB	6.12	Yes	****	<0.0001
NED 170 °C WPD vs. Acid 4% WPB	2.41	Yes	***	0.0002
NED 170 °C WPD vs. Acid 3% WPD	1.89	Yes	**	0.0083
NED 190 °C WPD vs. Acid 1% WPB	6.93	Yes	****	<0.0001
NED 190 °C WPD vs. Acid 2% WPB	1.87	Yes	**	0.0095
NED 190 °C WPD vs. Acid 3% WPB	1.72	Yes	*	0.026
NED 190 °C WPD vs. Acid 4% WPB	3.22	Yes	****	<0.0001
NED 190 °C WPD vs. Acid 3% WPD	2.7	Yes	****	<0.0001

Table A2. Cont.

Dunn's Multiple Comparisons Test	Mean Rank Diff.	Significant?	Summary	Adjusted p Value
NED 200 °C WPD vs. Acid 1% WPB	8.41	Yes	****	<0.0001
NED 200 °C WPD vs. Acid 2% WPB	3.35	Yes	****	<0.0001
NED 200 °C WPD vs. Acid 3% WPB	3.2	Yes	****	<0.0001
NED 200 °C WPD vs. Acid 4% WPB	4.7	Yes	****	<0.0001
NED 200 °C WPD vs. Acid 1% WPD	2.58	Yes	****	<0.0001
NED 200 °C WPD vs. Acid 2% WPD	2.38	Yes	***	0.0002
NED 200 °C WPD vs. Acid 3% WPD	4.18	Yes	****	<0.0001
NED 200 °C WPD vs. Acid 4% WPD	2.08	Yes	**	0.0021
Acid 1% WPB vs. Acid 2% WPB	−5.06	Yes	****	<0.0001
Acid 1% WPB vs. Acid 3% WPB	−5.21	Yes	****	<0.0001
Acid 1% WPB vs. Acid 4% WPB	−3.71	Yes	****	<0.0001
Acid 1% WPB vs. Acid 1% WPD	−5.83	Yes	****	<0.0001
Acid 1% WPB vs. Acid 2% WPD	−6.03	Yes	****	<0.0001
Acid 1% WPB vs. Acid 3% WPD	−4.23	Yes	****	<0.0001
Acid 1% WPB vs. Acid 4% WPD	−6.33	Yes	****	<0.0001
Acid 4% WPB vs. Acid 1% WPD	−2.12	Yes	**	0.0015
Acid 4% WPB vs. Acid 2% WPD	−2.32	Yes	***	0.0003
Acid 4% WPB vs. Acid 4% WPD	−2.62	Yes	****	<0.0001
Acid 2% WPD vs. Acid 3% WPD	1.8	Yes	*	0.0154
Acid 3% WPD vs. Acid 4% WPD	−2.1	Yes	**	0.0018

* $p \leq 0.05$, ** $p \leq 0.01$, *** $p \leq 0.001$, **** $p \leq 0.0001$.

Table A3. Two-way ANOVA, Tukey's multiple comparisons test, for ethanol yields.

Dunn's Multiple Comparisons Test	Mean Rank Diff.	Significant?	Summary	Adjusted p Value
NED 150 °C WPB vs. NED 200 °C WPB	−2.02	Yes	**	0.0032
NED 150 °C WPB vs. NED 150 °C WPD	−4.12	Yes	****	<0.0001
NED 150 °C WPB vs. NED 170 °C WPD	−3.28	Yes	****	<0.0001
NED 150 °C WPB vs. NED 190 °C WPD	−3.69	Yes	****	<0.0001
NED 150 °C WPB vs. NED 200 °C WPD	−4.33	Yes	****	<0.0001
NED 150 °C WPB vs. Acid 2% WPB	−2.16	Yes	**	0.0011
NED 150 °C WPB vs. Acid 3% WPB	−1.9	Yes	**	0.0077
NED 150 °C WPB vs. Acid 1% WPD	−2.67	Yes	****	<0.0001
NED 150 °C WPB vs. Acid 2% WPD	−2.87	Yes	****	<0.0001
NED 150 °C WPB vs. Acid 3% WPD	−1.87	Yes	**	0.0095
NED 150 °C WPB vs. Acid 4% WPD	−3.07	Yes	****	<0.0001
NED 170 °C WPB vs. NED 150 °C WPD	−3.06	Yes	****	<0.0001
NED 170 °C WPB vs. NED 170 °C WPD	−2.22	Yes	***	0.0007
NED 170 °C WPB vs. NED 190 °C WPD	−2.63	Yes	****	<0.0001
NED 170 °C WPB vs. NED 200 °C WPD	−3.27	Yes	****	<0.0001
NED 170 °C WPB vs. Acid 2% WPD	−1.81	Yes	*	0.0143
NED 170 °C WPB vs. Acid 4% WPD	−2.01	Yes	**	0.0035
NED 190 °C WPB vs. NED 150 °C WPD	−2.62	Yes	****	<0.0001
NED 190 °C WPB vs. NED 170 °C WPD	−1.78	Yes	*	0.0176
NED 190 °C WPB vs. NED 190 °C WPD	−2.19	Yes	***	0.0009
NED 190 °C WPB vs. NED 200 °C WPD	−2.83	Yes	****	<0.0001
NED 190 °C WPB vs. Acid 1% WPB	1.73	Yes	*	0.0244

Table A3. Cont.

Dunn's Multiple Comparisons Test	Mean Rank Diff.	Significant?	Summary	Adjusted p Value
NED 200 °C WPB vs. NED 150 °C WPD	−2.1	Yes	**	0.0018
NED 200 °C WPB vs. NED 190 °C WPD	−1.67	Yes	*	0.0357
NED 200 °C WPB vs. NED 200 °C WPD	−2.31	Yes	***	0.0003
NED 200 °C WPB vs. Acid 1% WPB	2.25	Yes	***	0.0006
NED 150 °C WPD vs. Acid 1% WPB	4.35	Yes	****	<0.0001
NED 150 °C WPD vs. Acid 2% WPB	1.96	Yes	**	0.005
NED 150 °C WPD vs. Acid 3% WPB	2.22	Yes	***	0.0007
NED 150 °C WPD vs. Acid 4% WPB	2.92	Yes	****	<0.0001
NED 150 °C WPD vs. Acid 3% WPD	2.25	Yes	***	0.0006
NED 170 °C WPD vs. Acid 1% WPB	3.51	Yes	****	<0.0001
NED 170 °C WPD vs. Acid 4% WPB	2.08	Yes	**	0.0021
NED 190 °C WPD vs. Acid 1% WPB	3.92	Yes	****	<0.0001
NED 190 °C WPD vs. Acid 3% WPB	1.79	Yes	*	0.0164
NED 190 °C WPD vs. Acid 4% WPB	2.49	Yes	****	<0.0001
NED 190 °C WPD vs. Acid 3% WPD	1.82	Yes	*	0.0134
NED 200 °C WPD vs. Acid 1% WPB	4.56	Yes	****	<0.0001
NED 200 °C WPD vs. Acid 2% WPB	2.17	Yes	**	0.001
NED 200 °C WPD vs. Acid 3% WPB	2.43	Yes	***	0.0001
NED 200 °C WPD vs. Acid 4% WPB	3.13	Yes	****	<0.0001
NED 200 °C WPD vs. Acid 1% WPD	1.66	Yes	*	0.038
NED 200 °C WPD vs. Acid 3% WPD	2.46	Yes	***	0.0001
Acid 1% WPB vs. Acid 2% WPB	−2.39	Yes	***	0.0002
Acid 1% WPB vs. Acid 3% WPB	−2.13	Yes	**	0.0014
Acid 1% WPB vs. Acid 1% WPD	−2.9	Yes	****	<0.0001
Acid 1% WPB vs. Acid 2% WPD	−3.1	Yes	****	<0.0001
Acid 1% WPB vs. Acid 3% WPD	−2.1	Yes	**	0.0018
Acid 1% WPB vs. Acid 4% WPD	−3.3	Yes	****	<0.0001
Acid 4% WPB vs. Acid 2% WPD	−1.67	Yes	*	0.0357
Acid 4% WPB vs. Acid 4% WPD	−1.87	Yes	**	0.0095

* $p \leq 0.05$, ** $p \leq 0.01$, *** $p \leq 0.001$, **** $p \leq 0.0001$.

Table A4. One-way ANOVA, Dunn's multiple comparisons test, for biomethane results.

Dunn's Multiple Comparisons Test	Mean Rank Diff.	Significant?	Summary	Adjusted p Value
Untreated WPD vs. NED 200 °C WPD	−225.6	Yes	****	<0.0001
Untreated WPD vs. Acid 2% WPD	−235.7	Yes	****	<0.0001
Untreated WPD vs. Acid 3% WPD	−223.3	Yes	****	<0.0001
Untreated WPD vs. NED 150 °C WPB	−176.6	Yes	**	0.006
Untreated WPD vs. NED 190 °C WPB	−164.1	Yes	*	0.0204
Untreated WPD vs. Acid 4% WPB	−159.7	Yes	*	0.0307
NED 150 °C WPD vs. Acid 2% WPB	195.6	Yes	***	0.0008
NED 170 °C WPD vs. Acid 2% WPB	186.5	Yes	**	0.0022
NED 190 °C WPD vs. Acid 1% WPB	167.2	Yes	*	0.0152
NED 190 °C WPD vs. Acid 2% WPB	223.6	Yes	****	<0.0001
NED 200 °C WPD vs. Acid 1% WPD	182.5	Yes	**	0.0033
NED 200 °C WPD vs. Untreated WPB	173	Yes	**	0.0087
NED 200 °C WPD vs. NED 170 °C WPB	233.3	Yes	****	<0.0001
NED 200 °C WPD vs. NED 200 °C WPB	211.9	Yes	***	0.0001
NED 200 °C WPD vs. Acid 1% WPB	273.6	Yes	****	<0.0001
NED 200 °C WPD vs. Acid 2% WPB	330.1	Yes	****	<0.0001

Table A4. *Cont.*

Dunn's Multiple Comparisons Test	Mean Rank Diff.	Significant?	Summary	Adjusted p Value
Acid 1% WPD vs. Acid 2% WPD	−192.6	Yes	**	0.0013
Acid 1% WPD vs. Acid 3% WPD	−180.2	Yes	**	0.0048
Acid 2% WPD vs. Untreated WPB	183.1	Yes	**	0.0036
Acid 2% WPD vs. NED 170 °C WPB	243.4	Yes	****	<0.0001
Acid 2% WPD vs. NED 200 °C WPB	222	Yes	****	<0.0001
Acid 2% WPD vs. Acid 1% WPB	283.7	Yes	****	<0.0001
Acid 2% WPD vs. Acid 2% WPB	340.2	Yes	****	<0.0001
Acid 3% WPD vs. Untreated WPB	170.7	Yes	*	0.0122
Acid 3% WPD vs. NED 170 °C WPB	231	Yes	****	<0.0001
Acid 3% WPD vs. NED 200 °C WPB	209.6	Yes	***	0.0002
Acid 3% WPD vs. Acid 1% WPB	271.3	Yes	****	<0.0001
Acid 3% WPD vs. Acid 2% WPB	327.8	Yes	****	<0.0001
Acid 4% WPD vs. NED 170 °C WPB	156.9	Yes	*	0.0439
Acid 4% WPD vs. Acid 1% WPB	197.2	Yes	***	0.0008
Acid 4% WPD vs. Acid 2% WPB	253.7	Yes	****	<0.0001
Untreated WPB vs. Acid 2% WPB	157.1	Yes	*	0.0432
NED 150 °C WPB vs. NED 170 °C WPB	184.3	Yes	**	0.0031
NED 150 °C WPB vs. NED 200 °C WPB	162.9	Yes	*	0.0255
NED 150 °C WPB vs. Acid 1% WPB	224.6	Yes	****	<0.0001
NED 150 °C WPB vs. Acid 2% WPB	281.1	Yes	****	<0.0001
NED 170 °C WPB vs. NED 190 °C WPB	−171.8	Yes	*	0.011
NED 170 °C WPB vs. Acid 4% WPB	−167.4	Yes	*	0.0168
NED 190 °C WPB vs. Acid 1% WPB	212.1	Yes	***	0.0001
NED 190 °C WPB vs. Acid 2% WPB	268.6	Yes	****	<0.0001
Acid 1% WPB vs. Acid 3% WPB	−181.5	Yes	**	0.0042
Acid 1% WPB vs. Acid 4% WPB	−207.7	Yes	***	0.0002
Acid 2% WPB vs. Acid 3% WPB	−237.9	Yes	****	<0.0001
Acid 2% WPB vs. Acid 4% WPB	−264.1	Yes	****	<0.0001

* $p \leq 0.05$, ** $p \leq 0.01$, *** $p \leq 0.001$, **** $p \leq 0.0001$.

References

1. Dabe, S.; Prasad, P.; Vaidya, A.; Purohit, H. Technological pathways for bioenergy generation from municipal solid waste: Renewable energy option. *Environ. Prog. Sustain. Energy* **2019**, *38*, 654–671. [CrossRef]
2. Bakraoui, M.; Hazzi, M.; Karouach, F.; Ouhammou, B.; El Bari, H. Experimental biogas production from recycled pulp and paper wastewater by biofilm technology. *Biotechnol. Lett.* **2019**, *41*, 1299–1307. [CrossRef] [PubMed]
3. Sanchis-Sebastiá, M.; Gomis-Fons, J.; Galbe, M.; Wallberg, O. Techno-Economic Evaluation of Biorefineries Based on Low-Value Feedstocks Using the BioSTEAM Software: A Case Study for Animal Bedding. *Processes* **2020**, *8*, 904. [CrossRef]
4. Brown, A.; Waldheim, L.; Landälv, I.; Saddler, J.; Ebadian, M.; McMillan, J.D.; Bonomi, A.; Klein, B. Advanced Biofuels—Potential for Cost Reduction. *IEA Bioenergy* **2020**, 88.
5. Ameli, M.; Mansour, S.; Ahmadi-Javid, A. A simulation-optimization model for sustainable product design and efficient end-of-life management based on individual producer responsibility. *Resour. Conserv. Recycl.* **2019**, *140*, 246–258. [CrossRef]
6. Luján-Ornelas, C.; del Mancebo, C.; Sternenfels, U.; Güereca, L.P. Life cycle assessment of Mexican polymer and high-durability cotton paper banknotes. *Sci. Total Environ.* **2018**, *630*, 409–421.
7. Yehia, A.; Yassin, K.; Eid, A. Recycling of shredded currency waste of Egyptian Central Bank for making good-quality papers. *Sep. Sci.Technol.* **2017**, *53*, 1–7. [CrossRef]
8. Sunil, S.; Panchmal, G.; Shenoy, R.; Kumar, V.; Jodalli, P.; Somaraj, V. Assessment of microbial contamination of indian currency notes in circulation. *J. Indian Assoc. Public Health Dent.* **2020**, *18*, 179–182. [CrossRef]
9. Sheikh, M.M.; Kim, C.H.; Park, H.J.; Kim, S.H.; Kim, G.C.; Lee, J.Y.; Sim, S.W.; Kim, J.W. Alkaline pretreatment improves saccharification and ethanol yield from waste money bills. *Biosci. Biotechnol. Biochem.* **2013**, *77*, 1397–1402. [CrossRef]
10. Zhang, Q.; Khan, M.U.; Lin, X.; Yi, W.; Lei, H. Green-composites produced from waste residue in pulp and paper industry: A sustainable way to manage industrial wastes. *J. Clean. Prod.* **2020**, *262*, 121251. [CrossRef]

11. Parra, R.; Vargas-Radillo, J.; Arzate, F.; Quiñones, J.; Aguilar, B.; Valdovinos, E.; Casillas, R. Ultrasonic treatment for deinking of laser paper using two frequencies, 25 and 45 khz (Destintado de papel de impresión láser mediante ultrasonido con dos frecuencias, 25 y 45 kHz). *Rev. Mex. Cienc. For.* **2015**, *6*, 126–141.
12. Carré, B.; Magnin, L.; Galland, G. Deinking difficulties related to ink formulation, printing process, and type of paper. *Tappi J.* **2000**, *83*, 60.
13. Ferguson, L.D. Deinking chemistry Part. 1. *Tappi J.* **1992**, *75*, 75–83.
14. Raud, M.; Kikas, T.; Sippula, O.; Shurpali, N.J. Potentials and challenges in lignocellulosic biofuel production technology. *Renew. Sustain. Energy Rev.* **2019**, *111*, 44–56. [CrossRef]
15. Rocha-Meneses, L.; Raud, M.; Orupõld, K.; Kikas, T. Second-generation bioethanol production: A review of strategies for waste valorisation. *Agron. Res.* **2017**, *15*, 830–847.
16. Rooni, V.; Raud, M.; Kikas, T. Technical solutions used in different pretreatments of lignocellulosic biomass: A review. *Agron. Res.* **2017**, *15*, 848–858.
17. Megala, S.; Rekha, B.; Saravanathamizhan, R. Chemical and non-chemical pre-treatment techniques for bio ethanol production from biomass. *Int. J. Energy Water Resour.* **2020**, *4*, 199–204. [CrossRef]
18. Raud, M.; Orupõld, K.; Rocha-Meneses, L.; Rooni, V.; Träss, O.; Kikas, T. Biomass Pretreatment with the Szego Mill™ for Bioethanol and Biogas Production. *Processes* **2020**, *8*, 1327. [CrossRef]
19. Shirkavand, E.; Baroutian, S.; Gapes, D.J.; Young, B.R. Combination of fungal and physicochemical processes for lignocellulosic biomass pretreatment—A review. *Renew. Sustain. Energy Rev.* **2016**, *54*, 217–234. [CrossRef]
20. Rocha-Meneses, L.; Raud, M.; Orupõld, K.; Kikas, T. Potential of bioethanol production waste for methane recovery. *Energy* **2019**, *173*, 133–139. [CrossRef]
21. Raud, M.; Olt, J.; Kikas, T. N2 explosive decompression pretreatment of biomass for lignocellulosic ethanol production. *Biomass Bioenergy* **2016**, *90*, 1–6. [CrossRef]
22. Rodriguez, C.; Alaswad, A.; El-Hassan, Z.; Olabi, A.G. Mechanical pretreatment of waste paper for biogas production. *Waste Manag.* **2017**, *68*, 157–164. [CrossRef] [PubMed]
23. Manyi-Loh, C.E.; Mamphweli, S.N.; Meyer, E.L.; Okoh, A.I.; Makaka, G.; Simon, M. Microbial anaerobic digestion (bio-digesters) as an approach to the decontamination of animal wastes in pollution control and the generation of renewable energy. *Int. J. Environ. Res. Public Health* **2013**, *10*, 4390–4417. [CrossRef]
24. Rocha-Meneses, L.; Ivanova, A.; Atouguia, G.; Avila, I.; Raud, M.; Orupold, K.; Kikas, T. The effect of flue gas explosive decompression pretreatment on methane recovery from bioethanol production waste. *Industri Crops Prod.* **2019**, *127*, 66–72. [CrossRef]
25. Test Method TAPPI/ANSI T 456 om-15. *Tensile Breaking Strength of Water-Saturated Paper and Paperboard ("Wet Tensile Strength"). Standard Method*; Technical Association of the Pulp and Paper Industry: Peachtree Corners, GA, USA, 2015.
26. Aghmashhadi, O.Y.; Asadpour, G.; Garmaroody, E.R.; Zabihzadeh, M.; Rocha-Meneses, L.; Kikas, T. The Effect of Deinking Process. *on Bioethanol Production from Waste Banknote Paper. Processes* **2020**, *8*, 1563.
27. Chatrath, H.; Durge, R. Repulping of Waste Paper Containing High Wet Strength. *IUP J. Sci. Technol.* **2011**, *7*.
28. Rocha Meneses, L.; Orupõld, K.; Kikas, T. Potential of bioethanol production waste for methane recovery (short version). In Proceedings of the 11th International Conference on Sustainable Energy & Environmental Protection, Paisley, UK, 8–11 May 2018.
29. Angelidaki, I.; Treu, L.; Tsapekos, P.; Luo, G.; Campanaro, S.; Wenzel, H.; Kougias, P.G. Biogas upgrading and utilization: Current status and perspectives. *Biotechnol. Adv.* **2018**, *36*, 452–466. [CrossRef]
30. Raud, M.; Rooni, V.; Kikas, T. The Efficiency of Nitrogen and Flue Gas as Operating Gases in Explosive Decompression Pretreatment. *Energies* **2018**, *11*, 2074. [CrossRef]
31. Yousef, S.; Eimontas, J.; Striūgas, N.; Trofimov, E.; Hamdy, M.; Abdelnaby, M.A. Conversion of end-of-life cotton banknotes into liquid fuel using mini-pyrolysis plant. *J. Clean. Prod.* **2020**, *267*, 121612. [CrossRef]
32. Rocha-Meneses, L.; Otor, O.F.; Bonturi, N.; Orupõld, K.; Kikas, T. Bioenergy Yields from Sequential Bioethanol and Biomethane Production: An Optimized Process Flow. *Sustainability* **2020**, *12*, 272. [CrossRef]
33. Rocha-Meneses, L.; Ferreira, J.A.; Bonturi, N.; Orupõld, K.; Kikas, T. Enhancing Bioenergy Yields from Sequential Bioethanol and Biomethane Production by Means of Solid–Liquid Separation of the Substrates. *Energies* **2019**, *12*, 3683. [CrossRef]
34. Jeihanipour, A.; Taherzadeh, M.J. Ethanol production from cotton-based waste textiles. *Bioresour. Technol.* **2009**, *100*, 1007–1010. [CrossRef] [PubMed]
35. Montgomery, L.F.R.; Bochmann, G. Pretreatment of Feedstock for Enhanced Biogas Production. In *IEA Bioenergy Task 37—Energy from Biogas*; Baxter, D., Ed.; IEA Bioenergy: Paris, France, 2014; ISBN 978-1-910154-05-2.
36. Phuttaro, C.; Sawatdeenarunat, C.; Surendra, K.C.; Boonsawang, P.; Chaiprapat, S.; Khanal, S.K. Anaerobic digestion of hydrothermally-pretreated lignocellulosic biomass: Influence of pretreatment temperatures, inhibitors and soluble organics on methane yield. *Bioresour. Technol.* **2019**, *284*, 128–138. [CrossRef] [PubMed]
37. Pilli, S.; Yan, S.; Tyagi, R.D.; Surampalli, R.Y. Thermal Pretreatment of Sewage Sludge to Enhance Anaerobic Digestion: A Review. *Crit. Rev. Environ. Sci. Technol.* **2015**, *45*, 669–702. [CrossRef]
38. Sjulander, N.; Kikas, T. Origin, Impact and Control of Lignocellulosic Inhibitors in Bioethanol Production—A Review. *Energies* **2020**, *13*, 4751. [CrossRef]

39. Sarto, S.; Hildayati, R.; Syaichurrozi, I. Effect of chemical pretreatment using sulfuric acid on biogas production from water hyacinth and kinetics. *Renew. Energy* **2019**, *132*, 335–350. [CrossRef]
40. Patinvoh, R.J.; Osadolor, O.A.; Chandolias, K.; Sárvári Horváth, I.; Taherzadeh, M.J. Innovative pretreatment strategies for biogas production. *Bioresour. Technol.* **2017**, *224*, 13–24. [CrossRef]
41. Venturin, B.; Frumi Camargo, A.; Scapini, T.; Mulinari, J.; Bonatto, C.; Bazoti, S.; Pereira Siqueira, D.; Maria Colla, L.; Alves, S.L.; Paulo Bender, J.; et al. Effect of pretreatments on corn stalk chemical properties for biogas production purposes. *Bioresour. Technol.* **2018**, *266*, 116–124. [CrossRef]
42. Rabie, S.; Rostom, M.; Gaber, M.; Enshasy, H. Treatment of Egyptian wastepaper using flotation technique for recovery of fillers and minerals. *Int. J. Sci. Technol. Res.* **2020**, *9*, 3077–3083.
43. Syaichurrozi, I.; Villta, P.K.; Nabilah, N.; Rusdi, R. Effect of sulfuric acid pretreatment on biogas production from *Salvinia molesta*. *J. Environ. Chem. Eng.* **2019**, *7*, 102857. [CrossRef]
44. Antonopoulou, G.; Vayenas, D.; Lyberatos, G. Biogas Production from Physicochemically Pretreated Grass Lawn Waste: Comparison of Different Process. *Schemes. Molecules* **2020**, *25*, 296. [CrossRef] [PubMed]
45. Avila, R.; Carrero, E.; Crivillés, E.; Mercader, M.; Vicent, T.; Blánquez, P. Effects of low temperature thermal pretreatments insolubility and co-digestion of waste activated sludge and microalgae mixtures. *Algal Res.* **2020**, *50*, 101965. [CrossRef]
46. Wang, L.; Templer, R.; Murphy, R.J. A Life Cycle Assessment (LCA) comparison of three management options for waste papers: Bioethanol production, recycling and incineration with energy recovery. *Bioresour. Technol.* **2012**, *120*, 89–98. [CrossRef] [PubMed]

MDPI
St. Alban-Anlage 66
4052 Basel
Switzerland
Tel. +41 61 683 77 34
Fax +41 61 302 89 18
www.mdpi.com

Polymers Editorial Office
E-mail: polymers@mdpi.com
www.mdpi.com/journal/polymers

www.ingramcontent.com/pod-product-compliance
Lightning Source LLC
LaVergne TN
LVHW070703100526
838202LV00013B/1024